HADRONIC SHOWER
SIMULATION WORKSHOP

To learn more about AIP Conference Proceedings,
including the Conference Proceedings Series, please visit the webpage
http://proceedings.aip.org/proceedings

HADRONIC SHOWER SIMULATION WORKSHOP

Batavia, Illinois   6 – 8 September 2006

EDITORS
Michael Albrow
Rajendran Raja
Fermilab
Batavia, Illinois

SPONSORING ORGANIZATIONS
Universities Research Association, Inc.
Department of Energy

Melville, New York, 2007
AIP CONFERENCE PROCEEDINGS ■ VOLUME 896

Editors

Michael Albrow
Rajendran Raja

Fermilab
MS122, P. O. Box 500
Batavia, Illinois   60510
U.S.A.

E-mail: albrow@fnal.gov
          raja@fnal.gov

L.C. Catalog Card No. 2007922639
ISBN 978-0-7354-0401-4
ISSN 0094-243X

Printed in the United States of America

CONTENT

HADRONIC SHOWER SIMULATION WORKSHOP

September 6th – 8th 2006
Fermi National Accelerator Laboratory

The goal of the workshop was to bring together world experts in the field of hadronic shower development, to establish a collaborative effort that will lead to a better understanding and simulation of hadronic showers relevant to hadron calorimetry at the ILC and LHC, neutrino fluxes and atmospheric showers. The workshop evaluated existing event generator and transport codes. Before the workshop we carried out a bench marking analysis, whereby various hadronic shower simulation scenarios were submitted to code developers and the results were presented at the workshop. The workshop identified the shortcomings of existing hadronic shower simulations and we home to develop a collaborative plan to improve our understanding of showers. The workshop brought out the need to acquire new data to improve shower models.

Local Organizing Committee

M. Albrow (Fermilab)
D. Chakraborty (NIU)
M. Demarteau (Fermilab)
D. Elvira (Fermilab)
J. Link (Virginia Tech.)
S. Magill (ANL)
A.Para (Fermilab)
R. Raja (Fermilab) (Chair)
C.Sazama (Fermilab)

International Organizing Committee

J.Apostolakis (CERN)(Chair)
S.Dytman(U.Pittsburgh)
A.Ferrari(CERN)
A.Heikkinen(Helsinki Inst. Of Physics)
P.Loch(U. Arizona)
S.Mashnik(LANL)
G.McKinney(LANL)
M.Messier(Indiana U.)
N.Mokhov(Fermilab)
K.Niita(RIST)
A.Ribon(CERN)
M.Thomson(Cambridge U.)
R.Wigmans(Texas Tech)
D.Wright(SLAC)

PREFACE

This is the first of a series of workshops dedicated to the problem of simulating hadronic showers in matter. During Sep 6-8, 2006, approximately 90 people convened at Fermilab to describe and discuss the triumphs and tribulations associated with the simulation of hadronic showers. We heard from the experts who author and maintain widely used simulation programs such as GEANT4, FLUKA, MARS, MCNPX, and PHITS. We had overview talks describing the structure and algorithms used in the various programs followed by talks describing attempts to validate the programs by comparison to data. We also had talks from the physics user community that makes extensive use of these programs. These included the neutrino experiments (MINOS, MiniBoone), the collider experiments (ATLAS, CMS) and the cosmic ray experiments.

Before the workshop, we set out to run a series of benchmarks to compare the performance of various programs in specific situations. These benchmarks were only partially completed at the time of the workshop but are mostly completed at the present time. The benchmarks reveal that much work needs to be done to further improve the simulation codes, although the agreement is impressive in some situations where there has been prior tuning with existing data. The need for more particle production data was brought out clearly during the workshop.

The workshop included a general lab-wide colloquium highlighting hadronic shower simulation in calorimeters and concluded with a Fermilab Wine and Cheese talk summarizing the workshop.

The workshop was an enjoyable and informative experience for the participants. It highlighted the need to improve significantly the state of simulations further and the benefits for the high energy community (and beyond) that would accrue as a result. It was decided to hold a second of these workshops in approximately 18-24 months in Europe.

The organizers are grateful for the financial support provided by the Universities Research Association and the U.S. Department of Energy in organizing the workshop.

Michael Albrow and Rajendran Raja(Editors)

An Overview of the Geant4 Toolkit

John Apostolakis* and Dennis H. Wright†

*CERN (European Organization for Nuclear Research), Geneva, Switzerland
†Stanford Linear Accelerator Center, Menlo Park, California, USA

(For the Geant4 collaboration)

Abstract. Geant4 is a toolkit for the simulation of the transport of radiation through matter. With a flexible kernel and choices between different physics modeling choices, it has been tailored to the requirements of a wide range of applications.

With the toolkit a user can describe a setup's or detector's geometry and materials, navigate inside it, simulate the physical interactions using a choice of physics engines, underlying physics cross-sections and models, visualise and store results.

Physics models describing electromagnetic and hadronic interactions are provided, as are decays and processes for optical photons. Several models, with different precision and performance are available for many processes. The toolkit includes coherent physics model configurations, which are called physics lists. Users can choose an existing physics list or create their own, depending on their requirements and the application area. A clear structure and readable code, enable the user to investigate the origin of physics results.

Application areas include detector simulation and background simulation in High Energy Physics experiments, simulation of accelerator setups, studies in medical imaging and treatment, and the study of the effects of solar radiation on spacecraft instruments.

Keywords: simulation, Geant4
PACS: 24.10.Lx, 02.70.Uu, 01.30.Cc

The Geant4 toolkit [1, 2] can be seen as three sets of modules. The 'kernel' initializes and manages run configurations, handles particles, tracks and materials, and enables the simulation of processes. 'Physics processes' is the largest module, and includes the cross sections and physics models. 'Interfaces' provides visualisation, input/output and the facility for the user to steer the application.

Geant4 is implemented in C++, utilizing Object Oriented technology to create a structure of class categories that provide key capabilities. The kernel includes the geometry, field propagation, particles, materials, tracking and run and event management.

The implementation of the kernel and physics models follow customised coding standards and other de-facto standards to promote readable, transparent code. This provides the ability for the user to trace the origin of physical or geometrical effects and investigate their relation to final, observable results.

THE KERNEL

The kernel provides all non-physics functionality required to run a simulation, and enables the simulation user to manage most aspects of simulation behavior. Mandatory for the user are actions to attach a set of physics models to a particle type (for example an

CP896, *Hadronic Shower Simulation Workshop*
edited by M. Albrow and R. Raja
© 2007 American Institute of Physics 978-0-7354-0401-4/07/$23.00

electron), to create a geometrical model describing a particular setup, and to configure a particle beam or source.

The kernel also provides hooks with which the user can choose the output required from measuring instruments or in radiation-sensitive parts. It also enables the use of visualization, which can show both the setup and event interactions using various graphics visualization systems.

A Geant4 'run' is a configuration of the particle source, the detector or setup's geometry and the chosen set of physics processes. Each user must configure all these aspects for a given simulation 'run'. Finer control of the run configuration is also provided to enable an experiment to embed a simulation using Geant4 into its framework in a simple, maintainable way.

The user can prioritise tracks by any criteria, including particle type and energy. Also using the event module, it is possible to store user-selected information for later use, and to obtain the primary particles from an event generator. These can include 'pre-assigned' decay products, used if the track does not interact earlier.

The tracking engine in Geant4 is general, and utilizes a list of processes customized for each particle type, including ions. The lists of processes for all types create a configuration of physics processes, known as 'physics' lists. Different physics lists allow choice between accuracy for a particular use case and computing performance.

Overall the kernel provides or configures all key simulation capabilities, including the detector descriptions and its external field, the creation and handling of particles, materials and tracks, and the management of runs and events.

This section provides an overview of one key part, the geometry module. Further details on kernel functionality can be found in Ref. [3].

Geometry

The geometry modeler of Geant4 enables a user to describe the volumes in a setup or detector, and provides the ability to navigate in the resulting geometry model. A variety of shapes are available: simple Constructed Solid Geometry (CSG) including boxes, tetrahedra and cylindrical and spherical shells, the Boundary Represented Solids (BREPS) with connected regular polyhedra, and a selection of additional specific shapes from ellipsoids to twisted trapezoids. The user can also create their own shape, either by combining Boolean operations on other solids or by writing an extension for their own customised solid.

The user can then create a volume hierarchy utilising these. A solid can be given material, visualization and field attributes, to create a 'logical' volume. A logical volume, when placed at a fixed location inside another parent volume, creates a 'physical volume'. Several sub-volumes (each one potentially containing its own sub-volumes) can be embedded inside one parent volume. In this way a volume hierarchy is created.

Volumes can contain an arbitrary number of sub-volumes – and it is also possible to create a 'flat' geometry with all sub-volumes at one level. Volume hierarchies are very powerful, and have been used to describe complex setups, such as large HEP detectors. Utilising several levels and a few thousand logical and physical volumes, LHC experiments have described their detectors which contain millions of distinct volumes.

The tasks of locating a point inside a flat or hierarchical geometry and of computing the distance to the next volume boundary are the key responsibilities of the Navigator. At the start of a simulation run the geometry cashes key information about the positioning of volumes, in order to optimise the location of a point and the step computation.

Advanced capabilities for creating geometry models enable users to create complex setups more easily and to change them during run time. These include the abilities to:

- create a mirror image reflection of a hierarchy, including all daughter volumes,
- modify the geometry, changing the setup for each run within one computer job, and
- have different geometrical views of one setup ('ghost geometries'); each one can have one or more uses: biasing, scoring, fast parameterization, or readout.

The above abilities are provided by the geometry module, and coordinated via the simulation state, which is managed by the Run module.

A single volume can represent several copies, each displaced or rotated, in order to save memory in case there are many repetitions. Replica and the more general division volumes slice their parent along an axis. Parameterized volumes allow the user to define the solid type, the position and rotation of each sub-volumes using methods of a user class.

The geometry module also enables a user to label different volumes and subtrees as belonging to a single geometrical region. A number of properties can be assigned to a region, including production thresholds, minimum step size and other user limits.

A geometry description mark-up language, GDML[4], has been created to describe, in XML, the volumes and material composition of a setup. A module for interfacing with Geant4 reads and writes geometry models from/into XML text files. Replicas and the full set of Geant4 solids are supported. It is possible also to store and recreate the contents of parametrised volumes.

For reasons of performance and simplicity, Geant4 performs few checks for mal-formed geometries during tracking. An option enables simple checking of the user's geometry model during navigation in an event.

It is a major challenge for users to create a consistent description of the geometry of a setup, without volumes that overlap or protrude from their parent volume. Different tools in Geant4 help users to identify overlaps, enabling them to correct their geometries. They include new optional checks during each volume's construction, a geometry verification sub-module to check assembled pieces, and the pioneering DAVID tool [5] which intersects the graphical representations of volumes. The tools have adjustable intersection tolerances.

Field

Charged particles in Geant4 are affected by external electromagnetic fields. Intersections of a track's curved trajectory with geometry boundaries are calculated approximately to a user specified precision. A 'global' field and optional 'local' fields are available. The user can adjust a number of parameters to obtain the desired accuracy or performance. These parameters may be chosen globally or on a per volume basis.

PHYSICS CHOICES AND CONFIGURATIONS

Geant4 offers modeling options for many physics processes, with different accuracy, computing requirements and strengths and weaknesses in describing particular physical aspects. Information on these modeling options, for electromagnetic and hadronic physics processes, is provided below, and detailed descriptions are available in the Geant4 Physics Reference Manual [6] .

The physics models and cross sections available must be assembled into a consistent set which covers the physics relevant to an application. Appropriate trade-offs between accuracy and computing performance must be made. This set of physics models is called a 'physics list' in Geant4.

Choosing a physics configuration, appropriate to one's use case, is ultimately the responsibility of the user. As part of the toolkit, Geant4 provides a number of physics list configurations. Several have been used and validated, to the extent possible, for typical use cases. Others are simply provided on an as-is basis, and as starting points for other use cases. They demonstrate the use of key physics capabilities, and are categorised in their applicability for selected use cases, including HEP calorimetry, tracking detectors, and dosimetry. Validation of key physics lists is undertaken in collaboration with HEP experiments [7]. Regression testing of these lists is also performed for each Geant4 release. It is possible for users to choose one of the physics lists provided, create their own list, or use existing lists as starting points. A number of other physics lists are supported as well as those provided with the Geant4 examples.

User software built on top of Geant4 may have its own configuration of Geant4 physics. For example, the GATE project [9], created for medical applications, provides configurations using both Standard and Low-energy EM packages. The GRAS software [10], created for studies of space radiation effects, also provides alternative physics configurations.

ELECTROMAGNETIC PHYSICS

Electromagnetic interactions of photons and charged particles with matter are implemented in two electromagnetic packages. The Standard EM package [12, 6] is tailored for performant simulation and for applications where secondaries only above 1 KeV are important. For applications where lower energy secondaries, below 1 KeV, and additional effects, such as atomic relaxation and shell effects, are important, the low energy EM package is provided.

The 'Standard' Electromagnetic package models the EM interactions of particles with energies between 1 keV and 10 PeV. It includes all relevant processes, including ionization, bremsstrahlung, and gamma conversion. The emission and transport of optical photons [6] is handled separately. This enables a complete simulation of electromagnetic interactions, starting from the beam transport through to the final detector. It was initially created to perform high quality Monte Carlo production for HEP experiments. Improvements and new models have also made it suitable for applications in space, medicine, and other domains [10, 9].

The package has been used for large-scale simulation production in HEP experiments like BaBar [11] for more than four years, and in several LHC experiments [7] since early 2004. To provide stability of the simulation results for long-term production, a verification suite was created covering simplified setups and simple calorimeters with typical materials. These are used to regularly monitor physics and computing performance.

Electromagnetic process implementations are provided for lepton, hadron and ion ionisation, Bremsstrahlung, Compton scattering, polarized Compton scattering, photo-electric effect, gamma conversion to e^+e^- and muons, muon ionization, muon bremsstrahlung, e^+e^- pair production by muons, and positron annihilation to muon pairs or hadrons. Photons are produced by Cerenkov radiation, scintillation, synchrotron radiation, and transition radiation processes.

The Low-energy Electromagnetic package [6] provides alternative models of electromagnetic interactions which take atomic effects into account in more detail. This enables very detailed simulation of particle transport in a media, but requires significantly more CPU resources. Developments in the Low-energy package and its applications are beyond the scope of this report.

Energy Loss

Energy loss processes in Geant4 cover the processes of ionization and bremsstrahlung. They are modeled with a combination continuous and discrete components [1]. High energy transfer interactions are simulated as discrete actions. Low energy transfers are treated together as a continuous process of energy loss.

The threshold for production of secondaries is set as the range of the produced electron or gamma. Thus the energy depends on the material. By default, and for early Geant4 versions[1] there is single threshold for each type of secondary particle (e.g. for electrons).

Fluctuations of the energy loss are sampled using the Lassila-Perrini-Urban approach [13].

The mean value of the restricted energy loss accounts for the production cut of secondary (δ-electrons or gammas) below a threshold energy value T_{cut} for the current material. This energy T_{cut} corresponds to the energy for which the range of the secondary particles in the material equal the user chosen 'cut-in-range'. Thus, energy transfers below ε_{lim} are considered as continuous energy loss. The restricted cross section of each process is calculated at initialization. It is sampled during each step using a new integral approach to calculate the restricted mean energy loss.

The step size is limited according to a smooth function of the range of the particle at the current energy, and a minimum step size parameter. At large energies this becomes a fraction $\alpha = 0.2$ of the current range. The minimum step value parameter $\rho = 1$ mm was chosen equal to the default Geant4 cut in range value.

Default values for transport of electrons and hadrons have been tested extensively in uniform media and standard HEP applications. The user can tune the stepping function parameters for studies or specialised applications.

Cross sections for the ionisation process of hadrons and ions have been improved to include Barkas, Bloch and Mott corrections and the nuclear stopping power[16]. These corrections are relatively small for hadrons, but significant for ions. The stopping powers is now in agreement with the NIST data inside the systematic uncertainty at all energy range (2% for incident proton energy above 1 MeV).

An optional algorithm [15] generates low energy δ-rays, below the production threshold for the current volume, during steps which are near the boundaries of volumes. This can lead to improved stability when varying cuts, enabling improved performance while maintaining physics results.

A new algorithm implements an integral approach for sampling an interaction length. In previous Geant4 versions, as in Geant3, the interaction length for the energy loss process was defined from the cross section at the initial energy of the step [1]. In the new integral algorithm, the energy dependence of the electromagnetic cross sections is utilised when sampling the interaction probability. The method proposed in [14] is used.

Multiple Scattering

Multiple scattering, in particular of electrons, is an important process for many application areas. Important recent development is been carried out in the treatment of the multiple scattering to improve its accuracy.

The Geant4 multiple scattering model [19] has been based on Lewis's approach [17]. Originally there was no limitation on the step size were imposed by this process. This provides results in a good agreement with the data for high energy hadrons and electrons [19, 21].

However a significant dependence was observed for the deposited energy on the production cuts (and/or maximum step size), in sampling calorimeters and benchmark setups. The analysis of the energy deposition of a 1 MeV photon beam inside a low density cavity [18] reported a departure from theoretical expectations when using Geant4 with the default parameters. To reach the necessary precision it was necessary to impose a small step size.

To address the causes of these effects a revision of the multiple scattering process was undertaken[20]. Improvements introduced in Geant4 8.0 included sampling in a correlated manner the scattering angle and the lateral displacement, limiting the step size using geometrical information to ensure at least two steps in each volume, and additional recalculations of the distance to the closest boundary. These provide results which are more stable to variations of cuts or user-imposed maximum step size, at the cost of additional CPU cycles. The extra cost ranges from 10% to 100% at the same cut value, depending on the cuts and geometry. With the improved stability it is possible, in many setups, to utilise larger values of the production cuts, recovering computing performance.

Recent Modifications

A refined design enables developers and advanced users to extend models and create alternative implementations for a physics process. The design separates management functions from physics models, providing generic abstract classes. Within a single process several physics models can be used for a given particle type, and the model selection can be made depending on the particle energy.

Since Geant4 release 5.1 a design iteration across the toolkit provided the possibility to change the threshold production cuts (for gammas, electrons, and positrons) by geometrical region. This enables the simulation of large HEP detector to undertake very high precision tracking of particles inside vertex detectors and for muons, while opting for high CPU performance in calorimeters and muon identifiers.

HADRONIC PHYSICS

Geant4 includes cross sections and physics models for hadronic interactions from thermal energies (for neutrons) to hundreds of GeV. For many regimes, a choice of physics models is available, enabling a user to choose between more precision and better CPU performance.

At high energies ($E > \sim 20$ GeV) the Quark-Gluon String (QGS) model and a Fritiof-like String model (FTF) provide theory-driven interaction models of the initial projectile-nucleon collision.

At energies below ~ 10 GeV, two cascade models are provided, one following the Bertini approach, and one, called the Binary cascade, which is more theory-based.

Each of the above models simulates the initial interaction within the nucleus, producing high energy secondaries and leaving the nucleus in a highly excited state. A number of models are available to perform the de-excitation. One is the Geant4 precompound model [22] which starts by decaying the excitons formed during the high energy interactions or cascades. It then calls one or more of the available fission, Fermi breakup, multifragmentation and evaporation routines. The precompound model is currently used as a "back-end" for the QGS and Binary cascade models.

The Bertini-like cascade employs its own evaporation model to de-excite the remnant nucleus.

The slowest remaining particles may also be re-absorbed by the nucleus. They may be treated by the Chiral Invariant Phase space (CHIPS) model, which may also be used as a back-end for the QGS model.

For low energy projectiles (< 20 MeV), high-precision neutron processes and photo-evaporation codes are provided, each of which rely on data libraries which are provided with Geant4.

For photo-nuclear and electro-nuclear interactions processes are provided that use the CHIPS model. Stopping particles are treated utilizing a CHIPS-based capture process.

Covering all long-lived particles at all energies are the Low Energy Parameterized (LEP) and High Energy Parameterized (HEP) models which have their origins in the GHEISHA hadronic package [23] which was used with Geant3. The GHEISHA Fortran code was cast into C++, re-engineered and split into the current high- and low-energy parts, and a number of corrections and improvements were undertaken. These processes utilize simplified descriptions of interaction mechanisms, with key quantities parameterized for speed. They cover all long-lived hadrons, and were designed to describe hadronic showers reasonably well. They are also intended to conserve energy and momentum on average but not event by event.

The LEP models are especially important because they cover the transition region between the cascade models and the string models (between 10 and 20 GeV), which for some particles is not covered by the theory-driven models.

The LEP and HEP models provide faster alternatives to the theory-driven models mentioned earlier. In addition they are utilized for long-lived hadrons (Ω, Σ), due to a lack of alternatives, and to transition between the cascade and string model regions of applicability.

The original universal elastic model is derived from GHEISHA and has been used in most physics lists. Specific models exist for coherent p-p and p-n elastic scattering[24]. A new process 'QElastic', including new cross-sections and final state models, has been developed for proton and neutron scattering on nuclei. New classes for cross sections for coherent hadron scattering from nuclei have been recently created.

In the following discussion, four of the most-used models (QGS, Bertini-like and Binary cascades, CHIPS) will be overviewed.

Models at high energies

The QGS model [25] is used in Geant4 to simulate the interactions of protons, neutrons, pions and kaons with nuclei, in the approximate energy range 15 GeV to 50 TeV. Other models are coupled to it to fragment and de-excite the damaged nucleus after the initial high energy interaction.

The model selects a target nucleon from a detailed, three-dimensional model of the nucleus, splits the projectile and target nucleon into quarks and di-quarks, and then forms and excites quark-gluon strings. Strings are stretched between partons of the projectile and target nucleon, using sampled parton densities. Hadronization proceeds by longitudinal string fragmentation in which the string is successively broken into a hadron and another string until the string mass becomes low enough to break into two hadrons.

The transverse momentum distribution is sampled from a Gaussian distribution, while the longitudinal momentum distribution is sampled from fragmentation functions. The amount of diffractive dissociation is chosen empirically.

This is the principal model for incident particles above 15 GeV in the QGSP physics lists, used by LHC experiments [7] for their main production simulation.

Bertini-like cascade

In the cascade energy range Geant4 has a Bertini-style cascade model which handles incident nucleons, pions, kaons and hyperons up to 10 GeV. It follows the INUCL implementation of N. Stepanov [26] and uses standard intra-nuclear cascade features of the Bertini approach [27].

Step-like nuclear density distributions and potentials are used. Since experimental cross sections and angular distributions are used, the model can be extended to any hadron for which sufficient experimental data are available.

The projectile (and induced secondaries) is transported along straight lines through the nuclear medium and interacts using the free hadron-nucleon total cross section. The nuclear medium is approximated by (up to three) concentric, constant-density shells. At shell boundaries a particle can be reflected or transmitted.

As cascade collisions occur, an excited residual nucleus is built up. An exciton decay routine is then used to de-excite the nucleus. For light, highly-excited nuclei Fermi breakup may occur, and a fission channel is available. A custom nuclear evaporation routine for neutrons and alphas is followed by gamma emission at the lowest excitation energies ($< 0.1 MeV$).

Binary cascade

The Geant4 Binary cascade model [28] is valid for incident protons and neutrons with $E_{kin} < 3$ GeV, pions with $E_{kin} < 1.5$ GeV, and light ions with $E_{kin} < 3$ GeV/A, but it can be used in some cases adequately up to 10 GeV.

The model is based on two-body to two-body, or two-body to one-body interactions within the target nucleus. Nucleon-nucleon scattering is done by resonance formation and decay. Elastic nucleon-nucleon scattering is also included. Meson-nucleon inelastic scattering, except for true absorption, is modeled as s-channel resonance excitation. Once resonances are formed they may interact with other nucleons or decay.

Target nucleons are sampled from a detailed three-dimensional nuclear model and the particle-particle collisions within it are approximated by free cross sections. The incident particle and subsequent secondaries propagate through the nucleus along curved paths which are calculated by numerical integration from the equation of motion.

When the cascade phase is finished, the Geant4 precompound model is used to de-excite the residual nucleus.

Chiral Invariant Phase Space Model

The Chiral Invariant Phase Space (CHIPS) model has its origins as an event generator [29, 30] first incorporated into Geant4 for treating anti-baryon-nucleon annihilation, the capture of negatively charged hadrons at rest, and gamma- and lepto-nuclear reactions. It is also used in some Geant4 models to handle the nuclear fragmentation part of nuclear de-excitation.

The basic building block of this model is the quasmon - a collection of massless free partons, which form hadronic systems. A critical temperature T_c relates the quasmon mass M_Q to the number of its partons n. Hadronization proceeds via quark fusion and quark exchange.

In the model, u, d, and s quarks are treated as massless and related by chiral symmetry. It can produce kaons, using a strangeness suppression parameter to adjust their multiplicity. The physical masses are used for strange hadrons that are produced.

Because the maximum energy from the primary quark parton contributes to the inclusive spectra, quark exchange or fusion can be considered as a one-dimensional process. It has been demonstrated experimentally by the fact that when the inclusive hadron spectra are plotted versus $k = (p + E_{kin})/2$, they not only have the same exponential slope but nearly coincide.

CONCLUSIONS

The Geant4 toolkit is used in Monte Carlo production for HEP experiments, and for space, medicine and radiation detector applications. The geometry modeler offers extensive capabilities for describing setups with complex shapes and many, repeated volumes. Navigation in these setups is efficient, for small and large numbers of volumes. The effect of fields on charged particles is handled using numerical integration. All essential electromagnetic processes are provided by the toolkit, and the upgrade of the multiple scattering model significantly improves the simulated electron transport across a boundary between different media. It results in improved stability in the energy deposition of sampling calorimeters. The hadronic models provided span the range of energies from thermal neutrons to TeV. String, cascade, precompound and evaporation models handle the interaction of nucleons, pions and kaons with matter. Parameterized models, derived from Gheisha, are utilized for other long-lived hadrons, and to transition between the cascade and string model regions of applicability.

REFERENCES

1. Geant4 Collaboration (S. Agostinelli et al.), Nucl. Instrum. Meth. **A506**, 250, (2003).
2. Geant4 Collaboration (J. Allison et al.), IEEE Trans. Nucl. Sci. **53**, 270 (2006).
3. J. Apostolakis, G. Cosmo, and M. Asai, in "The Monte Carlo Method: Versatility Unbounded In A Dynamic Computing World", Proceedings of the Monte Carlo 2005 Conference, Chattanooga, Tennessee, April 17-21, 2005, on CD-ROM, American Nuclear Society, (2005).
4. R. Chytracek, J. Mccormick, W. Pokorski, G. Santin, IEEE Trans.Nucl.Sci **53**, 2892 (2006)
5. S. Tanaka and K. Hashimoto, in Proceedings of the CHEP '98 Conference, Chicago, September 1998.
6. Geant4 Physics Reference Manual, http://cern.ch/geant4/UserDocumentation/UsersGuides/ PhysicsReferenceManual/html/
7. A.E. Kiryunin et al.(ATLAS collaboration), Nucl. Instrum. Methods Phys. **A** 560, 278, (2006); S. Abdoulline, et al. IEEE Nuclear Science Symposium Conference, Oct 2004, 2024 (2004) ; I. Belyaev et al., arXiV physics/0306035.
8. C. Alexaa et. al., CERN-LCGAPP-2004-10, July 2004
9. S. Jan et al., Phys. Med. Biol. **49**, 4543 (2004).
10. G. Santin, V. Ivanchenko, H. Evans, P. Nieminen and E. Daly, IEEE Trans.Nucl.Sci **52**, 2294 (2005).
11. R. Aubert et al., Phys.Rev.Lett. **95**, 142003 (2005), and **91**, 221802 (2003).
12. Geant4 Application Developers Guide, Chapter 5: Tracking and Physics. Available at http://geant4.web.cern.ch/geant4/support/userdocuments.shtml
13. K. Lassila-Perini and L. Urban, Nucl. Instr. Meth. **A362**, 416 (1995).
14. V.N. Ivanchenko, A.D. Bukin et al., in Proceedings MC1991, Detector and event simulation in high energy physics, Amsterdam, p. 79 (1991).
15. J. Apostolakis, S. Giani, M. Maire, L. Urban, CERN-OPEN-99-299 (1999).
16. A. Allisy et al., 1993, ICRU Report 49.
17. H.W. Lewis, Phys. Rev. **78** (1950) 526.
18. E. Poon, J. Seuntjens, F. Verhaegen, Phys. Med. Biol. **50**, 681 (2005).
19. L. Urban, CERN-OPEN-2002-070 (2002).
20. L. Urban, CERN-OPEN-2006-077 (2006).
21. H. Burkhardt et al., in Proceedings of the Monte Carlo 2005 Conference, *as above*.
22. Geant4 Physics Reference Manual, Chapter 29: Precompound Model Available at http://geant4.web.cern.ch/geant4/support/userdocuments.shtml
23. H. Fesefeld, "Simulation of hadronic showers, physics and applications", Technical Report PITHA 85-02, Aachen, Germany, Sept. 1985.
24. R.A. Arndt, I. I. Strakovsky, and R.L. Workman, Phys. Rev. **C 62**, 034005 (2000).
25. N.S. Amelin et al.,Phys. Rev. Lett. **67**, 1523 (1991); N.S. Amelin et al.,Nucl. Phys. **A544**, 463c (1992); L.V. Bravina et al., Nucl. Phys. **A566**, 461c (1994); L.V. Bravina et al., Phys. Lett. **B344**, 49 (1995).
26. N.V. Stepanov, ITEP Preprint ITEP-55, Moscow (1988).
27. M.P. Guthrie, R.G. Alsmiller and H.W. Bertini, Nucl. Instr. Meth. **66**, 29 (1968); H.W. Bertini and P. Guthrie, Nucl. Phys. **A169**, (1971).
28. G. Folger, V.N. Ivanchenko and J.-P. Wellisch, Eur. Phys. J. **A21**, 407 (2004).
29. M.V. Kossov, "Manual for the CHIPS event generator", KEK internal report 2000-17, Feb. 2001 H/R; P.V. Degtyarenko, M.V. Kossov and H.P. Wellisch, Eur. Phys. J. **A8** (2), 217 (2000), Eur. Phys. J. **A9** (2), 211 (2001), and Eur. Phys. J. **A9** (2), 221 (2001).
30. M.V. Kossov and L.M. Voronina, ITEP Preprint 165-84, Moscow (1984).

Low and High Energy Modeling in Geant4

Dennis H. Wright, Tatsumi Koi*, Gunter Folger, Vladimir Ivanchenko,
Mikhail Kossov, Nikolai Starkov†, Aatos Heikkinen** and
Hans-Peter Wellisch‡

*Stanford Linear Accelerator Center, Menlo Park, California, USA
†CERN, Geneva, Switzerland
**Helsinki Institute of Physics, Helsinki, Finland
‡Geneva, Switzerland

Abstract.
   Four of the most-used Geant4 hadronic models, the Quark-gluon string, Bertini-style cascade, Binary cascade and Chiral Invariant Phase Space, are discussed. These models cover high, medium and low energies, respectively, and represent a more theoretical approach to simulating hadronic interactions than do the Low Energy and High Energy Parameterized models. The four models together do not yet cover all particles for all energies, so the Low Energy and High Energy Parameterized models, among others, are used to fill the gaps.
   The validity range in energy and particle type of each model is presented, as is a discussion of the models' distinguishing features. The main modeling stages are also described qualitatively and areas for improvement are pointed out for each model.

Keywords: simulation, Geant4
PACS: 24.10.Lx, 02.70.Uu, 01.30.Cc

INTRODUCTION

In most Geant4 applications in high energy and nuclear physics, hadronic interactions are handled by four models which cover the high, medium and low energy domains. These are, respectively, the Quark-Gluon String, Bertini-style and Binary cascades, and Chiral Invariant Phase Space models. These models are detailed and theory-based (as opposed to parameterized) and explicitly conserve energy-momentum and most quantum numbers.

The above models handle most of the long-lived hadrons at almost all energies. Other models, which are not discussed here, are used to fill in the gaps in coverage. These include the High Precision neutron model for energies from thermal to 20 MeV, several elastic scattering models optimized for various energy ranges, and several types of nuclear de-excitation code, including fission, Fermi breakup and multifragmentation.

There are also the Low Energy Parameterized (LEP) and High Energy Parameterized (HEP) models which have their origins in the GHEISHA hadronic package [1] which was used with Geant3. The GHEISHA Fortran code was cast into C++, re-engineered and split into the current high- and low-energy parts. Like the GHEISHA code, these models are intended to be fast, cover all long-lived particles at all energies, and to describe hadronic showers reasonably well. They are also intended to conserve energy and momentum on average but not event by event.

CP896, *Hadronic Shower Simulation Workshop*
edited by M. Albrow and R. Raja
© 2007 American Institute of Physics 978-0-7354-0401-4/07/$23.00

In addition to hadron-nucleus interactions, Geant4 also offers lepto-nuclear and gamma-nuclear models and several ion-ion interaction models.

QUARK-GLUON STRING MODEL

The Quark-Gluon String (QGS) model [2] is used in Geant4 to simulate the interaction with nuclei of protons, neutrons, pions and kaons in the approximate energy range 20 GeV to 50 TeV. When coupled to gamma-nuclear models, QGS is also valid for incident high energy photons. Additional models are required to fragment and de-excite the damaged nucleus which remains after the initial high energy interaction.

Most of the QGS code is unique to Geant4, but theoretical guidance was taken from the Dubna QGS model of N.S. Amelin [3]. The model handles the selection of collision partners, splitting of the nucleons into quarks and di-quarks, the formation and excitation of quark-gluon strings, string hadronization and diffractive dissociation.

The modeling sequence begins by building a 3-D model of the target nucleus. Nucleon momenta are sampled using the Fermi gas model. The nuclear density is assumed to have a Woods-Saxon shape for all nuclei with $A \geq 17$. For lighter nuclei a harmonic oscillator shape is used. The momentum sampling is done in a correlated manner, with local phase space densities constrained by the Pauli principle and the sum of all nucleon momenta constrained to zero.

The large value of γ collapses the nucleus to a $2 - D$ disk. The impact parameters of all nuclei on the disk are then calculated. The hadron-nucleon collision probabilities can then be calculated using a quasi-eikonal model and gaussian density distributions for the hadron and nucleons. The Regge-Gribov [4] approach is used to determine the probability of an inelastic collision with the ith nucleon:

$$p_i(b_i,s) = \frac{1}{c}\left[1 - exp[-2u(b_i,s)]\right] = \sum_{n=1}^{\infty} p_{p^n}(b_i,s) \tag{1}$$

where

$$p_{p^n}(b_i,s) = \frac{1}{c}exp[-2u(b_i,s)]\frac{2u^2(b_i,s)}{n!} \tag{2}$$

is the probability of finding n cut pomerons in the collision. s as usual is the square of the CM energy, b_i is the impact parameter of the incident hadron with a nucleon, and c is the shower enhancement parameter described below.

$$u(b_i,s) = \frac{z(s)}{2}exp[b_i^2/4L(s)] \tag{3}$$

is the eikonal amplitude for hadron nucleon elastic scattering with pomeron exchange, where $z(s)$ is proportional to the pomeron-hadron coupling, and $L(s)$ contains the effective radius of the pomeron-hadron interaction region.

The initial interaction is assumed to proceed by pomeron exchange between the interacting hadrons. The pomeron parameters were determined by a fit to $N-N$, $\pi-N$, and $K-N$ collision data which include elastic, total and single diffraction cross sections.

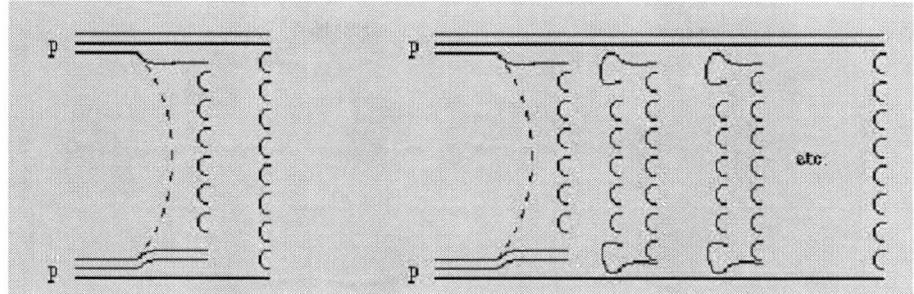

FIGURE 1. Diagram of quark-gluon string formation from cut pomeron. Internal vertical chains represent hadronizing strings stretched between sea quarks.

The resulting pomeron trajectory parameters are

$$\alpha'_P = 0.25 GeV^{-2},\qquad(4)$$

and

$$\alpha_P(0) = \begin{cases} 1.0808 & \text{for} \quad \pi\,,K \\ 0.9808 & \text{for} \quad p\,,n \end{cases}$$

The energy scale is set by

$$s_0 = \begin{cases} 3.0 GeV^2 & \text{for} \quad p\,,n \\ 1.5 GeV^2 & \text{for} \quad \pi \\ 2.3 GeV^2 & \text{for} \quad K. \end{cases}$$

The pomeron-hadron vertex parameters are:

$$\gamma_P^N = 6.56 GeV^{-2} \text{ and } R_P^{2N} = 3.56 GeV^{-2}.\qquad(5)$$

Strings are constructed from cut cylindrical pomerons and parton interaction leads to color coupling of the valence quarks. String formation follows the method of Capella [5] and Kaidalov [6] in which the parton densities are sampled for each participating hadron. This requires the quark structure functions of the hadrons. Parton pairs are combined into color singlets and sea quarks are included in the proportion $u : d : s = 1 : 1 : 0.27$. This is shown schematically in Fig. 1.

Hadronization proceeds by longitudinal string fragmentation. As each string is stretched between constituent partons, the number of breaks is sampled and $q - \bar{q}$ pairs are inserted at each break using the ratio $u : d : s : qq = 1 : 1 : 0.27 : 0.1$. The transverse momentum of each hadron is sampled from a gaussian distribution with $< P_t^2 >= 0.5 GeV^2$ while the longitudinal momentum is sampled from fragmentation functions native to the Geant4 QGS code.

The amount of diffractive dissociation is chosen empirically by the "shower enhancement" parameter c which is 1.4 for nucleons and 1.8 for pions:

$$p_{ij}^{diff} = \frac{c-1}{c}[p_{ij}^{tot}(b_{ij},s) - p_{ij}(b_{ij},s)].\qquad(6)$$

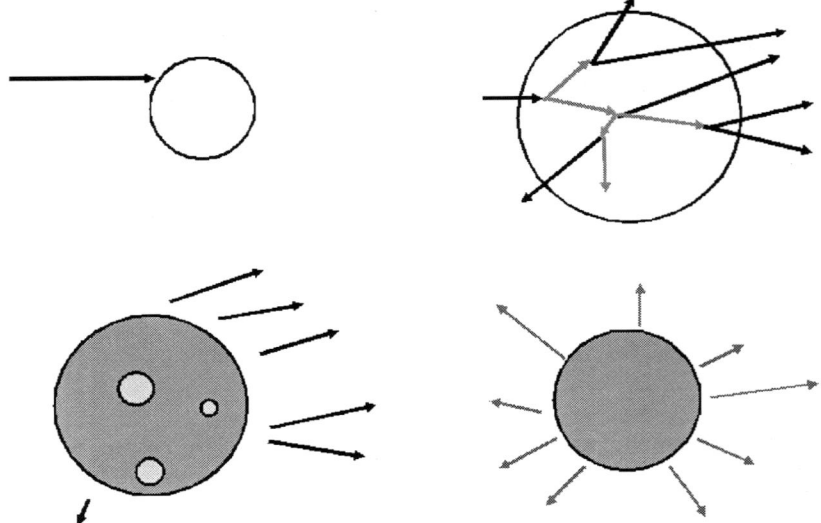

FIGURE 2. Schematic of cascade modeling sequence. Upper left: hadron incident upon target nucleus. Upper right: cascade development showing interactions within the nuclear medium and secondary hadrons leaving the nucleus. Lower left: high energy hadrons departing the nucleus, leaving it in a highly excited particle-hole state. Lower right: de-excited nucleus undergoes evaporation.

BERTINI-STYLE CASCADE

In the cascade energy range Geant4 has a Bertini-style cascade model which handles incident protons, neutrons, pions, kaons and hyperons. With the help of an internal pre-compound de-excitation handler, this code has been extended down to 0 initial energy. Its upper limit is roughly 10 GeV.

This implementation is a re-engineered version of Stepanov's INUCL code [7] and employs many of the standard intra-nuclear cascade features developed by Bertini [8]. Three of these are the use of

- classical scattering without matrix elements,
- free hadron-nucleon cross sections and angular distributions which are taken from experiment, and
- step-like nuclear density distributions and potentials.

The second feature, in principle, allows the model to be extended to any particle for which there are sufficient double-differential cross section measurements.

The modeling sequence is similar to many other cascade codes and is shown pictori-ally in Fig. 2.

The projectile enters the nucleus at a point sampled over the projected area of the nu-cleus. The projectile is then transported along straight lines through the nuclear medium and interacts according to the mean free path determined by the free hadron-nucleon total cross section. The nuclear medium is approximated by up to three concentric, constant-density shells. The initial nucleon momenta are distributed according to the Fermi gas model, and Pauli blocking is invoked for the nucleons. For the most part the projectile interacts with a single nucleon, but some nucleon-nucleon correlation is included by allowing pions to be absorbed on quasi-deuterons.

Each secondary from initial and subsequent interactions is also propagated in the nuclear potential until it interacts or leaves the nucleus. During propagation, particles may be reflected from, as well as transmitted through, the shell boundaries mentioned above. One drawback of the current model is that there is no Coulomb barrier implemented, thus the low energy proton spectrum is incorrectly modeled.

As cascade collisions occur, an excited residual nucleus is built up. This is done by forming particle-hole states based on the selection rules:

$$\Delta p = 0, \pm 1, \Delta h = 0, \pm 1, \Delta n = 0, \pm 2. \tag{7}$$

As mentioned above, the Bertini-style cascade has its own exciton routine which is used to collapse the particle-hole states and de-excite the residual nucleus. This routine is based on that of Griffin [9] and uses the Kalbach matrix elements [10] and nuclear level densities parameterized as functions of Z and A. The transition from cascade stage to exciton stage occurs when the secondary kinetic energy drops below either 20% of its original value, or seven times the nuclear binding energy.

For light, highly-excited nuclei Fermi breakup may occur, and fission is also possible.

In the final stage, nuclear evaporation occurs as long as the excitation energy is large enough to remove a neutron or alpha from the nucleus. Gamma emission then occurs at energies below 0.1 MeV

BINARY CASCADE

A more theoretically motivated alternative to the Bertini-style cascade is the Geant4 Binary cascade model [11]. A hybrid between a classical cascade and a full quantum-molecular dynamics model, it is native to Geant4 and based in part on Amelin's kinetic model [12]. It is nominally valid for incident protons and neutrons with $0 < KE < 3$ GeV, pions with $0 < KE < 1.5$ GeV, and light ions with $0 < KE < 3$ GeV/A. However, it works reasonably well up to 10 GeV when compared to the Bertini-style cascade.

The modeling sequence is in large part similar to that of other cascades (see Fig. 2), so only the characteristic details will be discussed here.

A detailed $3 - D$ model of the nucleus is used, placing nucleons in space according to Woods-Saxon-shaped nuclear densities, and in momentum according to the Fermi gas model. The nucleon momentum is taken into account when evaluating cross sections and collision probabilities. An optical potential is included to simulate the collective effect of the nucleus on the nucleons participating in the reaction. The incident particle and subsequent secondaries are then propagated through the nucleus along curved paths by numerical integration of the equation of motion in the potential.

Nucleon-nucleon scattering is handled by t-channel resonance formation and decay. The excitation cross sections are derived from $p - p$ scattering using isospin invariance and the corresponding Clebsch-Gordan coefficients. Elastic nucleon-nucleon scattering is also included. Meson-nucleon inelastic scattering, except for true absorption, is modeled as s-channel resonance excitation. Here, the Breit-Wigner form is used for the cross sections.

Once resonances are formed, they may interact or decay. At present the binary cascade model takes into account 25 strong resonances: 10 delta resonances from 1232 MeV to 1950 MeV, and 15 nucleon resonances from 1440 MeV to 2250 MeV. It is the mass of the highest included resonances which currently limits the upper energy of the model's validity. Nominal PDG branching ratios are used for resonance decay and the masses are sampled from the Breit-Wigner shape. The imaginary part of the R-matrix is calculated using free two-body cross sections from experimental data and parameterizations. For resonance re-scattering, the solution of the BUU equation is used.

Other features included in the binary cascade are:

- a Coulomb barrier for charged hadrons,
- for nucleon-nucleon elastic scattering, the use of angular distributions taken from the Arndt phase shift analysis of experimental data [13], and
- true pion absorption modeled as s-wave absorption on quasi-deuterons.

Cascade models are generally not valid for energies below a few tens of MeV. For the binary model, the cascade stops when the mean energy of all scattered particles is below an A-dependent cut, which varies from 18 to 9 MeV. Below this energy, the properties of the residual nucleus and exciton system, which are built up during the cascade, are passed to the Geant4 precompound model [14] which handles the nuclear de-excitation. When the primary particle is below 45 MeV, the cascade is not initiated; instead control is passed directly to the precompound model.

CHIRAL INVARIANT PHASE SPACE MODEL

The Chiral Invariant Phase Space (CHIPS) model began as an event generator [15] and was incorporated into Geant4 as a novel way of treating the capture of negatively charged hadrons at rest, anti-baryon-nucleon interactions, gamma- and lepto-nuclear reactions. It is also used in some Geant4 models to handle the nuclear fragmentation part of nuclear de-excitation.

CHIPS is based on a few fundamental concepts:

- the quasmon - an ensemble of massless partons uniformly distributed in invariant phase space. This is a $3-D$ bubble of quark-parton plasma and can be any excited hadron system or ground state hadron
- critical temperature T_c - a model parameter which relates the quasmon mass M_Q to the number of its partons:

$$M_Q^2 = 4n(n-1)T_c^2 \rightarrow M_Q \simeq 2nT_c \tag{8}$$

$$T_c = 180 - 200 MeV \tag{9}$$

- quark fusion hadronization - two quark-partons may combine to form an on-shell hadron,

- quark exchange hadronization - quarks from quasmon and neighboring nucleon may trade places.

The model treats u,d, and s quarks symmetrically, in that they are all assumed to be massless. It can produce kaons, but to get kaon multiplicities correct, a strangeness suppression parameter is required, as is an η suppression parameter. The real s-quark mass is only taken into account when producing strange hadrons.

Another important assumption of the model is that quark fusion occurs in one dimension, that is, the fusing partons have exactly opposite directions. This is born out by the fact that only secondary hadrons which get the maximum energy from the primary quark parton contribute to the inclusive spectra. It is demonstrated experimentally by the fact that when the inclusive hadron spectra are plotted versus $k = \frac{p+KE}{2}$, they not only have the same exponential slope but nearly coincide [16].

The modeling sequence for CHIPS simulation varies somewhat according to the application. To illustrate the method, the example of proton-anti-proton annihilation in vacuum is discussed first. A simplified picture of this process is shown in Fig. 3. In this case there is no quark exchange with neighboring nucleons. The number of quark partons is given by the mass M of the system as in Eq. 8. These are spread uniformly over the phase space with spectrum

$$dW/kdk \propto (1 - 2k/M)^{n-3}. \tag{10}$$

Here, k is the parton momentum.

Then quark fusion is simulated by calculating the probability that two quark-partons in the quasmon combine to produce the effective mass of the outgoing hadron. This is done by:

- sampling the momentum k in three dimensions,
- obtaining the second quark momentum q from the spectrum of $n-1$ quarks, and
- integrating over q with the mass shell constraint for the outgoing hadron.

Next the type of final state hadron must be determined. The probability that a hadron of spin s_h and a given quark content is produced is given by

$$P = (2s_h + 1)z^{n-3}C_q \tag{11}$$

where C_q is the number of ways a hadron h can be made from the set of quarks in the quasmon, and z^{n-3} is a kinematic factor that comes from the momentum selection. The first hadron is thus produced and escapes the quasmon.

Next the remaining quasmon mass is sampled, based on its original mass M and the emitted hadron mass. Quark fusion as described above is then repeated with the residual reduced quasmon mass and parton content. The hadronization process ends when a minimum quasmon mass m_{min} is reached. Its value is determined by the quark content in the final quasmon. The final quasmon may decay into two hadrons or a hadron and a resonance. As mentioned above, kaon multiplicity is regulated by the s-suppression parameter, $s/u = 0.1$. The η and η' suppression is regulated by an η suppression factor of 0.3.

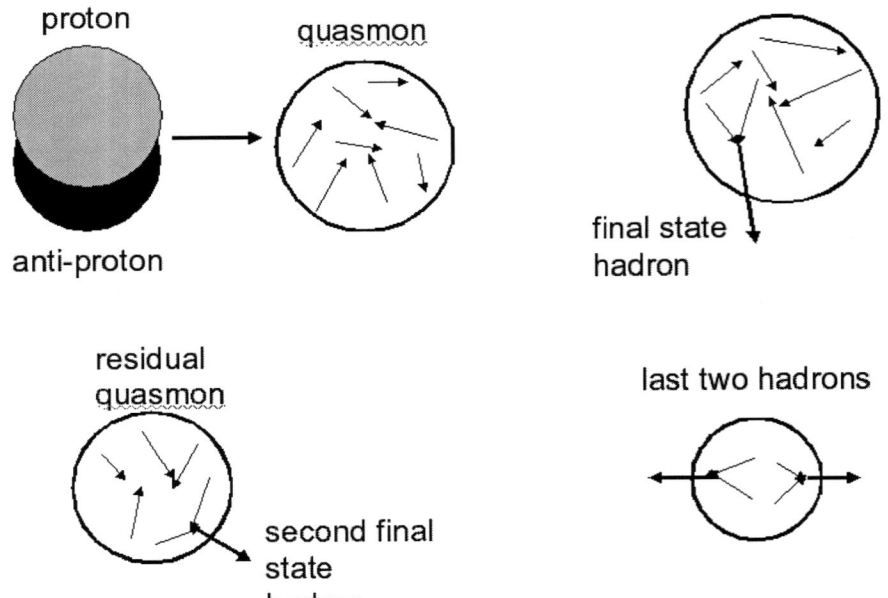

FIGURE 3. CHIPS modeling sequence for proton- anti-proton annihilation. Top left: proton and anti-proton merge, forming a quasmon. Top right: partons within the quasmon fuse to form an on-shell hadron which escapes. Lower left: hadronization continues from the residual quasmon. Lower right: hadronization ends when residual quasmon mass reaches a lower limit. Then it decays into on-shell hadrons.

The more complicated process of pion capture in a nucleus has added features. In this case the pion may capture on a nucleon or a cluster of nucleons. The resulting quasmon thus has a large mass and many partons. The capture probability is proportional to the number of clusters in the nucleus, which in turn is determined by three clusterization parameters. Because the quasmon is formed in nuclear matter, quark exchange with neighboring nuclei may occur. This gives rise naturally to correlation of the final state hadrons. Fusion can also happen but only between quarks and diquarks, hence mesons cannot be created.

Hadrons, once created, may escape the nucleus or be stopped by the Coulomb barrier. As in the vacuum case, hadronization continues until the residual quasmon mass reaches its lower limit m_{min}. In nuclear matter it is at this point that the nuclear evaporation processes take over. If the residual nucleus is far from stability, fast emission of nucleons and alphas is performed to avoid making short-lived isotopes.

AREAS FOR MODEL IMPROVEMENT

Validation of these models is underway for many energies and particle types. Areas of improvement have been indicated for each of the above models. For the QGS model, it is known that the sampling of p_T is too simple, leading to incorrect diffraction and not enough π^- suppression in proton scattering. Also, cross sections internal to the model are currently being improved.

One obvious improvement for the Bertini-style cascade is to install a Coulomb barrier, which would improve the behavior of low energy protons. In general the energy region between 10 and 60 GeV poses a problem in that the cascade models are not intended for such high energies, while the QGS model should not really be applied below 20-30 GeV. Here the Low Energy Parameterized models can fill in, but a new model is likely to be needed in this region.

The CHIPS model is likely to be very versatile in its application and may be extended to medium and even high energies in some cases. However, it was designed as a final state generator and not intended for projectile interaction with the target nucleus. Significant development is required for extensions beyond its current uses.

ACKNOWLEDGMENTS

This work was supported by the Stanford Linear Accelerator Center (SLAC) through a grant from the U.S. Department of Energy. It was also supported by the European Organization for Nuclear Research (CERN) and the Helsinki Institute of Physics.

REFERENCES

1. H. Fesefeld, "Simulation of hadronic showers, physics and applications", Technical Report PITHA 85-02, Aachen, Germany, Sept. 1985.
2. G. Folger and J.-P. Wellisch, "String Parton Models in Geant4", CHEP03, La Jolla, California, March 2003, nucl-th/0306007.
3. N.S. Amelin et al.,Phys. Rev. Lett. **67**, 1523 (1991); N.S. Amelin et al.,Nucl. Phys. A544, 463c (1992); L.V. Bravina et al., Nucl. Phys. A566, 461c (1994); L.V. Bravina et al., Phys. Lett. **B344**, 49 (1995).
4. M. Baker and K.A. Ter-Martirosyan, Phys. Rep. 28C, 1 (1976).
5. A. Capella, A. Krzywicki, Phys. Rev. **D18**, 4120 (1978).
6. A.B. Kaidalov, K.A. Ter-Martirosyan, Phys. Lett. **B117**, 247 (1982).
7. N.V. Stepanov, ITEP Preprint ITEP-55, Moscow (1988).
8. M.P. Guthrie, R.G. Alsmiller and H.W. Bertini, Nucl. Instr. Meth. 66, 29 (1968); H.W. Bertini and P. Guthrie, Results from Medium-Energy Intranuclear-Cascade Calculation, Nucl. Phys.A169, (1971).
9. J. J. Griffin, Statistical Model of Intermediate Structure, Physical Review Letters 17, 478 (1966); J. J. Griffin, Statistical Model of Intermediate Structure, Physics Letters 24B, 1 (1967).
10. C. Kalbach, Z. Physik A 287, 319 (1978).
11. G. Folger, V.N. Ivanchenko and J.-P. Wellisch, Eur. Phys. Jour. A21, 407 (2004).
12. N.S. Amelin, K.K. Gudima, V.D. Toneev, Sov. J. Nucl. Phys. **51** (1990) 327; N.S. Amelin, JINR Report **P2-86-56** (1986).
13. R.A. Arndt, SAID phase shift analysis program, George Washington University Data Analysis Center.

14. Geant4 Physics Reference Manual, Section IV, Chapter 28,
    http://geant4.web.cern.ch/geant4/UserDocumentation/UsersGuides/
    PhysicsReferenceManual/html/PhysicsReferenceManual.html
15. M.V. Kossov, "Manual for the CHIPS event generator", KEK internal report 2000-17, Feb. 2001 H/R;
    P.V. Degtyarenko, M.V. Kossov and H.P. Wellisch, Eur. Phys. J. **A8**, 217 (2000); P.V. Degtyarenko,
    M.V. Kossov and H.P. Wellisch, Eur. Phys. J. **A9** (2001); P.V. Degtyarenko, M.V. Kossov and H.P.
    Wellisch, Eur. Phys. J. **A9** (2001).
16. M.V. Kossov and L.M. Voronina, ITEP Preprint 165-84, Moscow (1984).

Validation of Hadronic Models in Geant4

Tatsumi Koi, Dennis H. Wright*, Gunter Folger, Vladimir Ivantchenko, Mikhail Kossov, Nikolai Starkov†, Aatos Heikkinen**, Pete Truscott , Fan Lei††, and Hans-Peter Wellisch‡

Stanford Linear Accelerator Center, Menlo Park, California, USA
†CERN, Geneva, Switzerland
**Helsinki Institute of Physics, Helsinki, Finland
††QinetiQ, Farnborough, UK
‡Geneva, Switzerland

Abstract. Geant4 is a software toolkit for the simulation of the passage of particles through matter. It has abundant hadronic models from thermal neutron interactions to ultra relativistic hadrons. An overview of validations in Geant4 hadronic physics is presented based on thin-target measurements. In most cases, good agreement is available between Monte Carlo prediction and experimental data; however, several problems have been detected which require some improvement in the models.

Keywords: simulation, Geant4
PACS: 24.10.Lx, 02.70.Uu, 01.30.Cc

INTRODUCTION

Abundant hadronic processes are available in Geant4 from thermal energy neutron interactions to energetic particle interactions available only at large accelerators or in cosmic rays. In some cases, there are multiple models for a given interaction so that users can select among them according to their requirements, in terms of application, precision and computing time. To aid user selection, validations of Geant4 hadronic models will be shown in this paper and detailed explanations about several selected models are available in another presentation [1] at this conference. Thin target experimental data are mainly used, because they allow a clean and detailed study of single hadronic interactions. Another type of validation uses calorimeter test-beam data with the complete detector setup. In this case, the observables are the convolution of many interactions; therefore, whole functions of Geant4 are validated at once. However, for the validation of each hadronic model, comparison to thin-target measurements is more suitable. The validations of Geant4 against calorimeter test-beam data were presented by other authors at this conference.

The coverage of Geant4 hadronic physics is quite wide and the number of hadronic models is also large. Due to space limitations however, not all models are represented and only one or a few validation plots are presented for each one. In the following sections, we will show validation results of Geant4 hadronic physics beginning with low energy models that include capture and pre-compound models and proceeding to high energy.

CP896, *Hadronic Shower Simulation Workshop*
edited by M. Albrow and R. Raja
© 2007 American Institute of Physics 978-0-7354-0401-4/07/$23.00

Then we will treat special topics such as gamma-nuclear, low energy neutron and ion-ion interactions.

LOW ENERGY INTERACTIONS

In this section, "low energy" refers to models in which the reaction energy is lower than the validity range of the Intra- Nuclear Cascade models. In addition to hadron scattering from a target nucleus, capture on the nucleus is also treated. The Chiral Invariant Phase Space (CHIPS) model [2] shows the best agreement to capture data. Fig. 1 shows comparisons of CHIPS model results with the experimental spectra [3, 4] of secondary particles and nuclear fragments from negative pion capture on ^{12}C nucleus. They are plotted as a function of $k = (p + E_{kin}) / 2$, where E_{kin} and p are the kinetic energy and the momentum of the secondary particles. Good agreement is observed in all energy ranges for all types of secondary particles. Other models are also available for this interaction, however they are not as detailed.

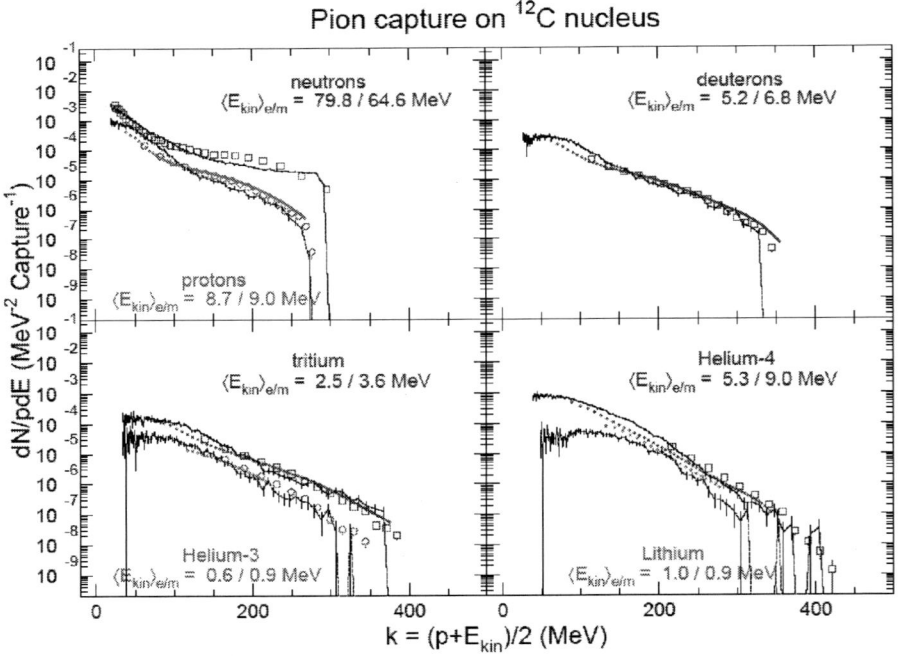

FIGURE 1. Comparison of the CHIPS model results with experimental data on proton, neutron, and nuclear fragment production in the capture of negative pions on C. The experimental neutron spectrum is taken from papers [3, 4]. The model calculations are shown by the two corresponding solid lines. The average kinetic energy carried away by each nuclear fragment is shown in the panels by the two numbers: first is the average calculated using the experimental data shown; second is the model result.

The pre-compound model, which can be divided into an exciton part and an evaporation part, is often used for hadron-nucleus interactions in this energy region.

Several pre-compound models are available in Geant4. Most of them are currently integrated within specific models such as the Beritini-style cascade [5], CHIPS model and Low Energy Parameterization (LEP) model which is based on the GHEISHA model of Geant3. There is also an independently implemented pre-compound model, which may be used by itself or coupled to Binary cascade [6] and/or high energy models of Geant4. Validations of the independently implemented pre-compound model are given in Fig. 2. The plots show neutron production cross sections resulting from protons bombarding tin and bismuth targets. Secondary neutrons created in the exciton and evaporation parts are presented separately in the plots. There is good agreement for both reactions.

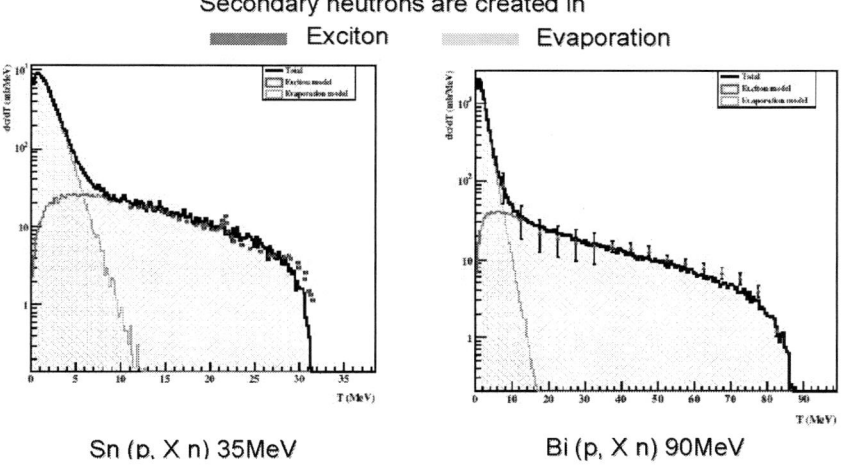

FIGURE 2. Comparisons of inclusive neutron production cross sections resulting from protons bombarding Sn and Bi targets [7]. The predicted flux is divided into neutrons generated by the exciton model (pink line) and the evaporation model (green line).

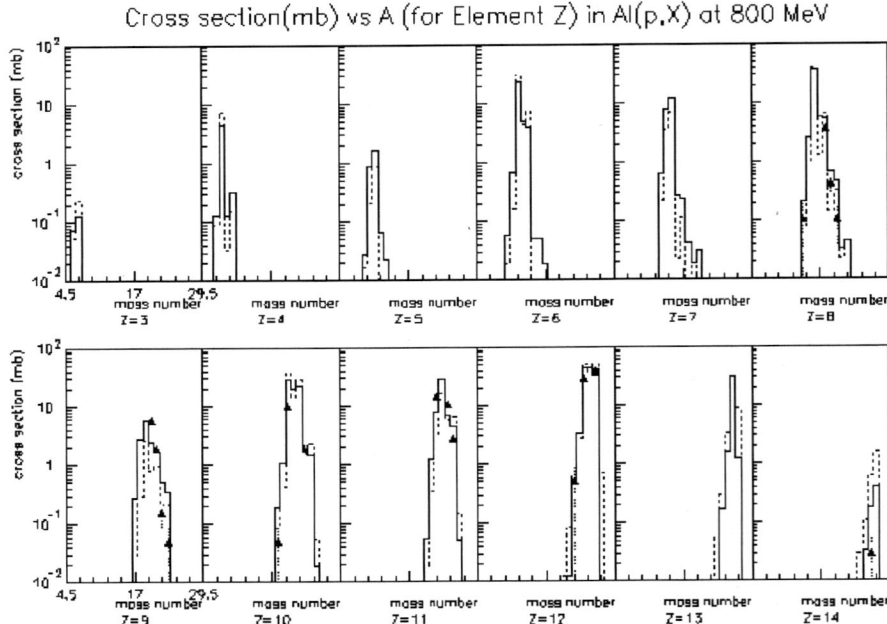

FIGURE 3. Comparison of isotope production resulting from 800MeV proton bombarding aluminum among Geant4 models and data [8]. Histogram – prediction of Bertini cascade (solid line) and Binary cascade (dashed line). Triangle –experimental data. Intermediate energy interactions

Isotope production by nuclear interactions is also treated in this section, because the pre-compound model usually plays the most important role in the isotope distribution from highly excited nuclei. Fig. 3 shows the validation of isotope production resulting from protons bombarding aluminum. Monte Carlo predictions are obtained from the Bertini and Binary cascade models. As already mentioned, different pre-compound models are used within each of these cascade models. The plot shows that the two pre-compound models work equally well.

INTERMEDIATE ENERGY INTERACTIONS

Intra-Nuclear Cascade models are generally used for simulation of hadron interactions above a few tens of MeV. The upper limits for these are model dependent. As mentioned before, Geant4 has two cascade models, namely a "Bertini-like cascade" and a "Binary cascade". The upper limit of the former model is roughly 10 GeV and the later is about 3 GeV. LEP models are also available for this energy region. A verification suite has been created in this energy region [9] which includes a large set of experimental data from thin-target scattering experiments. Fig. 4 is an example of a plot obtained using the suite. It shows the validation of Binary and Bertini cascade models for double differential neutron production cross sections. In most cases good agreement is found, however relatively large disagreements exist at the most forward scattering angles in the Bertini

24

cascade. In the Binary cascade, agreement at forward angles is better than at backward angles.

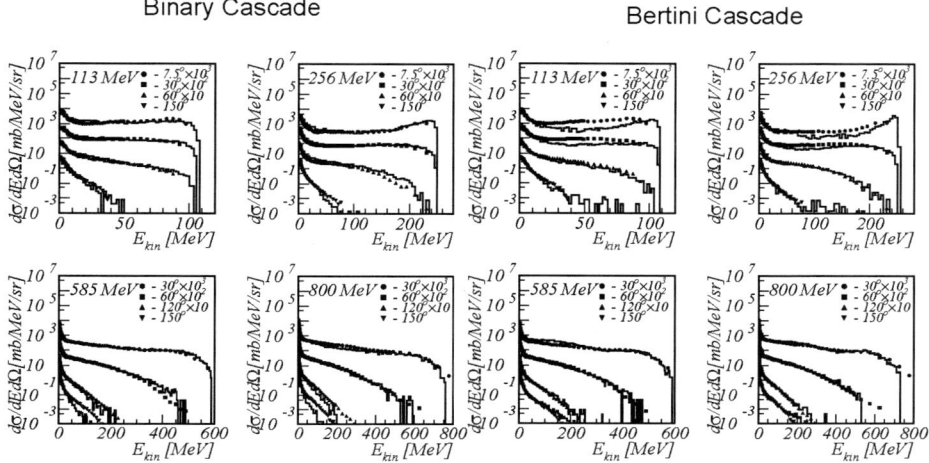

FIGURE 4. Double differential cross-section for neutrons produced in proton scattering off aluminum. Histograms –Monte Carlo predictions. Points-Data [10].

HIGH ENERGY INTERACTIONS

The Geant4 has two parton string models, namely the Quark-Gluon String (QGS) and Fritiof (FTF). These models are able to calculate hadronic interactions with reaction energies above a total center-of-mass energy of 5 GeV. Between the upper limit of our cascade models and the lower limit of these models, currently Geant4 has only the LEP model. The High Energy Parameterized (HEP) model (also based on GHEISHA) is available for hadronic interactions above 20GeV. Fig. 5 is a validation plot of transverse momentum (Pt) and rapidity distributions of secondary pions resulting from pion-Magnesium reactions at a laboratory momentum of 320 GeV/c. In the Pt distribution, the QGS model underestimates the data around the high Pt region. The rapidity distribution of the HEP model probably overestimates the peak and has an unphysical dip around 2. The peak position of QGS model is shifted slightly to the lower side.

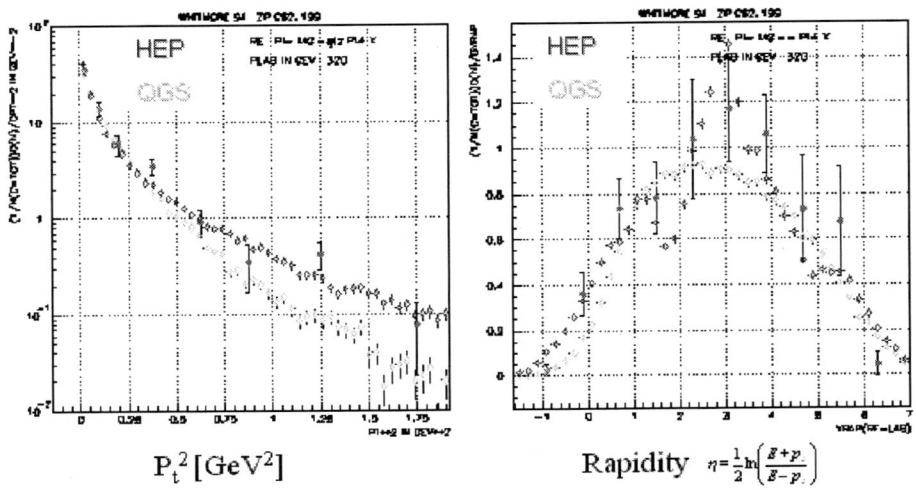

$$P_t^2 \, [GeV^2] \qquad\qquad \text{Rapidity} \quad \eta = \tfrac{1}{2}\ln\!\left(\frac{E+p_z}{E-p_z}\right)$$

FIGURE 5. Comparison of Pt and rapidity distributions of secondary pions resulting from pion-Magnesium reactions at laboratory momentum of 320 GeV/c. Open blue diamonds – prediction of HEP model. Open green diamonds – predictions of QGS model. Red box – experimental data[11].

PHOTON-NUCLEAR INTERACTIONS

The CHIPS model in Geant4 can also apply to the calculation of complicated mechanisms of interaction between photons and hadrons in nuclear matter. Fig. 6 shows comparison of the CHIPS model results with the experimental data [12, 13]. Proton yields from about 60 MeV gamma-rays bombarding ^{40}Ca are shown in the figure. The invariant inclusive cross sections ($d\sigma/p_p dE_p d\Omega_p$) of secondary protons are plotted as a function of k = (p + E_{kin}) / 2, where E_{kin} and p are the kinetic energy and the momentum of the secondary protons. The angular dependence of the proton yield in photoproduction is reproduced quite well. The Low Energy Parameterized (LEP) model is also available for this interaction, however it is not as detailed.

^{40}Ca(γ,p) spectral cross section

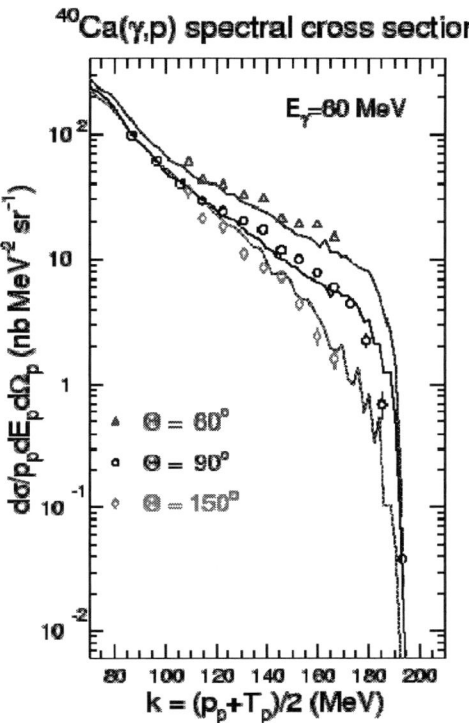

FIGURE 6. Comparison of CHIPS model with experimental data on proton production in photo-nuclear reactions on Ca at 59 - 65MeV. Open circles, triangles and diamonds represent the experimental proton spectra. Lines show the results of the corresponding CHIPS model calculation.

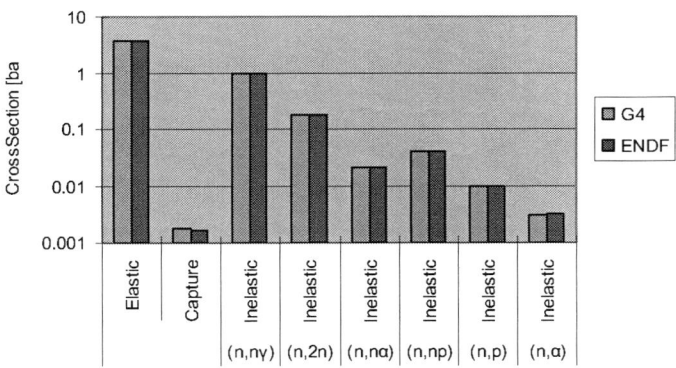

FIGURE 7. Comparison of channel cross sections between simulation results and reference data of ENDF/B-VI release8. Simulation results derived from thin-target calculation with Neutron HP models.

ION INTERACTIONS

Geant4 has several models for nucleus-nucleus interactions. The Binary Light Ion Cascade model is an extension of the Binary cascade model. It can handle ion interactions up to 3 GeV/n when the mass number of the projectile and/or target nucleus is less than 12. This limitation in mass number of the reaction system is not strict and in some cases the model can predict the final state well beyond the limitation [16]. Fig. 8 shows the neutron yield from 400MeV/n ^{56}Fe beams bombarding thick carbon and aluminum targets. Reactions between ^{56}Fe and Al nuclei are beyond the limitation while reactions between ^{56}Fe and C nuclei are within the limitation. However, judging from the validation plots, no significant difference in agreement can be seen between the two reactions.

The Abrasion-Ablation model is a C++ implementation of NUCFRG2 [17] physics and is also available in Geant4. This is a simplified macroscopic model for nucleus-nucleus interactions up to 10 GeV/n.

There is another kind of nucleus-nucleus interaction, namely electromagnetic dissociation. It is the liberation of nucleons or nuclear fragments as a result of the electromagnetic field by exchange of virtual photons. The EM Dissociation model of Geant4 can handle this interaction. Table 1 shows a validation of the EM Dissociation model for ultra relativistic ion projectiles in nuclear emulsion. Monte Carlo predictions are within experimental error bars in most cases.

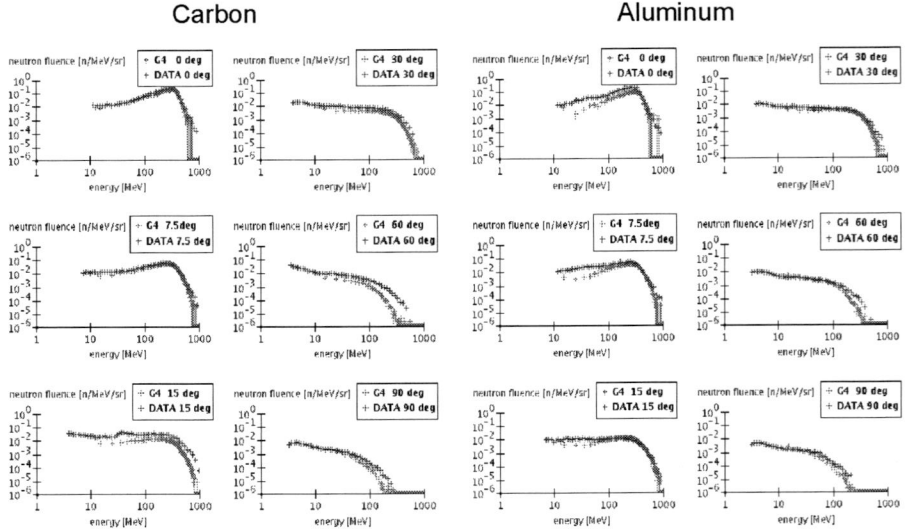

FIGURE 8. Comparison between data [18] and Geant4 simulation of double differential neutron production for a Fe beam of 400 MeV/n on thick Carbon and Aluminum targets.

TABLE 1. Comparison of cross-sections for ED interactions in silver in which projectile produces a single proton Experimental data are as reported by Jilany [19]

Projectile	Energy [GeV/nuc]	Product from ED	G4EM Dissociation [mbarn]	Experiment [mbarn]
Mg-24	3.7	Na-23+p	124 ± 2	154 ± 31
Si-28	3.7	Al-27+p	107 ± 1	186 ± 56
	14.5	Al-27+p	216 ± 2	165 ± 24
				128 ± 33
O-16	200	N-15+p	331 ± 2	293 ± 39
				342 ± 22

CONCLUSION

We have shown validations of low energy neutron, pre-compound, cascade, and high energy models. These are the most important for hadronic shower shapes and their agreement with data is good for most cases. However, relatively large disagreements are shown in the Pt distribution of the QGS model and the secondary neutron distribution at forward angles in the Bertini-like cascade model. These indicate that improvements in the diffractive part of the QGS model, and in the nuclear model of the Bertini-like cascade are required. The CHIPS model in Geant4 shows good agreement with data both for hadron capture and photonuclear interactions. The Binary Light Ion Cascade model has good agreement even for heavy ion collisions which are beyond its limitation in mass number, but improvement is needed in the correlation of participant nucleons and the transition to the pre-compound model. An extensive validation suite was prepared for the cascade energy region. Similar validation suites for other energy regions are required.

ACKNOWLEDGMENTS

This work was supported by the Stanford Linear Accelerator Center (SLAC) through a grant from the U.S. Department of Energy. It was also supported by the European Organization for Nuclear Research (CERN) and the Helsinki Institute of Physics.

REFERENCES

1. D. H. Wright, T. Koi, G. Folger, V. Ivanchenko, M. Kossov. N. Starkov. A. Heikkinen and H. P. Wellisch. in these proceedings of Hadronic Shower Simulation Workshop, Batavia, Illinois, 2006
2. M.V. Kossov, "Manual for the CHIPS event generator", KEK internal report 2000-17, Feb. 2001 H/R; P.V. Degtyarenko, M.V. Kossov and H.P. Wellisch, *Eur. Phys. J.* **A8**, 217 (2000); P.V. Degtyarenko, M.V. Kossov and H.P. Wellisch, *Eur. Phys. J.* **A9** (2001); P.V. Degtyarenko, M.V. Kossov and H.P. Wellisch, *Eur. Phys. J.* **A9** (2001).
3. G. Mechtersheimer et al., *Nucl. Phys.* **A324**, 379 (1979)
4. R. Madey, T. Villaithong, B. D. Anderson, J. N. Knudson, T. R. Witten, A. R. Baldwin, and F. M. Waterman, *Phys. Rev.* **C25**, 3050 (1982)
5. A.Heikkinen, N.Stepanov and HP.Wellisch, In the Proceedings of 2003 Conference for Computing in High-Energy and Nuclear Physics (CHEP 03), La Jolla, California, 24-28 Mar 2003, pp MOMT008
6. G. Folger, V.N. Ivanchenko and J.P. Wellisch, *Eur. Phys. J.* **A21**, pp. 407, (2004)
7. A. M. Kalend, B. D. Anderson, A. R. Baldwin, R. Madey, J. W. Watson, C. C. Chang, H. D. Holmgren, R. W. Koontz, J. R. Wu, and H. Machner, *Phys. Rev.* **C 28**, 105 (1983)

8. H. Vonach, A. Pavlik, A. Wallner, M. Drosg, R. C. Haight, D. M. Drake, and S. Chiba, *Phys. Rev.* **C55**, 2458 (1997)

9. V. Ivanchenko, G.Folger, J. P. Wellisch, T. Koi and D. H. Wright In the Proceedings of 2003 Conference for Computing in High-Energy and Nuclear Physics (CHEP 03), La Jolla, California, 24-28 Mar 2003, pp MOMT009

10. M. M. Meier et al., "Differential neutron production cross sections and neutron yields from stopping-length targets for 113-MeV protons" *Nucl. Scien. Engin.* **102**, 310, (1989); M. M. Meier et al., "Differential neutron production cross sections for 256-MeV protons" *Nucl. Scien. Engin.* **110**, 289, (1992); W. B. Amian et al., "Differential neutron production cross sections for 597-MeV protons" *Nucl. Scien. Engin.* **115**, 1, (1993) W. B. Amian et al., "Differential neutron production cross sections for 800-MeV protons" *Nucl. Scien. Engin.* **112**, 78, (1992)

11. J. J. Whitmore, F. Persi, W. S. Toothacker, P. A. Elcombe, J. C. Hill, W. W. Neale, W. D. Walker, W. Kowald, P. Lucas, L. Voyvodic, R. Ammar, D. Coppage, R. Davis, J. Gress, S. Kanekal, N. Kwak, J. M. Bishop, N. N. Biswas, N. M. Cason, V. P. Kenney, M. C. K. Mattingly, R. C. Ruchti and W. D. Shephard, *Z. Phys.* **C62** 199-227 (1994)

12. D. Ryckbosh, L. van Horrebeke, R. Van de Vyver, F. De Smet, J. O. Adler, D. Nilsson, B. Schröder and R. Zorro, *Phys. Rev.* **C42**, 444 (1990)

13. P.D. Harty et al. (unpublished); (private communication). Cited in the reference: J. Ryckebusch, L. Machenil, M. Vanderhaeghen, V. Van der Sluys and M. Waroquier, *Phys. Rev.* **C49**, 2704 (1994).

14. J.P. Wellisch. *"THE NEUTRON_HP NEUTRON TRANSPORT CODE"* from Monte Carlo 2005 Topical Meeting ISBN: 0-89448-695-0 Chattanooga, Tennessee, April 17-21, 2005

15. ENDF/B-VI: Cross Section Evaluation Working Group, ENDF/B-VI Summary Document, Report BNL-NCS-17541 (ENDF-201) (1991), edited by P.F. Rose, National Nuclear Data Center, Brookhaven National Laboratory, Upton, NY, USA.

16 T. Koi et al., *"ION TRANSPORT SIMULATION USING GEANT4 HADRONIC PHYSICS"* from Monte Carlo 2005 Topical Meeting ISBN:0-89448-695-0 Chattanooga, Tennessee, April 17-21, 2005

17. J. W. Wilson, R. K. Tripathi, F. A. Cucinotta, J. L. Shinn, F. F. Badavi, S. Y. Chun, J. W. Norbury, C. J. Zeitlin, L. Heilbronn, and J. Miller, *"NUCFRG2: An Evaluation of the Semiempirical Nuclear Fragmentation Database."* NASA TP-3533, 1995.

18 T. Kurosawa, N. Nakao, T. Nakamura, H. Iwase, H. Sato, Y. Uwamino, and A. Fukumura, *Phys. Rev.* **C62**, 044615 (2000).

19. M.A. Jilany; *Nuclear Physics* **A705** 477–493 (2002).

The FLUKA code: description and benchmarking

G. Battistoni*, F. Cerutti†, A. Fassò**, A. Ferrari†, S. Muraro*, J. Ranft‡, S. Roesler† and P.R. Sala*

*INFN sezione di Milano, via Celoria 16, I-20133 Milano, Italy
†CERN, CH-1211 GENEVE 23, Switzerland
**SLAC, Stanford, USA
‡Siegen University, Germany

Abstract. The physics model implemented inside the FLUKA code are briefly described, with emphasis on hadronic interactions. Examples of the capabilities of the code are presented including basic (thin target) and complex benchmarks.

Keywords: Monte Carlo simulations, calorimetry, nuclear interactions
PACS: 02.70.Uu; 24.10.Lx; 87.53.Wz; 29.40.Vj

INTRODUCTION

The FLUKA code [1] is a general purpose Monte Carlo code for the interaction and transport of hadrons, heavy ions, and electromagnetic particles from few keV (or thermal energies for neutrons) to cosmic ray energies in whichever material. It is built and maintained with the aim of including the best possible physical models in terms of completeness and precision. In this "microscopic" approach, each step has sound physical basis. Performances are optimized comparing with particle production data at single interaction level. No tuning whatsoever on "integral" data, like calorimeter resolutions, thick target yields etc, is performed. Therefore, final predictions are obtained with minimal free parameters, fixed for all energies and target/projectile combinations. Results in complex cases as well as scaling laws and properties come out naturally from the underlying physical models and the basic conservation laws are fulfilled "a priori". Moreover, the microscopic approach preserves correlations within interactions and among the shower components, and it provides predictions where no experimental data is directly available.

When needed, powerful biasing techniques are available to reduce computing time.

Descriptions of FLUKA models and extensive benchmarking can be found in the literature (see the web page, www.fluka.org). The development and maintenance of FLUKA are performed in the framework of an INFN–CERN agreement.

Old versions of the FLUKA hadronic models are still used by other codes. The 1993 version of the FLUKA hadronic physics [2], excluding the intermediate energy model, was interfaced to GEANT3 [3], and activated both by the so-called GEANT-FLUKA package at all energies, and by the GEANT-CALOR package for interactions above 5-10 GeV. The 1990 version has been made available to LAHET [4] and MCNPX [5] users. No upgrade of these interfaces has been performed and they have to be considered

CP896, *Hadronic Shower Simulation Workshop*
edited by M. Albrow and R. Raja
© 2007 American Institute of Physics 978-0-7354-0401-4/07/$23.00

obsolete; their results are not representative of the present performances of the FLUKA hadronic models. The CORSIKA simulation package [6] also includes the FLUKA hadronic routines as an option [7], but in this case the interface is kept up to date.

E.M. and muon transport in FLUKA

For historical reasons, FLUKA is best known for its hadron event generators, but since more than 17 years FLUKA can handle with similar or better accuracy electromagnetic effects [8]. Briefly, the energy range covered by this sector of FLUKA is very wide: the program can transport photons and electrons over about 12 energy decades, from 1 PeV down to 1 keV. The e.m. part is fully coupled with the hadron sector, including the low energy (i.e. < 20 MeV) neutrons. The simulation of the electromagnetic cascade in FLUKA is very accurate, including the Landau-Pomeranchuk-Migdal effect and a special treatment of the tip of the bremsstrahlung spectrum. Electron pairs and bremsstrahlung are sampled from the proper double differential energy-angular distributions improving the common practice of using average angles. In a similar way, the three-dimensional shape of the e.m. cascades is reproduced in detail by a rigorous sampling of correlated energy and angles in decay, scattering, and multiple Coulomb scattering.

Recently, since the FLUKA2005.6 version, the need for an external cross section preprocessor has been eliminated, integrating all the needed functionality into the initialization stage. At the same time, data from the EPDL97 [9] photon cross section library have become the source for pair production, photoelectric and total coherent cross-section tabulations, as well as for atomic form factor data.

Bremsstrahlung and direct pair production by muons are modeled according to state-of-the-art theoretical description and have been checked against experimental data [10, 11]. Muon photonuclear interactions are also modeled.

CHARGED PARTICLE TRANSPORT

Transport of charged particles is performed through an original Multiple Coulomb scattering algorithm [12], supplemented by an optional single scattering method. The treatment of ionization energy loss is based on a statistical approach alternative to the standard Landau and Vavilov ones that provides a very good reproduction of average ionization and of fluctuations [13]. Multiple scattering with inclusion of nuclear form factors is applied also to heavy ion transport. Up-to-date effective charge parameterizations are employed, and straggling of ion energy loss is described in "normal" first Born approximation with inclusion of charge exchange effects.

The precise determination of ion range and ionization losses is of utmost importance in dosimetry and in therapeutical applications. For this reason, FLUKA is being heavily benchmarked [37] against models and experimental data concerning ions beams of interest for hadrotherapy. In fig.1 an example of very nice agreement between Bragg peak calculations and data is shown. The contribution of fragmented ions is also evident after the peak.

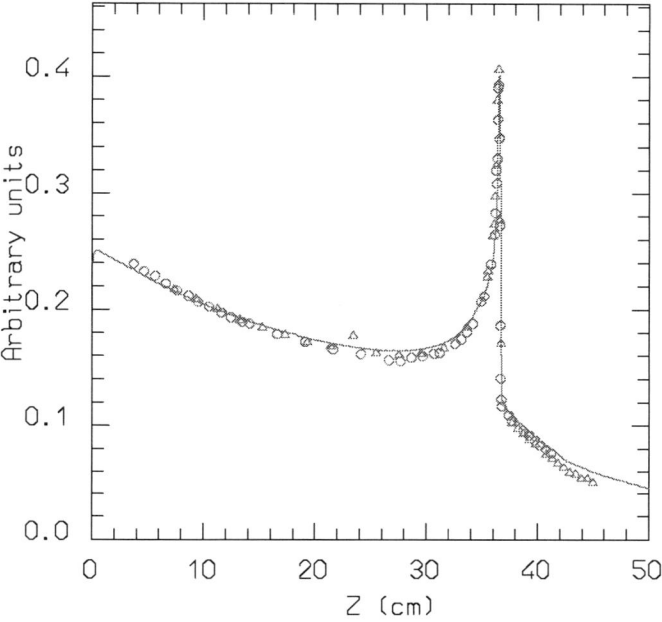

FIGURE 1. Dose versus depth distribution for 670 MeV/n ^{20}Ne ions on a water phantom. The symbols represent LBL (circles) and GSI (triangles) experimental data [38], the line is the prediction of FLUKA including the new BME interface. For the profile reproduction at large depths, nuclear interactions below 100 MeV/n play an important role.

FLUKA HADRONIC MODELS

A basic description of hadronic interactions in FLUKA and of their most recent developments can be found in [14, 15, 16]. Hadron-nucleon interactions at energies below a few GeV are simulated in FLUKA by the isobar model, through resonance production and decay, and by taking into account elastic, charge and strangeness exchange. Elementary hadron-hadron collisions at energies above a few GeV are described thanks to an implementation of the Dual Parton Model (DPM) [17], coupled to a hadronization scheme. This model allows a successful description of soft collision processes that cannot be addressed by perturbative QCD.

Hadron-hadron collisions are the main building blocks of hadron-nucleus collisions. Multiple collisions of each hadron with the nuclear constituents are taken into account by means of the Glauber-Gribov calculus [18, 19]. Particular efforts are devoted to the study of nuclear effects on hadron propagation. These are treated by the FLUKA nuclear interaction model called PEANUT [25, 26, 14, 16]. This model includes a Generalized IntraNuclear Cascade (GINC) with smooth transition to a pre-equilibrium stage performed with standard assumptions on exciton number or excitation energy.

GINC modeling in PEANUT is highly sophisticated. Different nuclear densities are

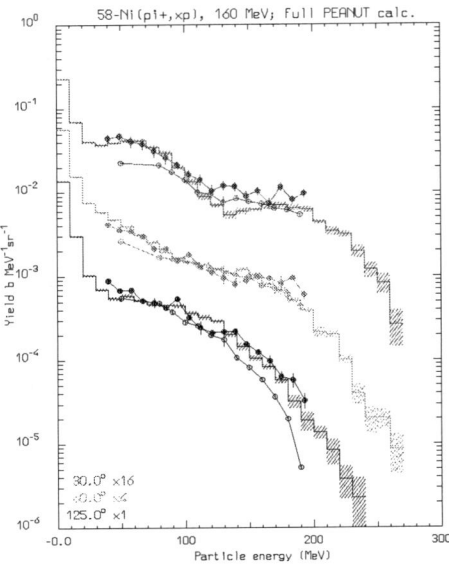

FIGURE 2. Emitted proton spectra at different angles, from 160 MeV π^+ on a nickel target. Histograms are FLUKA results, points are experimental data from [28, 29]. Note that proton spectra extend up to 300 MeV

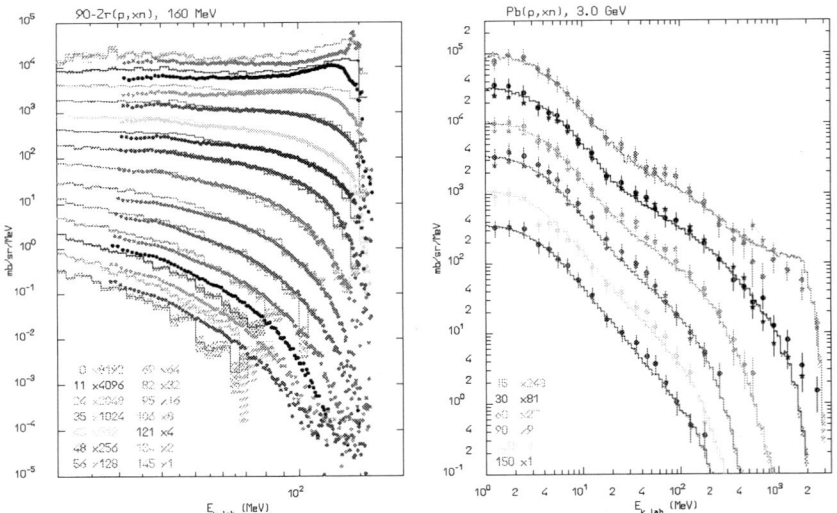

FIGURE 3. Emitted neutron spectra at different angles, from 160 MeV protons on Zr (left) and 3 GeV protons on Pb (right). Histograms are FLUKA results, points are experimental data from [30, 31].

adopted for neutrons and protons, Fermi motion is defined locally including wave packet-like uncertainty smearing, the curvature of particle trajectories due to the nuclear potential is taken into account, binding energies are obtained from mass tables and updated after each particle emission, energy-momentum conservation including the recoil of the residual nucleus is ensured. Quantum effects are explicitly included: Pauli blocking, formation zone, nucleon anti-symmetrization, nucleon-nucleon hard-core correlations, coherence length. Nuclear medium effects on the Δ resonance properties are accounted for when treating pion interactions[14, 26] and pion reinteractions in the nucleus (see fig.2 for an example)

The GINC step goes on until all nucleons are below a smooth threshold around 50 MeV, *and* all particles but nucleons (typically pions) have been emitted or absorbed. At the end of the GINC stage a few particles may have been emitted and the input configuration for the pre-equilibrium stage is characterized by the total number of protons and neutrons, by the number of particle-like excitons (nucleons excited above the Fermi level), and of hole-like excitons (holes created in the Fermi sea by the INC interactions), by the nucleus excitation energy and momentum. All the above quantities can be derived by proper counting of what occurred during the INC stage. The exciton formalism of FLUKA follows that of M. Blann and coworkers[20, 21, 22, 23], with some modifications:

- Inverse cross sections from systematics
- Correlation/formation zone/hardcore effects on reinteractions
- Constrained exciton state densities for the configurations 1p-1h, 2p-1h, 1p-2h, 2p-2h, 3p-1h and 3p-2h
- Energy dependent form for the single particle density g_x [24]
- Starting values for the position dependent parameters given by the point like ones as obtained out of the GINC stage.
- Angular distributions of emitted particles in the fast particle approximation

For further details see ref.[14].

PEANUT has proved to be a precise and reliable tool for intermediate energy hadron-nucleus reactions. Its "nuclear environment" is also used in the modelization of (real and virtual) photonuclear reactions, neutrino interactions, nucleon decays, muon captures.

Examples of PEANUT results on neutron production from low energy proton interactions are shown in fig.3. These benchmarks are of high relevance for, for instance, calorimetry. Indeed, even in showers initiated by high energy projectiles, most of the interactions occur at medium-low energies, and the amount of visible energy depends critically on the energy balance and neutron balance in low energy reactions.

Emission of energetic light fragments through the coalescence process is included all along the PEANUT reaction chain. This allows to reproduce the high energy tail of the light fragment spectra, as in fig.4

A major improvement carried out in the last months has been the extension of PEANUT to cover the whole energy range, replacing the simplified intranuclear cascade model that was used for projectile energies larger than 5 GeV. Results obtained with the latest FLUKA version are shown in figs.5, 6 and 7

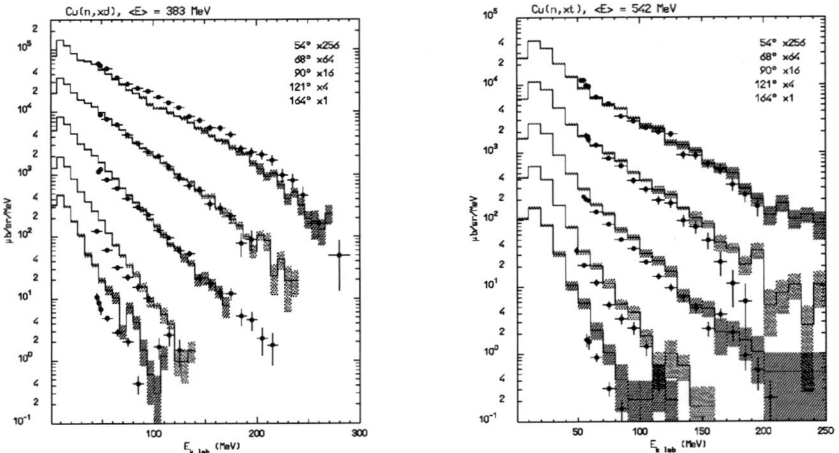

FIGURE 4. Deuteron (left) and triton (right) emission from 383 MeV and 542 MeV neutrons on Cu. respectively (exp data from [27])

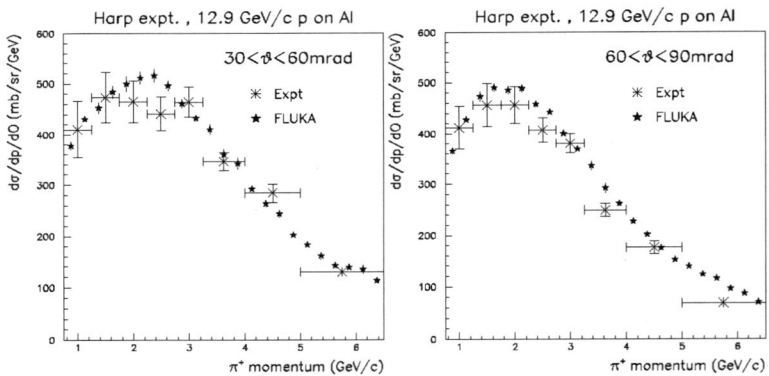

FIGURE 5. Computed π^+ double differential production cross section for 12.9 GeV/c protons on Aluminum for different angular ranges, compared with experimental data [32].

The final steps of the reaction include evaporation in competition with fission and gamma deexcitation. For light nuclei, a Fermi break-up model is implemented. These equilibrium processes are critical for a correct calculation of residual nuclei distributions. This topic is obviously important for activation and residual dose rate studies, it is also indirectly important for calorimetry: since the energy spent in breaking nuclear bonds is a major source of non-compensation and spread in energy deposition, a correct reproduction of residual nuclei distribution is a proof that binding energy losses are correctly taken into account. The FLUKA evaporation model, which is based on the Weisskopf-Ewing approach, has been continuously updated along the years, with the inclusion, for instance, of sub-barrier emission, full level density formula, analytic so-

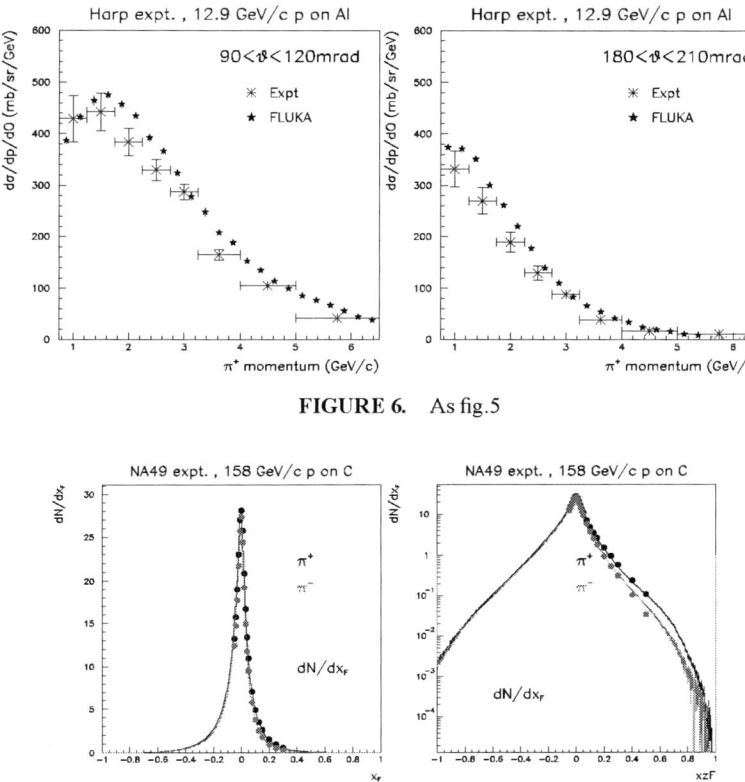

FIGURE 6.   As fig.5

FIGURE 7.   Feynman−x distributions for π^+ and π^- production for proton interactions on Carbon at 158 GeV/c , as measured by NA49 [39] (symbols) and predicted by FLUKA (histograms). Linear scale on the left, logarithmic scale on the right.

lution of the emission widths, evaporation of nuclear fragments up to $A \leq 24$. Recent improvements in the treatment of fission and in the adopted level densities were particularly effective for the description of residual nuclei production from heavy targets. An example of the present code capabilities is shown in fig.8. More complex benchmarks have been carried out at the CERF[41] facility at CERN. Samples of different materials have been irradiated with a mixed hadron field with broad energy spectrum. Comparison of activation and dose rate curves with FLUKA simulations [42] show very nice agreement, as for example in fig9.

LOW ENERGY NEUTRONS

Transport of neutrons with energies lower than 19.6 MeV is performed in FLUKA by a multigroup algorithm. The multi-group technique, widely used in low-energy

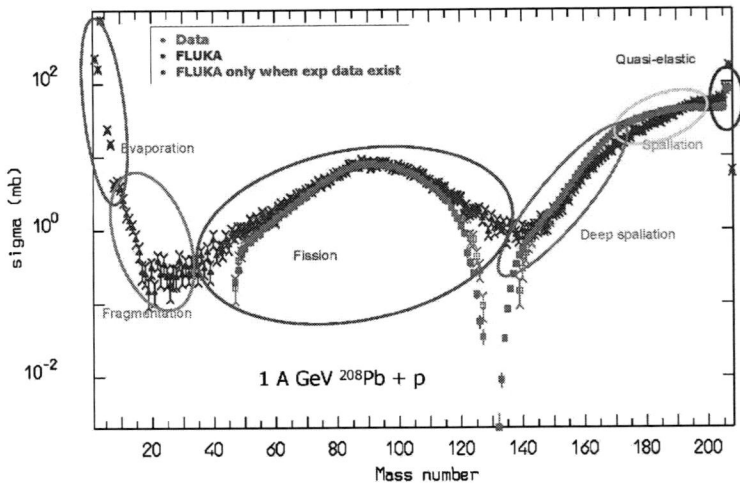

FIGURE 8. Residual nuclei production from 1 GeV protons on Lead. Data from [40]

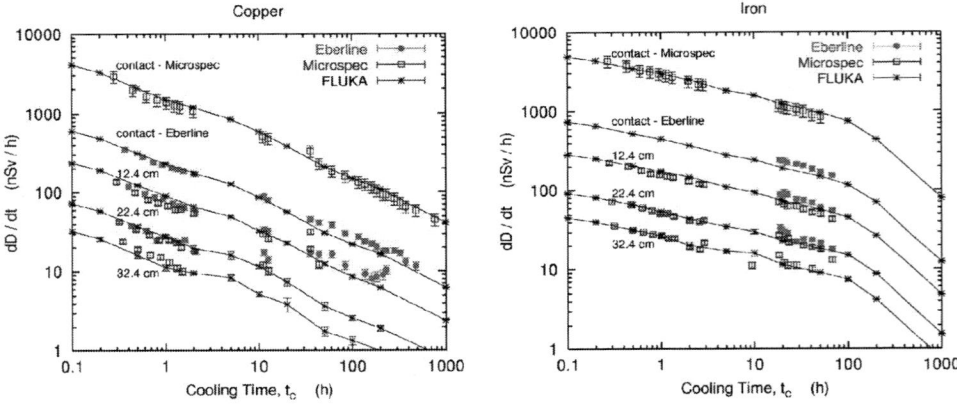

FIGURE 9. Dose rate as a function of cooling time for different distances between sample and detector. Left:Copper sample. Right: iron sample, both irradiated at the CERF facility at CERN. Adapted from [42]

neutron transport programs, consists in dividing the energy range of interest in a given number of intervals ("energy groups"). In the FLUKA cross-section library, the energy range is divided into 72 energy groups of approximately equal logarithmic width (one of which is thermal). The angular probabilities for inelastic scattering are obtained by a discretisation of a P5 Legendre polynomial expansion For a few isotopes only, neutron transport can be done also using continuous (pointwise) cross-sections. For ^{1}H, ^{6}Li and ^{10}B, it is applied as a user option (above 10 keV in ^{1}H, for all reactions in ^{6}Li, and only for the reaction ^{10}B(n,tγ)^{4}He in ^{10}B). For the reaction ^{14}N(n,p)^{14}C, pointwise neutron transport is always applied. In general, gamma generation by low-energy neutrons (*but not gamma transport*) is treated in the frame of a multigroup scheme too. A downscattering matrix provides the probability, for a neutron in a given energy group, to generate a photon in each of 22 gamma energy groups, covering the range 10 keV to 20 MeV. In all cases, the generated gammas are transported in the same way as all other photons in FLUKA.

NUCLEUS-NUCLEUS INTERACTIONS

Nucleus-nucleus interactions up to 10000 TeV/n are performed through interfaces with external generators. The interface with a modified version of rQMD-2.4 [34, 43] is used for energies below 5 GeV/n. The DPMJET-III [33] code is used from this energy up to the maximum supported one. It is worth mentioning that the external generators, as well as the ones under development, share the same evaporation/deexcitation stage developed for hadron-nucleus interactions. Examples of results can be found in [43, 44]

Work is in progress to complement and eventually substitute the rQMD interface with new QMD codes developed by the FLUKA collaboration[35]. A non-relativistic QMD model has already been interfaced to FLUKA and tests on thin and thick target data are in progress.

The implementation of a Boltzmann Master Equation (BME)[36] model for very low energies is in progress.

GEOMETRY

Transport in arbitrarily complex geometries, including magnetic field, can be accomplished using the FLUKA combinatorial geometry. A suitable voxel geometry module allows to model properly CT scans or other detailed 3D representations of human beings, typically for dosimetry or therapy planning purposes.

FLUGG : the GEANT4 geometry Interface

FLUGG[45] (FLUKA with GEANT4 Geometry) is an extension of FLUKA that uses the GEANT4 geometry package to build the geometry, find the particle locations and boundary interceptions. It provides a more flexible geometry than the default one and allows to run FLUKA using geometry inputs in the GEANT4 format. It has been tested on HP and Linux platforms for single level and multi-level geometries, for neutral and charged particles, including biasing options and magnetic field. An input user interface has been developed, while the output is in FLUKA format.

APPLICATION TO CALORIMETERS

The use of calorimeter data as a benchmark for Monte Carlo codes is a common bad habit. Calorimeters are complex objects, where many physical and instrumental effects are deeply entangled, and can easily mask the goodness or deficiencies of simulation models. Indeed, benchmarking and optimization of codes should be performed on clean data, like thin target or more complex benchmarks in controlled conditions. In this way, codes can then be used to optimize calorimeters design and understand their response. Of course, this does not prevent investigation on the Monte Carlo side in case of disagreement!

FLUKA has been used to simulate calorimeter response, in the framework of the ATLAS and ICARUS collaborations. A few examples and discussion are given in the following subsections.

The ATLAS tile calorimeter

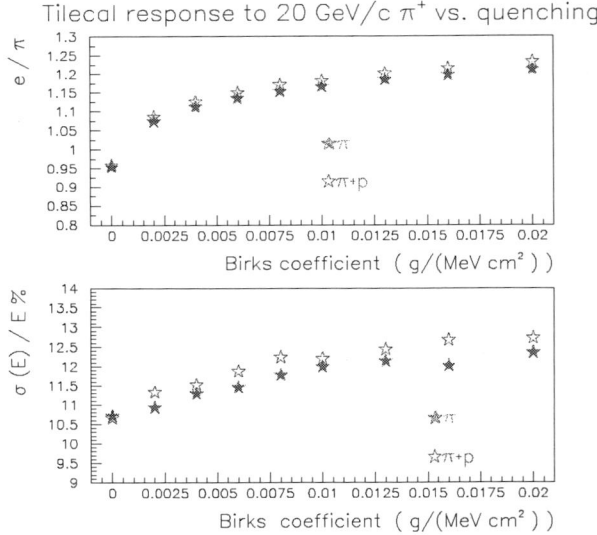

FIGURE 10. e/π response (top) and fractional energy resolution (bottom) of the tile calorimeter as a function of the Birks quenching parameter. Simulations for 20 GeV/c π^+ (full red stars) and 20 GeV/c mixed π^+ + proton beam (empty blue stars), at 20° incidence angle.

The ATLAS hadronic calorimeter is composed of scintillator tiles in an iron structure. Several test beams were carried over in the SPS secondary beam lines before the production of the final modules. The results reported here refer to the 1994 setup of the test modules, as reported in [46]. An ensemble of 5 modules was exposed to positron and positive pion beams, with momentum varying from 20 to 300 GeV/c at various angles of incidence. Simulations performed here refer to 20° incidence. The calorimeter geome-

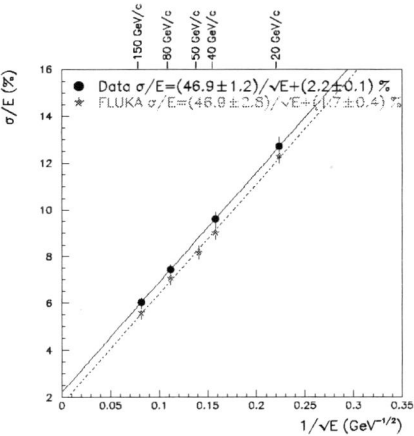

FIGURE 11. Fractional energy resolution as a function of beam energy, for a π^++proton beam at 20°. Experimental data (dots) from [46].

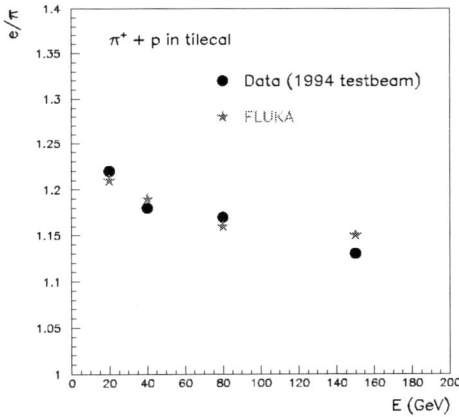

FIGURE 12. As in fig 11 for the relative response to electrons and hadrons.

try was faithfully reproduced, and instrumental effects were included in the calculation. Namely, photostatistics has been convoluted offline as well as random cell-to-cell miscalibration. Signal quenching in scintillator has been simulated on line at each energy deposition. The effect of signal quenching, being proportional to the specific energy loss, is to suppress the signal from slow hadrons and heavy particles. Therefore, it has a strong influence on compensation and energy resolution. As shown in figure10, the iron-scintillator combination would be even over-compensating in the absence of signal quenching, due to neutrons produced in the iron layers and scattering on hydrogen. The e/π signal ratio can vary of more than 10% when the quenching parameter is varied over a "reasonable" range of values. Variations of the same order affect the fractional energy

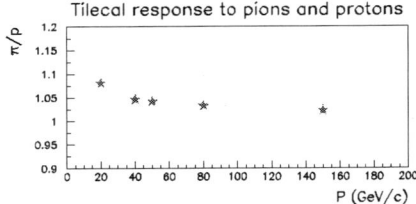

FIGURE 13. Simulated relative response to π^+ and protons of the tile calorimeter, as a function of energy. Beam at 20° incidence, tile configuration as in [46]

resolution. In the following, a value of $1.30 \cdot 10^{-2}$ $(MeV/(g/cm^2))^{-1}$ has been adopted for the Birks parameter. Simulations also included the proton contamination present in the positive pion beam. This contamination has been both evaluated by FLUKA simulations, and measured with Cerenkov counters during a later test beam[47]. The effect of the proton contamination for the 20 GeV beam is also shown in fig.10 for the 20 GeV/c beam. The effect on resolution is more than 6% for the adopted quenching parameter. The final results of simulations for energy resolution and e/π ratio are shown in figs.11 and 12. A very nice agreement with data is visible.

In pure beams, the simulated response to pions is higher than that to protons, as measured experimentally (see [47] with a different tile configuration) and decreases with energy (see fig.13). Energy resolution is worse for pions (about 10% higher) than for protons, also in agreement with [47].

The ATLAS electromagnetic calorimeter

The ATLAS electromagnetic (E.M.) calorimeter is a lead-liquid argon ionization chamber, with electrodes and absorbers shaped in accordion. Its performances were studied in many test-beams[48]. Comparisons of the response of the E.M. calorimeter to electrons and positrons with FLUKA simulations are very good, both for what concerns energy resolution and shower shape. The calculated energy resolution is $\frac{\sigma_{FLUKA}}{E} = \frac{9.2 \pm 0.3\%}{\sqrt{E}}$, vs an experimental one $\frac{\sigma_{Exp}}{E} = \frac{9.8 \pm 0.4\%}{\sqrt{E}}$. In fig. 14 the dependence of the calorimeter response to the beam impact position is shown. The azimuthal modulation due to the accordion cell structure is well reproduced.

The ATLAS combined calorimeter test beam

Prototypes of the ATLAS electromagnetic and hadronic calorimeters were tested together in 1994 and 1996 with pion beams. The experimental set-up is described in [49, 50], as well as comparisons with simulations. Simulated data were analyzed exactly in the same way as the experimental one, after event-by-event convolution of electronic noise. Calibration was done on an absolute scale. Results for e/π ratios and longitudinal

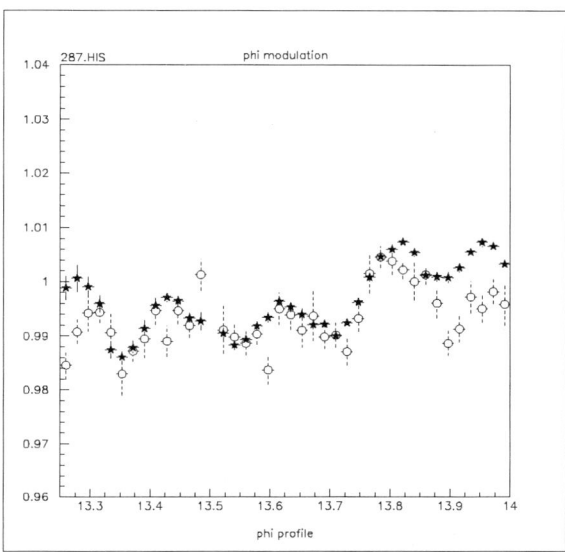

FIGURE 14. Modulation of the response of the accordion E.M. calorimeter to a 287 GeV electron beam, as a function of beam impact position. Stars are FLUKA simulations, dots are experimental data. Abscissa values are in ϕ cell units, one cell spanning about 2.5 cm

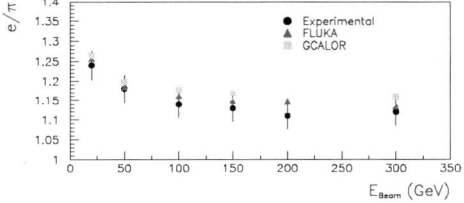

FIGURE 15. Experimental (dots) and simulated (triangles: FLUKA, squares:GCALOR) relative response to electrons and hadrons of the ATLAS combined calorimeter test beam, as a function of beam momentum (adapted from [49])

shower development are reproduced in figs.15 and fig.16. Energy resolution deserves a few comments, demonstrating that care must be taken in considering instrumental effects. In the 1994 test beam a significant discrepancy between measured and simulated energy resolution at 20 GeV/c was found (see fig.17). However, this discrepancy had nothing to do with the physics model used in the simulations, rather it was due to the incomplete knowledge of the beam line characteristics. Indeed, in the 1996 test beam, where the amount of dead materials upstream the detectors was better controlled, the resolution at 20 GeV/c was better and came very near to the simulated value.

In fig.17 two sets of simulation results are presented. The two sets differ only for the details of the algorithm chosen for the reconstruction of the preshower detector response. This detector, a thin lead-liquid argon layer in front of the E.M. calorimeter, was used

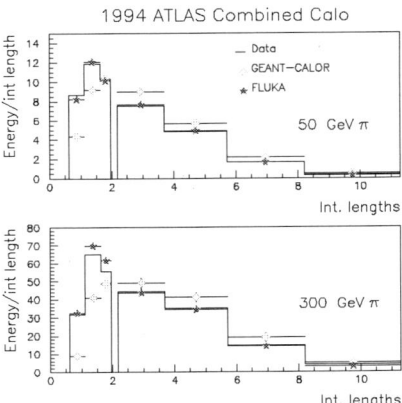

FIGURE 16. Longitudinal shower development in the ATLAS combined calorimeter set-up, for 100 GeV/c (top) and 300 GeV/c (bottom) pions. Adapted from [49]

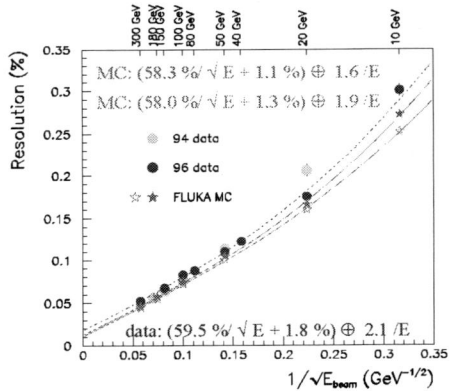

FIGURE 17. Fractional energy resolution as a function of beam momentum for the ATLAS combined calorimeter set-up. Two data sets (dots) correspond to two data takings[49, 50]. Stars: FLUKA simulations with two different preshower reconstruction algorithms.

to veto events with interactions upstream the detector. It is evident that the resolution at low beam energies is heavily affected by the presence of "dirty" events.

COSMIC RAY SHOWERS

The application of FLUKA to simulations of cosmic rays showers in the atmosphere began may years ago, with the initial aim to calculate atmospheric neutrino fluxes[56, 57, 58, 59]. This implied the development of more tools, namely:

- Primary spectra from $Z = 1$ to $Z = 28$ derived from NASA and updated to most recent measurements (the results of AMS [51] and BESS [52] for proton and helium) and modulated for a given date, according to solar modulation.
- A spherical representation of the earth geometry with the surrounding atmosphere up to 70 km a.s.l.
- The MSIS (Mass-Spectrometer-Incoherent-Scatter) [54] atmospheric model. The atmosphere is layered in 100 shells with a density scaling according to the chosen profile as a function of height.
- A solar modulation model as taken from [53].
- A geomagnetic model, whose degree of complexity can be varied, according to the difficulty of the problem, from a simple dipolar approximation to the spherical harmonic expansion of IGRF[55].

The first important result was that a full 3-dimensional simulation of neutrino fluxes[57] proves to be significantly different from the customary mono-dimensional approximation.

Data on different shower components have been used to benchmark the code[60, 61, 62]. Among the latest calculations[63], we show in fig18 comparisons with experimental muon fluxes at Mount Norikura ($\sim$2700 m a.s.l., 740g/cm^2, 11.2 GV) and at CERN.

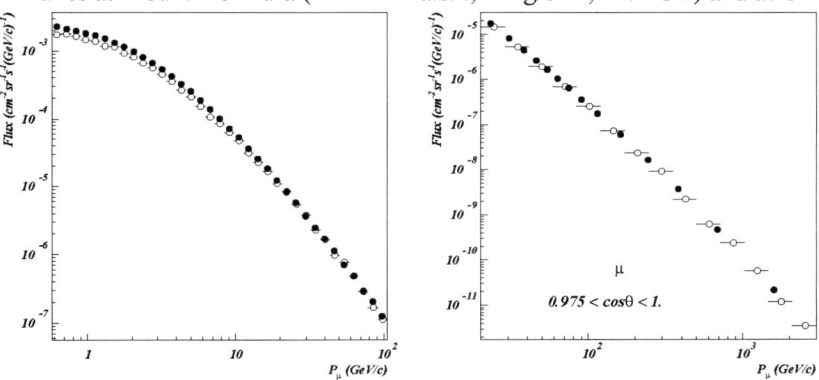

FIGURE 18. left: comparison of the simulated μ^- flux (open symbols) with the BESS[64] experimental data (full symbols) at Mt. Norikura (2700 m a.s.l.). Right: the comparison of simulated $\mu^\pm$ flux at sea level (open symbols) with the L3+C[65] experimental data ($0.975 < cos\theta_z < 1$, full symbols).

Reliable simulations of particle fluxes in the atmosphere are also needed for the evaluation of dose rates on aircrafts. Comparisons of FLUKA prediction with measured doses on commercial aircrafts can be found in [66, 67]. As an example, fig.19 shows the dose equivalent rate on two different aircraft routes. The striking difference between

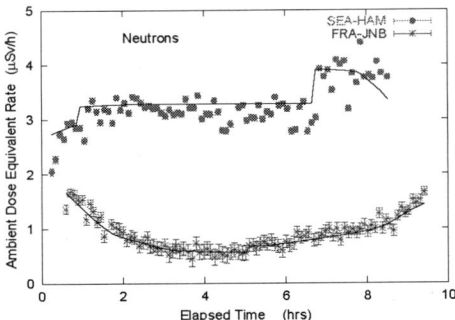

FIGURE 19. Ambient dose equivalent from neutrons measured during solar maximum on commercial flights from Seattle to Hamburg and from Frankfurt to Johannesburg, as function of time after take-off (symbols, exp. data, Lines: FLUKA).

polar and equatorial routes is evident, as well as the agreement between simulations and measurements.

NEUTRINO BEAMS AND INTERACTIONS

Many of the improvements to the hadronization model described in [15] were aimed to a better description of hadron production in the CERN Neutrino to GranSasso (CNGS) beam. All simulations for the CNGS facility, from energy deposition to neutrino spectra, are performed with an integrated simulation set-up based on FLUKA[68, 69].

Neutrino interactions are simulated in the framework of the PEANUT nuclear model. Quasi-elastic interactions are generated directly by FLUKA, while for DIS interactions the neutrino-nucleon generator NUX[70] has been interfaced. Work is in progress to develop an internal generator for DIS and resonant reactions.

Nuclear effects on Quasi-Elastic interactions has been tested by comparison with data from the ICARUS 50l prototype exposed to the CERN WANF neutrino beam[71]. In the analysis, "golden" events were selected, i.e. events with only one muon and one identified proton. The input spectrum for the FLUKA event simulation was the one calculated for the NOMAD experiment[72], normalized to the integrated experimental beam intensity. The expected number of quasi-elastic events resulted to be 400, of which 16% were selected as "golden". A 20% background came from DIS events passing the "golden" cuts. As a total, $80 \pm 9(\text{stat.}) \pm 13(\text{syst.})$ were expected, to be compared with 86 events observed. The effect of the nuclear environment on these events is visible in fig20, where the missing transverse momentum distribution is plotted. Dots are the experimental results, the various histograms correspond to simulated quantities. The dashed histogram shows the expected p_{Tmiss} distribution when only the Fermi motion of the target nucleons is taken into account. The distribution is broadened and acquires a high p_{Tmiss} tail when reinteractions inside the nucleus are correctly simulated (dotted histogram), and the agreement with data is further improved when the misidentified (hatched) DIS events are added, summing up to the full response (full line histo). It should be pointed out that the ability to preserve correlations among interaction products is an essential feature in order to reproduce the experimental selection cuts. This is one of the many examples where microscopic models cannot be substituted by data parametrizations.

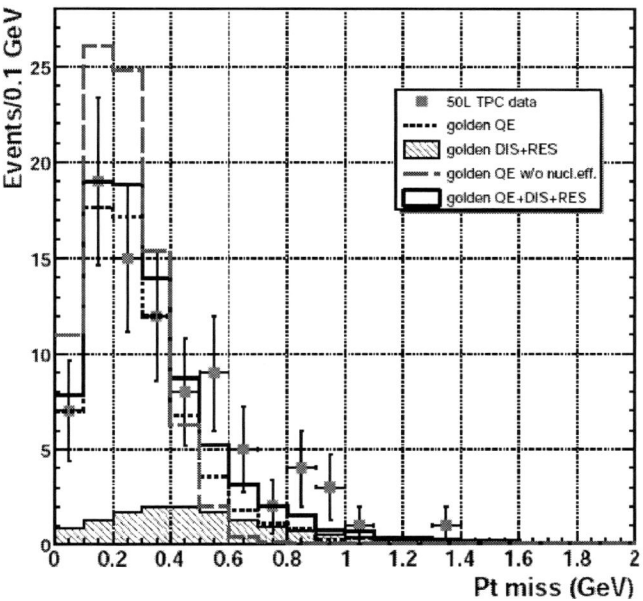

FIGURE 20. $p_{T miss}$ distribution of experimental and simulated events in the ICARUS 50l prototype (from [71]). See text.

CONCLUSIONS

FLUKA is a multiparticle transport and interaction Monte Carlo code, able to work both in analog and biased mode. Its physical models are continuously upgraded and benchmarked against experimental data. It has a wide range of applications, in particle physics but also in accelerator design and shielding, dosimetry, radiation protection, hadrotherapy. In particular, it has proven capabilities in the simulation of calorimeters and neutrino beams.

ACKNOWLEDGMENTS

This work has been carried out in the framework of the FLUKA collaboration. Part of this work was supported by Department of Energy contract DE-AC02-76SF00515.

REFERENCES

1. A. Ferrari, P.R. Sala, A. Fassò and J. Ranft, CERN 2005-10 (2005), INFN/TC_05/11, SLAC-R-773; A. Fassò et al., Computing in High Energy and Nuclear Physics 2003 Conference (CHEP2003), La Jolla, CA, USA, March 24-28, 2003, (paper MOMT005) eConf C0303241 (2003), arXiv:hep-ph/0306267.
2. A. Fassò, A. Ferrari, J. Ranft, and P.R. Sala, " Proc. of the IV Int. Conf. on Calorimetry in High Energy Physics", La Biodola, (Elba), A. Menzione and A. Scribano eds., World Scientific, p. 493 (1994).
3. R. Brun et al., "GEANT3", CERN DD/EE/84-1 (1987).
4. R.E. Prael and H. Lichtenstein, Los Alamos report **LA–UR–89–3014** (1989)
5. J. Hendricks et al., Los Alamos report **LA-UR-05-2675** (2005).
6. D. Heck et al., Forschungszentrum Karlsruhe Report FZKA 6019 (1998). `http://www-ik.fzk.de/corsika/Welcome.html`
7. D. Heck et al.Forschungszentrum Karlsruhe Report FZKA-6890ZC, Prepared for *28th International Cosmic Ray Conferences (ICRC 2003), Tsukuba, Japan, 31 Jul - 7 Aug 2003*
8. A. Fassò, A. Ferrari, and P.R. Sala, *Proceedings of the MonteCarlo 2000 Conference* A. Kling, F. Barão, M. Nakagawa, L. Távora, P. Vaz eds., Springer-Verlag Berlin, 159 (2001).
9. D.E. Cullen et al.,"EPDL97: the Evaluated Photon Data Library, '97 Version", UCRL–50400, Vol. 6, Rev. 5, 1997.
10. M. Antonelli *et al.*, *Proc. VI Int. Conf. on Calorimetry in High Energy Physics (Calor 96)*, Frascati Physics Series Vol. VI (1997), pp. 561-570
11. G. Battistoni et al., *Nucl. Instr. Meth.* **A394** , 136–145 (1997)
12. A. Ferrari et al., *Nuclear Instr. Meth.* **B71**, 412-426 (1992).
13. A. Fassò et al.in *Proceedings of SARE-3*, H. Hirayama ed., KEK report Proceedings 97-5, 32 (1997).
14. A. Ferrari, and P.R. Sala, *Proceedings of Workshop on Nuclear Reaction Data and Nuclear Reactors Physics, Design and Safety*, A. Gandini, G. Reffo eds., **2**, 424 (1998).
15. G. Collazuol, A. Ferrari, A. Guglielmi, P.R. Sala, *Nucl. Instr. Meth.* **A449**, 609 (2000).
16. G. Battistoni F. Cerutti, A. Ferrari et al., *Proc. 11th Int. Conf. on Nuclear Reaction Mechanisms*, Ric. Scient. ed Ed. Perm. Suppl. , E. Gadioli ed., **126**, 483 (2006).
17. A. Capella, U. Sukhatme, C.-I. Tan and J. Tran Thanh Van, *Phys. Rep* **236**, 225 (1994).
18. R.J. Glauber and G. Matthiae, *Nucl. Phys.* **B21**, 135 (1970).
19. V.N. Gribov, *Sov. Phys. JETP* **29**, 483 (1969).
20. M. Blann, *Phys. Rev. Lett.*, **27**, 337 (1971).
21. M. Blann, *Phys. Rev. Lett.*, **28**, 757 (1972).
22. M. Blann and H.K. Vonach, *Phys. Rev.* **C28**, 1475 (1983).
23. M. Blann, *Phys. Rev.* **C28**, 1648 (1983).
24. S. Shlomo, *Nucl. Phys.* **A539**, 17 (1992).
25. A. Ferrari and P. R. Sala, *Proc. MC93 Int. Conf. on Monte-Carlo Simulation in High-Energy and Nuclear Physics*, World Scientific ed., 277 (1994).
26. A. Fassò, A. Ferrari, J. Ranft, and P. R. Sala, *Proceedings of the "Specialists' Meeting on Shielding Aspects of Accelerators, Targets & Irradiation Facilities"*, p. 287-304 (1995).
27. J. Franz, P. Koncz, E. Rössle, *Nucl. Phys.* **A510**, 774 (1990).
28. Burger et al., *Phys. Rev.* **C41**, 2215 (1990)
29. McKeown et al., *Phys. Rev.* **C24**, 211 (1981)
30. Scobel et al., *Phys. Rev.* **C41** , 2010 (1990);
31. Ishibashi et al., *Nucl. Sci. Technol.* **32**, 827 (1995).
32. The HARP Collaboration, *Nucl. Phys.* **B732**, 1 (2006).
33. S. Roesler, R. Engel and J. Ranft, *Proceedings of the MonteCarlo 2000 Conference*, , A. Kling, F. Barão, M. Nakagawa, L. Távora, P. Vaz eds., Springer-Verlag Berlin, 1033 (2001).
34. H. Sorge, H. Stöcker, and W. Greiner, *Nucl. Phys.* **A498**, 567c (1989).
35. M.V. Garzelli et al., (2005) *Journ. of Phys. : Conference Series* **41** , 519 (2006)
36. F. Cerutti et al., *Journ. of Phys. : Conference Series* **41** , 212 (2006)
37. F. Sommerer, et al.*Physics in Medicine and Biology* **51**, 4385 (2006)
38. L. Sihver et al., *Jpn. J. Med. Phys.* **18**, 1 (1998).
39. C. Alt *et al.*, arXiv:hep-ex/0606028

40. T. Enqvist et al., *Nucl. Phys.* **A 686** 481 (2001).
41. C. Birattari et al., *Radiat. Prot. Dosim.* **76**, 135 (1988)
42. M. Brugger et al., *Radiat. Prot. Dosim.* **116**, 12-15 (2005).
43. V. Andersen et al., *Adv. Space Res.* **34**, 1302 (2004)
44. H. Aiginger, et al., *Adv. Space Res.* **35**, 214-222 (2005)
45. M. Campanella et al., ATL-SOFT-98-039 (1998), ATL-SOFT-99-004 (1999).
46. F. Ariztizabal et al., *Nucl. Instr. Meth.* **349**, 384 (1994)
47. CERN internal note ATL-TILECAL-2001-005
48. D.M. Gingrich et al., *Nucl. Instr. Meth.* *A364*, 290 (1995)
49. Z. Ajaltouni et al. *Nucl. Instr. Meth.* **A 387**, 333 (1997)
50. S. Akhmadaliev et al.*Nuclear Instruments & Methods A*, **449**, 461-477 (2000).
51. J. Alcaraz et al., *Phys. Lett.* **B490** (2000) 27; J. Alcaraz et al., *Phys. Lett.* **B494** (2000) 193.
52. T. Sanuki et al., *Astrophys. J.* **545** (2000) 1135; K. Abe et al., Phys. Lett. **B564** (2003) 8.
53. G.D. Badhwar, P.M. O Neill, *Adv. Space Res.* **17** n. 2 (1996) 7.
54. MSIS-E-00 Atmosphere Model. Available from: `http://nssdc.gsfc.nasa.gov/space/model/models/msis.html`
55. see `http://swdcwww.kugi.kyoto-u.ac.jp/igrf/index.html`
56. G. Battistoni et al.,*Nucl. Phys. B (Proc. Suppl.)*, **70**, 358 (1998).
57. G. Battistoni et al. ,*Astropart. Phys.*, **12**, 315-333 (2000).
58. G. Battistoni et al.,*Astropart. Phys.* **19**, 269 (2003)
59. G. Battistoni, A. Ferrari, T. Montaruli and P. R. Sala, *Astropart. Phys.* **23** (2005) 526.
60. G. Battistoni, A. Ferrari, T. Montaruli and P. R. Sala, *Astropart. Phys.* **17**, 477-488 (2002).
61. P. Zuccon et al., *Astropart. Phys.* **20**, 221 (2003).
62. S. Muraro, PhD thesis, Milan University (2006)
63. G. Battistoni et al., "Atmospheric muon simulation using the FLUKA MC Model" *proceedings of the NOW06 conference*, in press (2006)
64. T. Sanuki et al. *Physics Letters* **B54**, n. 3-4, (2002) 234;
65. P. Le Coultre, *Proc. of 29th International Cosmic Ray Conference* Pune (2005),00, 101-106 P. Le Coultre and the L3+C collaboration, *Nuclear Physics B - Proc. Supplements*, **151**, n.1 314 (2006).
66. A. Ferrari et al., *Rad. Prot. Dosim.* **93**, 101 (2001) G. Battistoni, A. Ferrari, M. Pelliccioni and R. Villari, *Rad. Prot. Dosim.* **112**, 331-343 (2004); *Adv. Space Res.* **36**, 1645-1652 (2005).
67. S. Roesler, W. Heinrich and H. Schraube, *Radiation Research* 149, 87 (1998); *Radiation Protection Dosimetry* **98**, 367-388 (2002)
68. A. Ferrari et al., CERN-AB-Note-2006-038 (2006).
69. A. Ferrari et al., *CNGS neutrino beam: from CERN to GranSasso*, presented at the NOW06 conference.
70. A. Rubbia "NUX- neutrino generator", *1st Workshop on Neutrino - Nucleus Interactions in the Few GeV Region (NuInt01)*, Tsukuba, Japan, 13-16 Dec 2001, `neutrino.kek.jp/nuint01/slide/Rubbia.1.pdf`
71. F. Arneodo et al.,*Phys. Rev. D 74*, 112001 (2006).
72. P. Astier et al., *Nucl. Instr. and Meth.* **A 515**, 800 (2003).

MARS15 Overview[1]

N.V. Mokhov[2] and S.I. Striganov

Fermilab, Batavia, IL 60510, U.S.A.

Abstract. MARS15 is a Monte Carlo code for inclusive and exclusive simulation of three-dimensional hadronic and electromagnetic cascades, muon, heavy-ion, and low-energy neutron transport in accelerator, detector, spacecraft, and shielding components in the energy range from a fraction of an electron volt up to 100 TeV. Main features of the code are described in this paper with a focus on recent developments and benchmarking. Newest developments concern inclusive and exclusive nuclear event generators, extended particle list in both modes, heavy-ion capability, electromagnetic interactions, enhanced geometry, tracking, histograming and residual dose modules, improved graphical-user interface, and other external interfaces.

Keywords: Hadrons, cascades, Monte Carlo, simulation
PACS: 13.85.-t, 24.10.Lx

INTRODUCTION

The MARS code system [1] is a set of Monte Carlo programs for detailed simulation of hadronic and electromagnetic cascades, muon, heavy-ion, and low-energy neutron transport in an arbitrary 3-D geometry of shielding, accelerator, detector, and spacecraft components with energy ranging from a fraction of an electron volt up to 100 TeV. It has been developed since 1974 at IHEP, SSCL and Fermilab. The current MARS15 version combines the well established theoretical models for strong, weak and electromagnetic interactions of hadrons, heavy ions and leptons with a system which can contain up to 10^5 objects, ranging in dimensions from microns to hundreds of kilometers in the same run. A setup can be made of up to 100 composite materials (those from 165 built-in or any user-defined), in presence of arbitrary 3-D magnetic and electric fields. A powerful user-friendly graphical-user interface (GUI) is used for visualization of geometry, materials, fields, particle trajectories, and results of calculations. MARS15 has five geometry options and flexible histograming options, can use MAD optics files as an input through a powerful MAD-MARS Beam Line Builder, and provides an MPI-based multiprocessing option, with various tagging, biasing and other variance reduction techniques.

There are quite substantial improvements in most of the code modules of the current version compared to those documented earlier. Some highlights are given below on the list of elementary particles and arbitrary heavy ions, their interaction cross-sections, inclusive and exclusive nuclear event generators, photohadron production, correlated

---

[1] Work supported by Fermi Research Alliance, LLC, under contract No. DE-AC02-07CH11359 with the U.S. Department of Energy.
[2] mokhov@fnal.gov

ionization energy loss and multiple Coulomb scattering, nuclide production, residual activation, and radiation damage (DPA). In particular, the details of a new model for leading baryon production and implementation of advanced versions of the Cascade-Exciton Model (CEM03) as well as the Los Alamos version of the Quark-Gluon String Model (LAQGSM03) are given.

Many people participated in the MARS code development over its 33-year history. Current contributors are N.V. Mokhov, S.I. Striganov, K.K. Gudima, C.C. James, M.A. Kostin, S.G. Mashnik, M.E. Monville, N. Nakao, I.L. Rakhno, and A.J. Sierk. Growing demand, diversity of applications and invaluable feedback from 300 MARS users worldwide motivate continuous developments of the code.

PARTICLES AND NUCLEAR CROSS-SECTIONS

A list of particles transported by MARS15 is extended to include neutral kaons, antineutrons and hyperons/antihyperons totaling in more than 40 species, all with their corresponding cross-sections, decay modes and electromagnetic processes. Arbitrary heavy ions with atomic mass A and charge Z are now fully treated by the code, including decays of unstable ions.

Total and elastic cross-sections of hadron-nucleon interactions for ordinary hadrons are described using corresponding fits to experimental data. Cross-sections for hyperon-nucleon interactions are described via the ordinary hadron cross-sections using the Additive Quark Model rules. At energies above 5 GeV, such an approach agrees well with data. At lower energies, the hyperon-nucleon cross-sections are very close to proton-nucleon ones. Hadron-nucleus total and inelastic cross-sections at energies above 5 GeV are calculated using the Glauber model. At lower energies, parameterizations to experimental data are used. For neutral kaons, cross-sections on both nucleon and nucleus targets are calculated using the relation based on an isospin and hypercharge conjugation. Total and inelastic cross-sections for heavy-ion nuclear interactions are based on the JINR model (see Ref. [2]). Photonuclear interaction cross-sections are described in great detail for all nuclei and energies from a few MeV up to 40 TeV using approximations from Ref. [3]. Fig. 1(a) gives an example of the neutron and heavy-ion cross-section descriptions in the code.

INCLUSIVE EVENT GENERATOR

The basic model for the original MARS program, introduced in 1974, came from Feynman's ideas concerning an inclusive approach to multiparticle reactions and *weighting* techniques. At each interaction vertex, a particle cascade tree can be constructed using only a fixed number of representative particles (the precise number and type depend on the specifics of the interaction), and each particle carries a statistical weight w, which is equal, in the simplest case, to the partial mean multiplicity of the particular event. Energy and momentum are conserved *on average* over a number of collisions. It was proved rigorously that such an estimate of the first moment of the distribution function is unbiased [5].

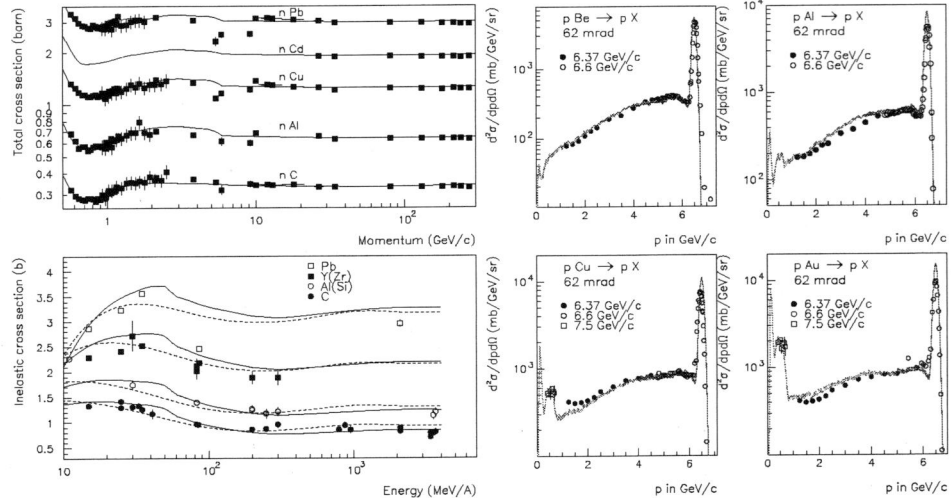

FIGURE 1. **Left:** Total neutron-nucleus cross-sections (top) and inelastic nuclear cross-sections of ^{12}C ions (bottom) as calculated in MARS15 (solid lines) *vs* data [4]; dashed lines are from a NASA model. **Right:** Invariant cross-sections for $pA \rightarrow pX$ reactions at 6.37 to 7.5 GeV/c on various nuclei in MARS15 (lines) *vs* data [6].

Inclusive particle production in nuclear interactions above 3 GeV is modeled in MARS15 using the following form for double differential distributions (taking $pA \rightarrow pX$ reaction as an example):

$$\frac{d^2N^{pA \rightarrow pX}}{dpd\Omega} = R^{pA \rightarrow pX}(A, E_0, p, p_\perp)\frac{d^2N^{pp \rightarrow pX}}{dpd\Omega} + F_{qel} + F_{ce}.$$

Differential cross-sections on a hydrogen target are described by a set of models and phenomenological formulae. For example, proton production in pp-collisions $\frac{d^2N^{pp \rightarrow pX}}{dpd\Omega}$, is described with a high accuracy in four kinematic regions of Feynman $x_F = |p_L^*/p_{max}^*|$:

1. Resonance region $x_F > 1 - 2.2/p_0$: a sum of five baryon resonances, Breit-Wigner formulae.
2. Diffractive dissociation region $1 - 2.2/p_0 < x_F < 0.9$: a triple-Reggeon formalism.
3. Fragmentation region $0.4 < x_F < 0.9$: a phenomenological model with a flat behavior on longitudinal and exponential one on transverse momenta.
4. Central region $0 < x_F < 0.4$: a fit to experimental data with normalizations at $x_F = 0$ and $x_F = 0.4$.

A nuclear modification factor $R^{pA \rightarrow pX}$ is known much better than the absolute yields, and its dependence on particle momenta $p_\perp$, p_0 and p is much weaker than for the

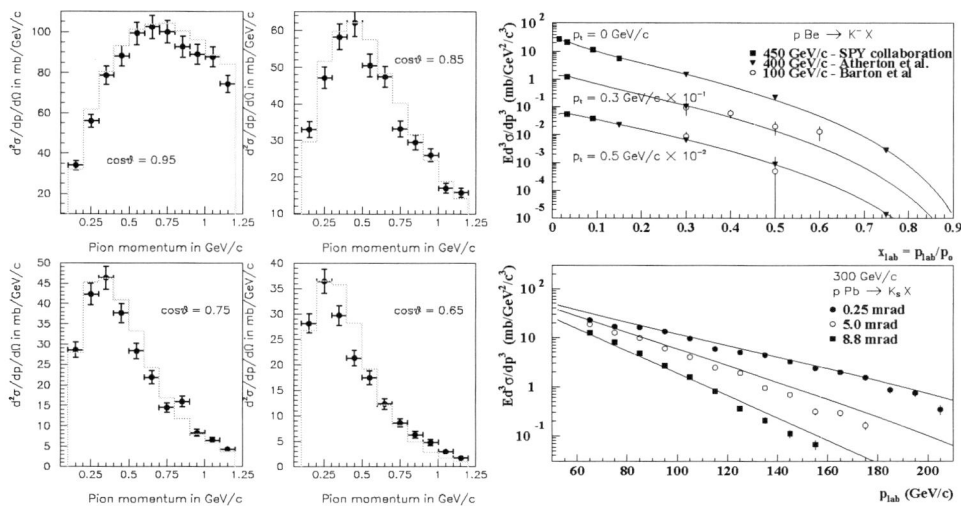

FIGURE 2. Invariant cross-sections for $pBe \rightarrow \pi^- X$ at 12.3 GeV/c [7] (left), $pBe \rightarrow K^- X$ at 100 to 450 GeV/c [8] and $pPb \rightarrow K_s X$ at 300 GeV/c [9] (right) vs MARS15; symbols are data and lines are calculations.

differential cross-sections themselves. For $pA \rightarrow pX$, it is presented in a factorized form

$$R^{pA \rightarrow pX} = F(p/p_0)(\frac{A}{2})^{\alpha p_T^2}(\frac{A}{9})^{\gamma(E_0)},$$

with a momentum dependence given by the Additive Quark Model. Appropriate formulae are used in the corresponding kinematic regions for meson production.

For all projectiles and secondary particles, quasi-elastic scattering (F_{qel}) and Fermi-motion are additionally simulated in this inclusive model and supplied with new phenomenological models for cascade/evaporation (F_{ce}), fission, and anti-nucleon production processes. Pion and charged kaon production phenomenological models have been improved. Recent additions to the inclusive event generator also include coupled nucleon / anti-nucleon production, neutral kaon production and heavy-ion nuclear interactions according to a superposition model. The later is used only for simplified non-exclusive modeling in a non-LAQGSM mode (see next section). A quality of the model for protons, pions and kaons is demonstrated in Fig. 1b, Fig. 2a and Fig. 2b, respectively.

EXCLUSIVE EVENT GENERATOR

A substantially improved Cascade-Exciton Model code, CEM03.01 [10] combined with the Fermi break-up model, the coalescence model, an improved version of the Generalized Evaporation-fission Model (GEM2), and a recent multi-fragmentation extension is used in MARS15 as a default for hadron-nucleus interactions below 5 GeV. It is a completely new, updated and modified version in comparison with its predecessors, not

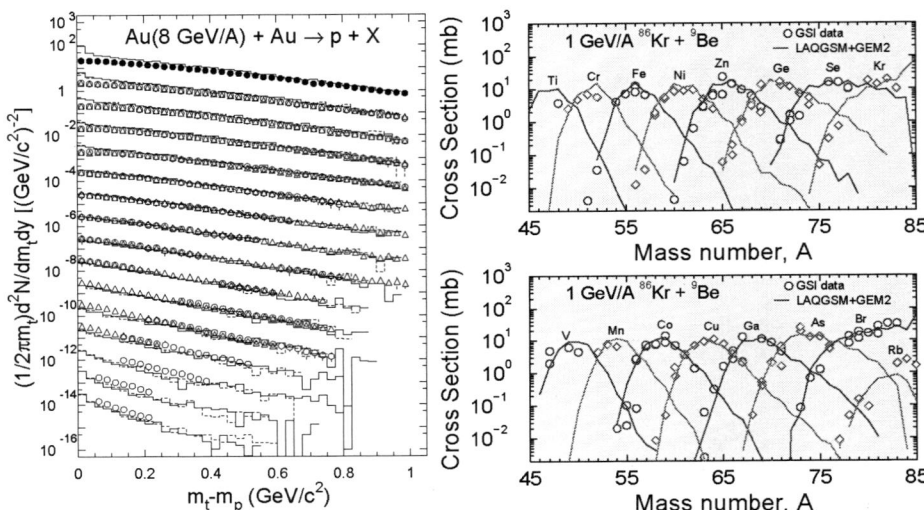

FIGURE 3. **Left:** Invariant proton yields per central $Au + Au$ collision at 8 GeV/A as calculated in MARS15 with LAQGSM03 (histograms) *vs* experiment [13]; solid lines and open circles is forward production, dashed lines and open triangles is backward production; mid-rapidity (upper set) is shown unscaled, while the 0.1 unit rapidity slices are scaled down by successive factors of 10. **Right:** Mass yield for $^{86}Kr + ^9Be$ as calculated with LAQGSM03 (lines) and measured in Ref. [14].

just an incremental improvement. CEM03.01 describes reactions induced by nucleons, pions and photons as a three-stage process: Intra-Nuclear Cascade (INC), followed by pre-equilibrium emission of particles during the de-excitation of the excited residual nuclei formed during the INC, followed by evaporation of particles from or fission of the compound nuclei. If the excited residual nucleus produced after the INC has a mass number $A < 13$, CEM03.01 uses a recently updated and improved version of the Fermi Break-up model to calculate its decay instead of considering a pre-equilibrium stage followed by evaporation from compound nuclei. CEM03.01 considers also coalescence of complex particles up to ^4He from energetic nucleons emitted during the INC.

The Los Alamos Quark-Gluon String Model code, LAQGSM03 [11], is implemented into MARS15 for particle and heavy-ion projectiles at 10 MeV/A to 800 GeV/A, including π^--capture and annihilation at rest. This provides a power of full theoretically consistent modeling of exclusive and inclusive distributions of secondary particles as well as spallation, fission, and fragmentation products. For quite some time, MARS has used the Dual-Parton Model code, DPMJET3 [12], for the very first vertex in a cascade tree. This is used in our numerous studies for the LHC 7×7 TeV collider and its detectors, and at very high energies up to 100 TeV. Fig. 3 shows results of MARS15/LAQGSM benchmarking at 1 and 8 GeV/A for proton and nuclide production in heavy-ion nuclear interactions.

ELASTIC SCATTERING AND LOW-ENERGY NEUTRONS

The elastic model at E<5 GeV is based on evaluated nuclear data from the LA-150 and ENDF/HE-VI libraries and from other sources (see Ref. [15]). For protons, the interference of nuclear and Coulomb elastic scattering is taken into account. At E>5 GeV, a simple analytical description used in the code for the coherent component of scattering cross-sections is quite consistent with data.

Once the energy of neutrons falls below 14 MeV during shower development, all subsequent neutron interactions are described using the appropriate MCNP4C [16] modules. Secondaries generated at this stage by neutrons – protons, photons and deuterons – are directed back to the MARS15 modules for a corresponding treatment. This implementation, along with algorithms developed for heavier recoils and photons from the thermal neutron capture on ^6Li and ^{10}B, allows the detailed description of corresponding effects in hydrogenous, borated and lithium-loaded materials. The interface includes several other modifications to the dynamically allocated storage, material handling, as well as an optional writing of low-energy neutrons and other particles to a file for processing at a later time.

ELECTROMAGNETIC PROCESSES, MUONS AND NEUTRINOS

The mean ionization energy loss for charged particles, except for heavy ions, is calculated using the Bethe formalism with the density correction. For heavy ions – as described in Ref. [2] – the Lindhard-Sorensen correction to the regular ionization logarithm and a Barkas term are taken into account. In addition, at low ion energies, the processes of electron capture and loss are accounted for by means of an effective ion charge. The effective charge, z_{eff}, is determined according to semi-empirical formulae and used instead of a bare ion charge. The projectile nuclear form-factor is also taken into account.

A further improved algorithm [17] for modeling correlated ionization energy loss and multiple Coulomb scattering is used in MARS15 for arbitrary mixtures. It takes into account arbitrary projectile and nuclear target charge distributions, exact kinematics of projectile-electron interactions, nuclear screening, and projectile-electron interactions. It accurately treats both soft and hard collisions. Calculated correlations between energy loss and scattering are quite substantial for low-Z targets. Radiative processes for single-charged particles and heavy ions – bremsstrahlung and direct pair production – are modeled directly [18].

Analog or inclusive simulation algorithms are used for prompt muon production (single muons in charmed meson decays, $\mu^+\mu^-$ pairs in vector meson decays, and the dimuon continuum), Bethe-Heitler $\mu^+\mu^-$ pairs and direct $e^+e^- \to \mu^+\mu^-$ annihilation. Neutrinos from meson and muon decays can be forced to interact with matter through all the possible mechanisms.

NUCLIDE INVENTORY, RESIDUAL DOSE AND DPA

Mass and charge number distributions (1D and 2D) are calculated in MARS15 for nuclides generated in arbitrary materials and regions of a setup studied. A further improved ω-factor based algorithm [19] to calculate residual dose rates in arbitrary composite materials for arbitrary irradiation and cooling times is used in MARS15. The algorithm distinguishes three major energy groups responsible for radionuclide production [1]: (1) above 20 MeV, (2) 1 to 20 MeV and (3) below 0.5 eV. Creation of the residual nuclides was pre-calculated with the FLUKA code. The emission rates of de-excitation photons are determined for irradiation time 12 hours$< T_i <$20 years and cooling time 1 sec$< T_c <$20 years. Corresponding dose rates on the outer surfaces are calculated from photon fluxes and related to the star density above 20 MeV (first group) and neutron fluxes in two other energy groups. A new algorithm [20] is used to calculate residual dose on very small objects as well as its attenuation in air around. Radiation damage to material – displacements per atom (DPA) – is calculated in MARS15 within a damage energy concept, taking into account recoil nuclei in elastic and inelastic nuclear interactions.

BIASING, VARIANCE REDUCTION AND TAGGING

Many processes in MARS15, such as electromagnetic showers, most hadron-nucleus interactions, decays of unstable particles, emission of synchrotron photons, photohadron production, and muon pair production, can be treated either analogously or inclusively with corresponding statistical weights. The choice of method is left for the user to decide, via the input settings. Other variance reduction techniques used in MARS15 are weight-window, splitting and Russian roulette, exponential transformation, probability scoring, and step/energy cutoffs. The goal here is to maximize computing efficiency t_0/t, where t is the CPU time needed to get a RMS error σ equal to the one in the reference method with the CPU time t_0 provided $\sigma < 20\%$. In addition to a standard multi-parameter bias setting in the code, a new user-friendly global bias control was recently introduced via a single card BIAS with parameters which define exclusive/inclusive switch, Russian Roulette and/or exponential transform control for six processes: decays of unstable particles, prompt muon production, Bethe-Heitler muon production, $e^+e^- \to \mu^+\mu^-$ annihilation, photo-nuclear reactions, and anti-proton production.

A further enhanced tagging module in MARS15 allows one to tag the origin of a given signal/tally: geometry, process and phase-space. It is invaluable in studying a source term and for sensitivity analysis. A user-friendly access is provided to the process ID at a scoring (histograming) stage, giving flags to 50 process types.

GEOMETRY AND MATERIALS

MARS15 provides a user with five geometry description options:

1. *Standard*: heterogeneous R-Z-ϕ cylinder.
2. *Non-standard*: arbitrary geometry defined by a user in a Fortran or C routine.

3. *Extended*: a set of contiguous or overlapping geometrical shapes, currently, boxes, spheres, cylinders, truncated cones, tetrahedra, and elliptical tubes. Each shape can be sub-divided into many sub-regions in each direction; arbitrary transformation matrices can be applied to any object.
4. *MCNP*: read in an input geometry description in the MCNP format.
5. *FLUKA*: read in an input geometry description in the FLUKA format.

Any or all of the five geometry options can be used simultaneously (co-exist) in a setup description. An arbitrary number of regions can be used, with a default maximum of 10^5. Volumes of all regions are auto-calculated for the predefined shapes. A short Monte-Carlo session of the code is used to calculate volumes of complex and overlapped regions. A corresponding output file provides calculated volumes with statistical errors and is directly linked to the main code.

The *extended* geometry provides exact crossing of particle tracks with the surfaces that prevents small regions within a large volume from being skipped over. In the other four options, boundary localization is based on an iterative algorithm and user needs to take care of appropriate region numbering, pilot steps and localization parameters.

A list of the built-in materials in MARS15 includes 165 elements and composites. Recently added are many kinds of steel, cast iron, mineral oil, gadolinium-loaded scintillator, etc. On top of that, customized composites can be defined in a user routine. The code does a separate treatment of gaseous and liquid states.

TRACKING AND HISTOGRAMING

Particle tracking – in a magnetic field with Coulomb scattering and energy loss for charged particles and decays for unstable particles – and histograming algorithms, which are already quite sophisticated, have further been refined to assure the highest accuracy and CPU performance. Slowing down pion and muon decays and nuclear capture are carefully modeled in a competition.

Performance of scoring algorithms for volume and surface detectors linked to the geometry was further improved. A further developed user-friendly flexible XYZ-histograming module allows scoring numerous distributions – total and partial particle fluxes, star density, total and partial energy deposition, DPA, temperature rise, prompt and residual dose rates, particle spectra, etc. – in boxes arbitrary positioned in a 3D system, independent of geometry description (mesh tally). Its performance and accuracy were also recently improved.

GUI AND MULTIPROCESSING

The existing Tcl/Tk-based 2D MARS-GUI-SLICE functionality was further improved and extended, which further extends the power of visualization of the modeled system: geometry, materials and magnetic field descriptions, simulated processes, and calculated results. New features include: selection of four GUI window sizes, online renormalization of histograms displayed, extended information on particle ID and histogram values at a point, switchable canvas aspect ratios, switchable format of a saved plot, etc. Arbitrary 3-D rotation of a slice is possible.

Since 2004, parallel processing is default in all CPU-hungry applications of MARS15 [1]; it is based on the Message Passing Interface (MPI) libraries. Parallelization is job-based, i.e. the processes, replicating the same geometry of the setup studied, run independently with different initial seeds. A unique master process – also running event histories – collects intermediate results from an arbitrary number of slaves and calculates the final results when a required total number of events has been processed. Intermediate results are sent to the master on its request generated in accordance with a scheduling mechanism. The performance scales almost linearly with the number of nodes used (up to tens of nodes at Fermilab clusters).

BEAM-LINE BUILDER

The MAD-MARS Beam-Line Builder (MMBLB) is the interface system [21, 22] to build beam-line and accelerator models in the MARS format. MMBLB reads in a MAD lattice file and puts the elements in the same order into MARS. Each element is assigned six functions: element type/name, geometry, materials, field, volume and initialization. MMBLB has been substantially extended [23]:

- The set of supported element types includes now almost all the elements supported by MAD.
- An arbitrary number of beam lines – arbitrary positioned and oriented – can be put in a MARS15 model.
- More sophisticated algorithms and new data structures enable more efficient searches through the beam line geometry.
- Tunnel geometry can now follow the beam line or be described independently of it.

MMBLB is heavily used in numerous MARS15 accelerator and machine-detector interface applications at Tevatron, LHC, J-PARC, and ILC. Fig. 4 shows just two recent examples of MARS15 use: muon fluxes in a 550-m region upstream of the LHC CMS detector due to 7-TeV proton interactions with residual gas, and photon tracks generated in the 220-m extraction line after 250×250-GeV e^+e^- collisions at the ILC IP. MMBLB and the GUI described above were used here to describe geometry, materials and magnetic fields, debug the model, perform MARS15 runs, and display results calculated.

BENCHMARKING

Debugging, validation and comparisons are done for the MARS code continuously once a new model or algorithm is implemented, a new problem is attacked, or a colleague provides interesting results from his/her code. Numerous verifications were performed at Fermilab over the last two decades in accelerator, shielding, targetry, and detector applications. Recent verifications of MARS15 by the authors and the MARS community were presented at the workshop [24, 25]. They prove the MARS15 code reliability, flexibility and robustness on microscopic and macroscopic levels.

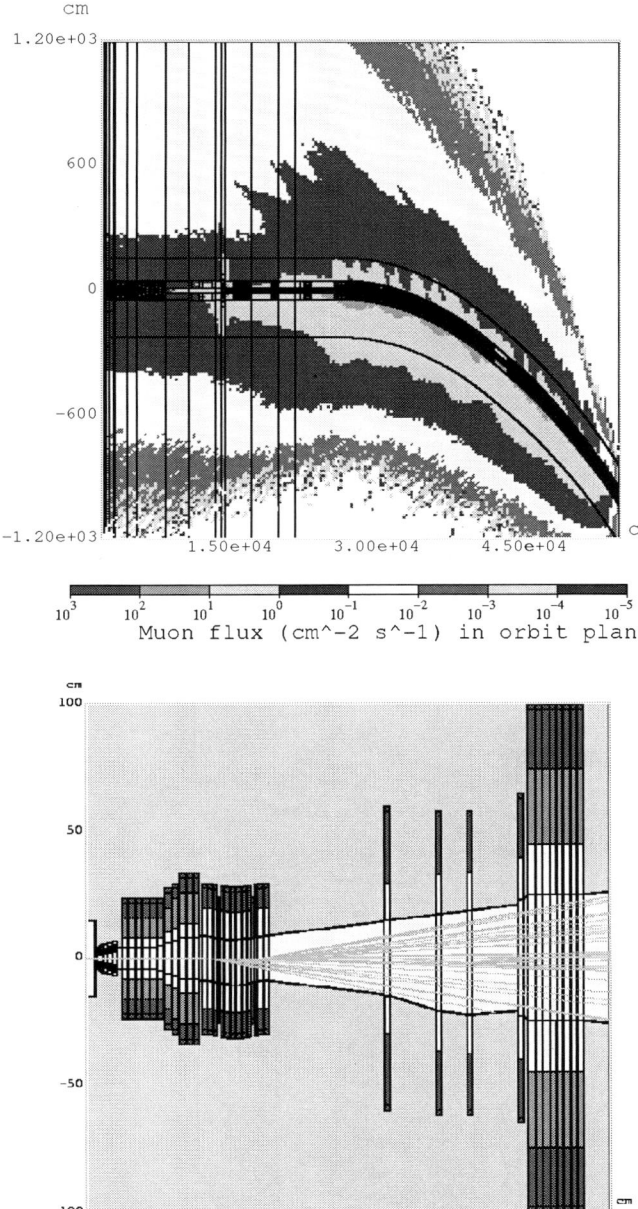

FIGURE 4. Muon flux isocontours in the LHC IP5 due to beam-gas interactions (top) and photons generated in the ILC 18-MW extraction line (bottom).

REFERENCES

1. N.V. Mokhov, *Fermilab-FN-628* (1995); N.V. Mokhov et al, *Rad. Prot. Dosimetry*, **116**, 99 (2005); N.V. Mokhov et al, *Int. Conf. on Nucl. Data Sci. Technology, AIP Conf. Proc.*, **769**, 1618 (2005); http://www-ap.fnal.gov/MARS/.
2. N.V. Mokhov et al, *Rad. Prot. Dosimetry*, **116**, 104 (2005).
3. M.V. Kossov, *Eur. Phys. J.*, **A14**, 377 (2002).
4. V.S. Barashenkov, "Cross-Sections of Particle and Nuclei Interactions with Nuclei", JINR, Dubna, Russia (1993).
5. , A.N. Kalinovskii, N.V. Mokhov and Yu.P. Nikitin, "Passage of High-Energy Particles through Matter", AIP, New York (1989).
6. A.S. Arefiev et al, ITEP-85-25 (1985); L.Z. Barabash et al, *Sov. J. Nucl. Phys.* **36**, 90 (1982); Yu.D. Bayukov et al, *Sov. J. Nucl. Phys.* **42**, 116 (1985).
7. I. Chemakin et al, *Phys. Rev.* , **C65:024904** (2002).
8. G. Ambrosini et al, *Eur. Phys. J.*, **C10**, 605 (1999); H.W. Atherton et al, CERN 80-07 (1980); D.S. Barton et al, *Phys. Rev.*, **D27**, 2580 (1983).
9. P.L. Skubic et al, *Phys. Rev.*, **D18**, 3115 (1978).
10. S.G. Mashnik et al, "CEM03.01 User Manual", LANL LA-UR-05-7321 (2005).
11. K.K. Gudima et al, *LA-UR-01-6804* (2001).
12. http://sroesler.home.cern.ch/sroesler/dpmjet3.html.
13. J.L. Klay et al, *Phys. Rev. Lett.*, **88**, 102301 (2002).
14. Y. Iwata et al, *Phys. Rev.*, **C64**, 054609 (2001).
15. I.L. Rakhno, N.V. Mokhov, E. Sukhovitski, S. Chiba, *ANS Topical Meeting on Accelerator Applications/Accelerator Driven Transmutation Technology Applications, AccApp/ADTTA '01*, Reno, Nevada, Omnipress CD ROM (2002).
16. J.F. Briesmeister, "MCNP - A General Monte Carlo N-Particle Transport Code, Version 4C", *Pub. LA-13709-M* (2000).
17. S.I. Striganov, *Rad. Prot. Dosimetry*, **116**, 293 (2005).
18. N.V. Mokhov, S.I. Striganov, in *AIP Conf. Proc.*, **372**, 234-256 (1996).
19. I.L. Rakhno et al, "Benchmarking Residual Dose Rates in a NuMI-like Environment", in [15].
20. N.V. Mokhov, E.I. Rakhno, I.L. Rakhno, "Residual Activation of Thin Accelerator Components", *Fermilab-FN-0788-AD* (2006).
21. D.N. Mokhov et al, "MAD Parsing and Conversion Code", *Fermilab-TM-2115* (2000).
22. O.E. Krivosheev et al, "A Lex-based MAD Parser and its Applications", in *Proc. 2001 Particle Accel. Conf.*, 3036-3038, Chicago, June 2001.
23. M.A. Kostin, O.E. Krivosheev, N.V. Mokhov, I.S. Tropin, "An Improved MAD-MARS Beam-Line Builder: User's Guide", *Fermilab-FN-0738-rev* (2004).
24. N.V. Mokhov, S.I. Striganov, "Recent Verifications in MARS15", talk at the HSSW06 workshop.
25. N.V. Mokhov, S.I. Striganov, "Hadronic Shower Code Inter-Comparison and Verification", in these Proceedings.

PHITS Overview

K. Niita[1], N. Matsuda[2], Y. Iwamoto[2], Y. Sakamoto[2],
H. Nakashima[2], T. Sato[2], H. Iwase[3], L. Sihver[4], D. Mancusi[4]

[1]*Research Organization for Information Science and Technology (RIST),
Tokai, Ibaraki, 319-1106, Japan,*
[2]*JAEA, Tokai, Ibaraki, 319-1195, Japan*
[3]*GSI, Darmstadt, Germany,*
[4]*Chalmers University of Technology, Gothenburg, Sweden*

Abstract. The paper presents a brief description of the models incorporated in PHITS and the present status of the code, showing some benchmarking tests of the PHITS code for accelerator facilities and space radiation.

Keywords: PHITS, Monte Carlo simulation, Heavy Ion.
PACS: 24.10.-i, 24.10.Lx

INTRODUCTION

A reliable and accurate particle and heavy ion transport code is an essential implement in the design study of accelerator facilities as well as for various applications such as radiotherapy, BNCT, spallation neutron sources, rare isotopes production, radiation protection, and space technology. We have therefore developed a multi-purpose particle and heavy ion transport Monte Carlo code system, PHITS [1] (Particle and Heavy Ion Transport code System), based on the NMTC/JAM code [2]. PHITS has three important ingredients which enable us to simulate (1) hadron-nucleus reactions with energies up to 200 GeV, (2) nucleus-nucleus collisions from 10 MeV/nucleon up to 100 GeV/nucleon, (3) transports of heavy ions, all hadrons including low energy neutrons down to 10^{-5} eV, and leptons. In this paper, we report a brief description of the models incorporated in the PHITS code, the present status of the code, and some benchmarking results.

OVERVIEW OF PHITS

In PHITS neutrons can be transported from thermal energies up to 200 GeV, and the same method as in the MCNP4C code [3] is employed for neutrons with energies below 20 MeV down to 1 meV based on the Evaluated Nuclear Data such as the ENDF-B/VI [4], JENDL-3.3 [5], and LA150 libraries [6]. Above 20 MeV, the simulation model JAM (Jet AA Microscopic Transport Model) developed by Nara et al. [7] is used. JAM is a hadronic cascade model, which explicitly treats all established hadronic states including resonances with explicit spin and isospin as well as their anti-particles. We have parameterized all hadron-hadron cross sections based on the resonance model and string model by fitting the available experimental data. Below the energy in the center-of-mass system (c.m.) $\sqrt{s}$ < 4 GeV, the inelastic hadron-hadron collisions are described by the resonance formations and their decays, and at higher energies, string formation and their fragmentation into hadrons are assumed. For protons and other hadrons, JAM is also used above 1 MeV up to 200 GeV, but for charged particles below 1 MeV only the ionization process is considered until the particles are stopped.

PHITS also uses Evaluated Nuclear Data for photon and electron transport below 1 GeV in the same manner as in the MCNP4C code [3] based on ITS version 3.0 code [8]. The energy range of electron and photon is restricted to the energy region 1 keV - 1 GeV at the present, but the extension of the maximum energy of these particles is in progress.

PHITS can also transport nuclei in materials. Below 10 MeV/n, only the ionization process for the nucleus transport is taken into account, but above 10 MeV/n the nucleus-nucleus collisions up to 100 GeV/n is described by the simulation model JQMD (JAERI Quantum Molecular Dynamics) developed by Niita et al. [9]. The JQMD code has been widely used to analyze various aspects of heavy ion reactions as well as of nucleon-induced reactions [10], and has shed light on several exciting topics in heavy ion physics, for example, the multifragmentation, the flow of the nuclear matter, and the energetic particle production [11]. In the QMD model, the nucleus is described as a self-binding system of nucleons, which are interacting with each other through the effective interactions in the framework of molecular dynamics. One can estimate the yields of emitted light particles, fragments and excited residual nuclei resulting from the heavy ion collision.

The QMD simulation, as well as the JAM simulation, describes the dynamical stage of the reactions. At the end of the dynamical stage, excited nuclei are created and must be forced to decay in a statistical way to get the final observed state. In PHITS the GEM model [12] (Generalized Evaporation Model) is employed for light particle evaporation and fission process of the excited residual nucleus.

When simulating the transport of charged particles and heavy ions, the knowledge of the magnetic field is sometimes necessary to estimate beam loss, heat deposition in the magnet, and beam spread. PHITS can provide dipole and quadrupole magnetic fields in any direction and any region of the setup geometry. In contrast to other beam transport codes, PHITS can simulate not only the trajectory of the charge particles in the field, but also the collisions and the ionization process at the same time. This is a great advantage of PHITS when designing high intensity proton and heavy ion accelerator facilities, where one must estimate radiation damage of the magnets and the surrounding materials and the radiation shielding, as well as perform trajectory calculations.

For the ionization process of the charged particles and nuclei, the SPAR code [13] is used for the average stopping power dE/dx, the first order of Moliere model for the angle straggling, and the Gaussian, Landau and Vavilov theories for the energy straggling around the average energy loss according to the charge density and velocity. In addition to the SPAR code, the ATIMA package [14] developed at GSI group has recently been implemented as an alternative code for the ionization process.

The total reaction cross section, or the life time of the particle for decay is an essential quantity in the determination of the mean free path of the transport particle. According to the mean free path, PHITS chooses the next collision point using the Monte Carlo method. It is therefore very important that reliable data of total non-elastic and elastic cross sections is used for the particle and heavy ion transport. In PHITS, the Evaluated Nuclear Data is employed for neutron-induced reactions below 20 MeV. For neutron-induced reactions above 20 MeV a parameterization is used [2]. As for the elastic cross sections, the Evaluated Nuclear Data is also used for neutron-induced reactions below 20 MeV, and a parameterization is used above 20 MeV [2]. Parameterizations are also used for proton induced reactions for all energies, and for the double differential cross sections of elastic nucleon-nucleus reactions [2]. Recently we have also adopted the NASA systematics developed by Tripathi et al., [15] for the total nucleus-nucleus reaction cross section, as an alternative to the Shen formula [16].

BENCHMARKING OF THE PHITS CODE

PHITS is used in a wide field of applications, e.g. semiconductor soft errors, BNCT, shielding around accelerators, radiotherapy, and radioprotection of personnel on space missions. In this section, we present validation of calculations compared with the experimental data.

Neutrons From Heavy Ion Reactions

The neutrons produced in heavy ion reactions are very important when designing shielding e.g. around heavy ion accelerators and spacecrafts because of their large attenuation length in shielding materials. Recently, secondary neutrons from heavy ion reactions have been systematically measured using thin and thick targets at HIMAC (Heavy Ion Medical Accelerator in Chiba) in Japan [17, 18]. In Fig. 1, the calculated neutron energy spectra using JQMD at different angles are shown together with measured values for the reaction of 400 MeV/n ^{12}C on a thin ^{208}Pb target. As can be seen the JQMD model used in PHITS reproduces the measured cross sections quite well. We have also compared the calculated results with measured performed by Kurosawa et al. [18]. These measurements were performed at HIMAC and include secondary neutrons produced from thick targets of C, Al, Cu, and Pb bombarded with various heavy ions from He to Xe in the energy region 1000 MeV/n to 800 MeV/n. In Fig. 2, two examples of the comparisons are shown.

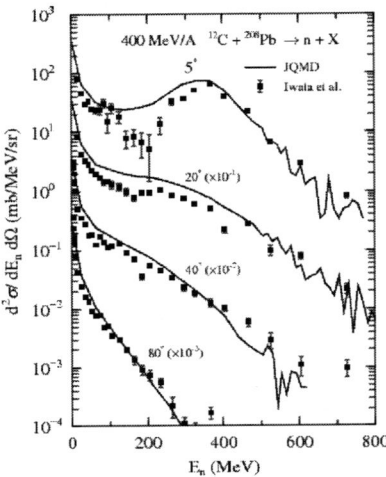

Figure 1. Calculated neutron energy spectra using JQMD at different angles is shown together with measured values [17] for the reaction of 400 MeV/n ^{12}C on a thin ^{208}Pb target

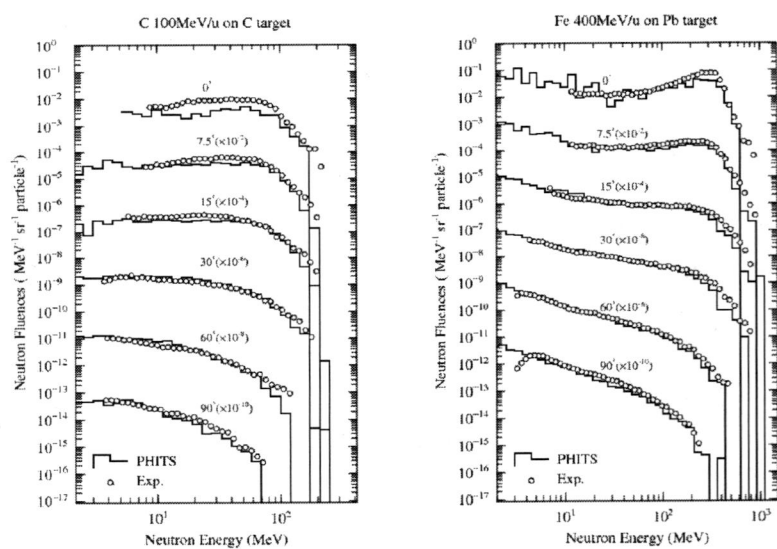

Figure 2. Comparison of the neutron double differential cross section calculated with PHITS and the measured data [18] for 1000 MeV/n C ion on C target (left panel) and 400 MeV/n Fe ion on Pb target (right panel).

Spallation Neutron Source

The J-PARC (Japan Proton Accelerator Research Complex) project at JAEA in Japan [19] is the main field where the NMTC/JAM code had been developed. This project includes a 400 MeV normal-conducting LINAC, a superconducting LINAC from 400 to 600 MeV, a 3 GeV synchrotron ring for 1 MW proton, a 50 GeV synchrotron ring for 0.75 MW proton beams. These are now under construction and aim to pursue frontier science in particle physics, nuclear physics, materials science, life sciences and nuclear technology, using a new proton accelerator complex with the highest beam power in the world. A reliable transport code was required for the shielding design and optimization study of each facility in J-PARC. The conditions which should be satisfied in the transport code are very severe, since the energy range is very wide, from 50 GeV down to meV neutrons in the material science facility. Furthermore the dimension of the system is very large (about several tens of meters), but the resolution of the calculation must be of the order of millimeters. First the NMTC/JAM [2] code was developed to satisfy these severe conditions, and then the code was upgraded to PHITS, which includes heavy ion transport.

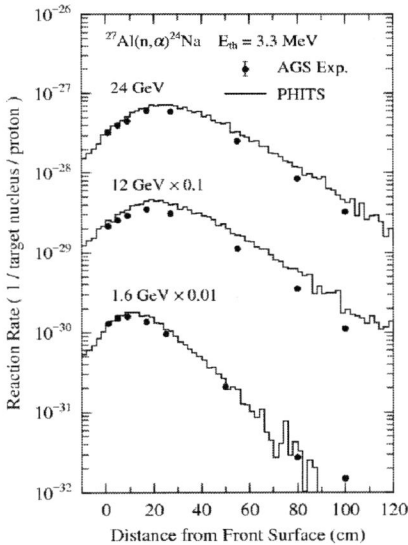

Figure 3. Distribution of the ^{27}Al(n,α)^{24}Na reaction rates along the cylindrical surface of a mercury target bombarded with 1.6, 12 and 24 GeV protons [20]. The solid histograms denote the results of PHITS.

For validation of the PHITS calculations for the neutron flux produced by the mercury spallation target, we have applied PHITS to the experiments under the ASTE (AGS Spallation Target Experiment) collaboration [20]. One of the experiments was carried out using a bare mercury target, which is a 20 cm diameter and 130 cm long cylinder, and detecting the reaction rate distributions along the cylindrical surface of the target by activation techniques at incident proton energies of 1.6, 12 and 24 GeV. Various activation detectors such as the ^{115}In(n,n')^{115m}In, ^{93}Nb(n,2n)^{92m}Nb and ^{209}Bi(n,xn) reactions with threshold energies ranging from 0.3 to 70.5 MeV were employed to obtain the reaction rate data for estimating spallation source neutron characteristics of the mercury target. Figure 3 shows the distribution of the

^{27}Al(n, α)^{24}Na reaction rates along the cylindrical surface of bare mercury target bombarded with 1.6, 12, and 24 GeV protons. The threshold of this reaction is 3.3 MeV, while the most effective neutron energy for this reaction is roughly 10 MeV. The results of PHITS, denoted by the solid histograms in these figures, reproduce the experimental distribution quite well for all positions and all energies. From these data, we have also derived the neutron flux spectra by the spectrum adjustment method. Figure 4 compares the adjusted and calculated neutron spectra at 17 cm from the front surface of the target. Over all the energy range, the PHITS results roughly trace the adjusted spectra.

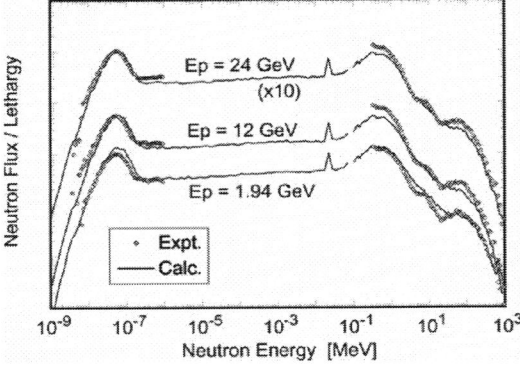

Figure 4. Comparison of the neutron spectra obtained by adjustment method from the experimental data [20] and by the PHITS calculation

Heat Deposition

One of the important issues for high intense proton and heavy ion accelerator facilities is a heat deposition to magnet components, target vessels and materials of beam dump. Precise knowledge of the heat deposition is required for designing them. Monte Carlo codes such as PHITS are usually used to estimate the heat deposition. There are, however, few data to validate the codes in connection with heat measurements. Recently, Ohnishi et. al. [21] measured the heat deposition of secondary particles emitted from a production target interacting with a primary proton beam at the KEK 12 GeV proton synchrotron by developing a "cryogenic calorimeter". The production target was a copper cylinder with dimensions of 30 mm in diameter and 30 mm in length, corresponding to a 0.2 interaction length. The target was remotely controlled so as to change its position from the center of the absorber along the beam direction. Figure 5 shows a comparison of the calculation results with the experimental data of the heat deposition. The results of PHITS and MCNPX well reproduce the experimental data, though the result of MARS underestimates the data. Ohnishi analyzed the discrepancies between the results of MARS and PHITS and concluded that the details of the distribution of secondary particles produced from the copper target at 12 GeV proton beam caused the differences of the heat [22]. We also checked the differences between PHITS and MCNPX, and found that the contributions of each secondary particle to the heat in the absorber are different between them, but finally the total heats are the same each other due to the cancellation of the differences. We therefore concluded that the heat estimation by these Monte Carlo codes still has an uncertainty about 20 % due to the difficulties with simulation models for secondary particle production around 10 GeV. Further studies are necessary to improve the accuracy.

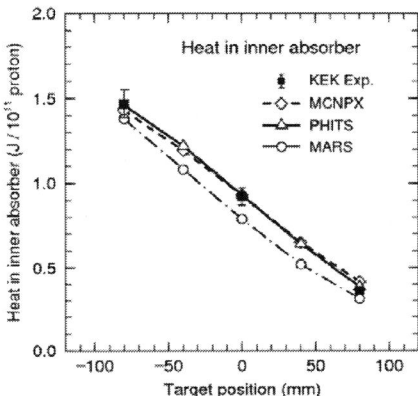

Figure 5. Comparison of the experimental heat [21] with the simulation results by MCNP, PHITS and MARS.

Space Radiation

Personnel on space missions are exposed to an enhanced level of radiation. Dose estimation and shielding is therefore of great importance in the planning of long-term space missions. The particle transport simulation inside the spacecraft is one of the key issues in the dose estimation, since the internal radiation fields are significantly disturbed from the external fields due to the cosmic-ray induced nuclear reactions occurred in the shielding materials of spacecraft. We have calculated the neutron and charge particle spectra inside the Space Shuttle [23]. As source particles, we considered almost all kinds of space radiations, the trapped and Galactic Cosmic Ray (GCR) protons, albedo neutrons from the earth atmosphere, and heavy ion with charges up to 28 and energy up to 100 GeV/u. The spectrum of albedo neutrons was also calculated by PHITS simulating the earth atmosphere based on the charge particles spectra outside of the earth. We have compared calculated neutron spectra in the imaginary vessel at the solar minimum with the orbit-averaged data measured by BBND (Bonner Ball Neutron Detector) at STS-89 [24]. An excellent agreement can be observed between the calculated and experimental results, particularly for neutron energies above 1 MeV which is very important in the evaluation of dose for astronauts.

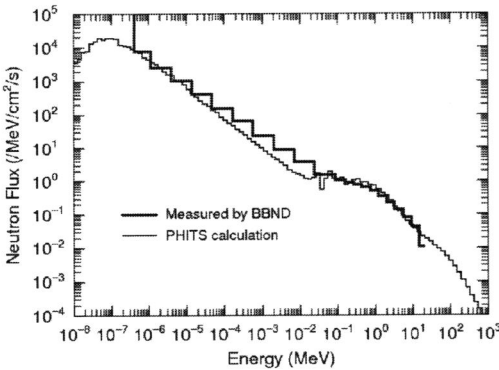

Figure 6. Comparison results of the calculated neutron spectra in the imaginary vessel at the solar minimum with the orbit-averaged data measured by BBND at STS-89 [24].

Atmospheric Cosmic Rays

In the last decade, radiation protection for aircrews against terrestrial cosmic-rays was one of the most intensively discussed dosimetric issues. Furthermore, increasing attention has been paid to the soft errors of semiconductor devices induced by the cosmic-rays even at the ground level, since the recent miniaturization of the devices causes a rapid decrease of their critical charges. These radiation effects are predominantly triggered by neutrons produced by nuclear reactions between the cosmic-rays and atmospheric components. Therefore, estimation of cosmic-ray neutron spectra in the atmosphere is an essential issue in the evaluation of the aircrew doses and the soft-error rates. A number of studies have been devoted to the estimation of the neutron spectra, *e.g.* [25-29]. However, cosmic-ray neutron spectra depend on the atmospheric depth, cut-off rigidity, solar modulation and local geometry in an intricate manner, and none of the existing models are able to reproduce the measured neutron spectra at any location and time with satisfactory accuracy.

We have therefore calculated the cosmic-ray neutron spectra by PHITS [30], utilizing the latest version of the nuclear data library JENDL-High-Energy File (JENDL/HE) [31, 32]. In the simulation, cosmic-rays with charges up to 28 (Ni) with their spectra calculated by CREME96 [33] were incident on the earth system represented by the concentric spherical shells, and the neutron spectra in each shell, *i.e.* a certain altitude range, were obtained. Based on a comprehensive analysis of the simulation results, we proposed analytical functions to predict the cosmic-ray neutron spectra at any global condition at the altitudes below 20 km, considering the local geometry effect. Figure 7 shows the neutron spectra obtained by PHITS and the analytical functions, in comparison with the several experimental data [34, 35]. In the figures, d denotes the atmospheric depth and r_c does the cut-off rigidity. It is evident from the figure that the neutron spectra of the experiments and the calculations are in excellent agreement. Furthermore, the analytical functions are substantially superior to the PHITS simulation in reproducing experimental data at lower energies, since they can consider the local geometry effect on the spectra even at high altitudes. Based on the analytical functions, we have made an Excel-based program (EXPACS) to predict the atmospheric cosmic-ray neutron spectra and released it from JAEA web site [36].

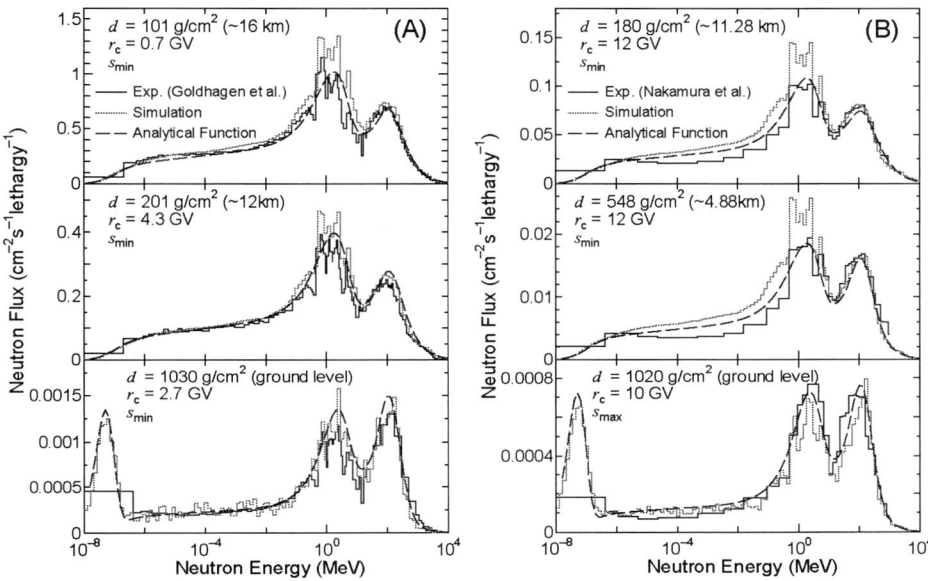

Figure 7. Calculated and experimental neutron spectra in the atmosphere for various global conditions. The panels (A) and (B) show the comparison with the data measured by Goldhagen et al. [34] and Nakamura et al. [35], respectively.

SUMMARY

We have briefly presented the models incorporated in the particle and heavy ion transport code PHITS and discussed the present status of the code. The results indicates that PHITS has a great ability of carrying out the radiation transport analysis of almost all accelerator facilities of protons and heavy ions within a wide energy range, as well as the possibility to be used in a wide field of applications, e.g. semiconductor soft errors, BNCT, shielding, radiotherapy, and radioprotection of personnel on space missions.

REFERENCES

1. H. Iwase, K. Niita and T. Nakamura, *J. Nucl. Sci. Technol.* **39**, 1142 (2002).
2. K. Niita, H. Takada, S. Meigo and Y. Ikeda, *Nucl. Instr. and Meth.* **B184**, 406 (2001).
3. J. F. Briesmeister, et al., "MCNP General Monte Carlo N-Particle Transport Code", Los Alamos National Laboratory report; *LA-12625-M* (1997).
4. V. McLane, et al., "ENDF/B-VI Summary Documentation", *BNL-NCS-17541* (1996).
5. Shibata, K., et al., "Japanese Evaluated Nuclear Data Library Version 3 Revision-3: JENDLE-3.3", *J. Nucl. Sci. Technol.* **39**, 1125 (2002).

6. M. B. Chadwick, et al., "LA150 Documentation of Cross Sections, Heating, and Damage", Los Alamos National Laboratory report; *LA-UR-99-1222* (1999).

7. Y. Nara, N. Otuka, A. Ohnishi, K. Niita and S. Chiba, *Phys, Rev.* **C61**, 024901 (2000).

8. J. A. Halbleib, et al., "ITS Version 3.0: The Integrated TIGER Series of Coupled Electron/Photon Monte Carlo Transport Codes", *SAND91-1634* (1992)

9. K. Niita, S. Chiba, T. Maruyama, H. Takada, T. Fukahori, Y. Nakahara, and A. Iwamoto, *Phys. Rev.* **C52**, 2620 (1995).

10. S. Chiba, O. Iwamoto, T. Fukahori, K. Niita, Toshiki Maruyama, Tomoyuki Maruyama, A. Iwamoto, *Phys. Rev.*, **C54**, 285 (1996), S. Chiba, M.B. Chadwick, K. Niita, Toshiki. Maruyama, Tomoyuki. Maruyama, A. Iwamoto, *Phys. Rev.*, **C53**, 1824 (1996).

11. J. Aichelin, *Phys. Rep.* **202**, 233 (1991).

12. S. Furihata, *Nucl. Instr. and Meth.* **B171**, 251 (2000).

13. T. W. Armstrong and K. C. Chandler, "A Fortran program for computing stopping powers and ranges for muons, charged pions, protons, and heavy ions". *ORNL-4869*, Oak Ridge National Laboratory, (1973).

14. C. Scheidenberger and H. Geissel, "Penetration of relativistic heavy ions through matter", *Nucl. Instr. Meth.* B136, 114 (1998).

15. R. K. Tripathi, F. A. Cucinotta, and J. W. Wilson, *Nucl. Instr. and Meth.* **B117**, 347 (1996); R. K. Tripathi, J. W. Wilson, and F. A. Cucinotta, *Nucl. Instr. and Meth.* **B129**, 11 (1997), R. K. Tripathi, F. A. Cucinotta, and J. W. Wilson, *Nucl. Instr. and Meth.* **B155**, 349 (1999).

16. W. Shen, B. Wang, J. Feng, W. Zhan, Y. Zhu and E. Feng, *Nucl. Phys.*, **A491**, 130 (1989).

17. Y. Iwata, et al., *Phys. Rev.*, **C64**, 054609 (2001).

18. T. Kurosawa, et al., *Nucl. Sci. and Eng.*, **132**, 30 (1999); *Journal of Nucl. Sci. and Technol.* **36-1**, 42 (1999); *Nucl. Instr. and Meth.* **A430**, 400 (1999); *Phys. Rev.* **C62**, 044615 (2000).

19. The Joint Project Team of JAERI and KEK, "The Joint Project for High Intensity Proton Accelerators'', *JAERI-Tech*, 2000-003, JAERI (2000).

20. H. Nakashima, et.al., "Research Activities on Nuetronics under ASTE Collaboration at AGS/BNL", *Proc. of Int. Conferece on Nuclear Data for Science and Technology*, Tsukuba, Japan, 2001.; "Current Status of the AGS Spallation Target Experiment", *Proc. of the 6th meeting of the Task Force on SATIF*, SLAC, 2002.

21. H. Ohnishi, et.al., *Nucl. Instr. and Meth.* A545, 88 (2005).

22. H. Ohnishi (private communication).

23. T. Sato, K. Niita, H. Iwase, H. Nakashima, Y. Yamaguchi and L. Sihver, *Radiat. Meas.*, **41**, 1142 (2006).

24. H. Matsumoto, T. Goka, K. koga, S. Iwai, T. Uehara, O. Sato and S. Takagi, *Radiat Meas.*, **33**, 321 (2001).

25. J. F. Ziegler, *IBM J. Res. Develop.* **42**, 117-139 (1998).

26. K. O'Brien, W. Friedberg, H. H. Sauer and D. F. Smart, *Environ. Int.* **22** (Suppl.1), S9-44 (1996).

27. S. Roesler, W. Heinrich and H. Schraube, *Radiat. Res.* **149**, 87 (1998).

28. A. Ferrari, M. Pelliccioni and T. Rancati, *Radiat. Prot. Dosim.* **93**, 101 (2001).

29. J. M. Clem, G. De Angelis, P. Goldhagen and J. W. Wilson, *Radiat. Prot. Dosim.* **110**, 423 (2004).

30. T. Sato and K. Niita, *Radiat. Res.* **166**, 544 (2006).

31. T. Fukahori, Y. Watanabe, N. Yoshizawa, F. Maekawa, S. Meigo, C. Konno, N. Yamano, A.Yu. Konobeyev and S. Chiba, *J. Nucl. Sci. Technol.*, *Suppl.* **2**, 25 (2002).

32. Y. Watanabe, et. al., "Nuclear data evaluations for JENDL high-energy file'' *Proc. of Int. Conference on Nuclear Data for Science and Technology*, Santa Fe, USA, Sep.26-Oct.1, 2004; AIP CP769, pp326 (2005).

33. A. J. Tylka, et. al., *IEEE Transactions on Nuclear Science* **44**, 2150 (1997).

34. P. Goldhagen, J. M. Clem and J. W. Wilson, *Radiat. Prot. Dosim.* **110**, 387-392 (2004).

35. T. Nakamura, T. Nunomiya, S. Abe, K. Terunuma and H. Suzuki, *J. Nucl. Sci. Technol.*, **42**, 843 (2005)

36. T. Sato, *http//www3.tokai-sc.jaea.go.jp/rphpwww/radiation-protection/expacs/expacs.html*.

Validation of Heavy Ion Transport Capabilities in PHITS

Reginald M. Ronningen

National Superconducting Cyclotron Laboratory, Michigan State University, East Lansing, MI 48824-1321, USA

Abstract. The performance of the Monte Carlo code system PHITS is validated for heavy ion transport capabilities by performing simulations and comparing results against experimental data from heavy ion reactions of benchmark quality. These data are from measurements of secondary neutron production cross sections in reactions of Xe at 400 MeV/u with lithium and lead targets, measurements of neutrons outside of thick concrete and iron shields, and measurements of isotope yields produced in the fragmentation of a 140 MeV/u ^{48}Ca beam on a beryllium target and on a tantalum target. A practical example that tests magnetic field capabilities is shown for a simulated ^{48}Ca beam at 500 MeV/u striking a lithium target to produce the rare isotope ^{44}Si, with ion transport through a fragmentation-reaction magnetic pre-separator. The results of this study show that PHITS performs reliably for the simulation of radiation fields that is necessary for designing safe, reliable and cost effective future high-powered heavy-ion accelerators in rare isotope beam facilities.

Keywords: PHITS;simulations;heavy ion transport;validation
PACS: 24.10.Lx; 25.70.-z; 25.70.Mn;

INTRODUCTION

Design studies at heavy ion accelerators in the past have been hampered by the lack of both reliable and wide-ranging data and appropriate computational tools. The next generation of rare isotope beam (RIB) facilities is expected to accelerate primary beams of stable ion species from hydrogen at through uranium to $\sim 0.5 - 1$ GeV/u with beam powers to 400 kW [1]. This beam power is up to two orders of magnitude greater than that handled at currently operating heavy ion RIB facilities.

One of the most difficult technical challenges in designing the accelerator and target areas to handle the beam power reliably and safely and to yield maximum performance relative to design requirement is to perform accurate simulations of heavy-ions interacting with matter. Due to complexities in describing heavy ion reactions, simulations of the radiation environment of the accelerator-target system are very challenging, as all appropriate reaction mechanisms must be taken into account to successfully describe heavy ion transport, interactions and resulting secondary radiation. Furthermore, such simulations are significantly more computationally intensive than for light ion transport.

For estimating the consequences of such intense and diverse radiation fields, it is important to know the reliability of currently available codes that transport heavy ions. However, such codes are not as mature as those developed for light-ion transport. In

CP896, *Hadronic Shower Simulation Workshop*
edited by M. Albrow and R. Raja

addition, up to now there have been very few high-quality benchmark data sets to test these newly developed codes. Fortunately, new high-quality data sets on heavy-ion-induced reactions are either now or soon becoming available. The recently published book by Nakamura and Heilbronn [2] references and summarizes many of the relevant experiments, and makes available the relevant data sets. These include studies of secondary neutron production at beam energies to be delivered by the next generation RIB facility. In addition, there are new high quality studies of rare isotope production via projectile fragmentation [3,4], and induced fission, fast fission, and spallation, *e.g.* [5,6].

The objective of this study is to validate the performance the general purpose Monte Carlo Particle and Heavy Ion Transport System PHITS [7] by performing simulations and comparing results against experimental data of benchmark quality. Quantitative tests of the theoretical bases and the uncertainties of the simulations of this and other applicable codes are essential for the safe, reliable and cost effective design of any future RIB facility.

CASE STUDIES USED FOR VALIDATING PHITS

The main objective of our study is to assess the performance of the PHITS code system [7] by performing simulations and comparing results against experimental data with benchmark quality. Simulations were performed and compared to benchmark data on the production of secondary neutrons from thin targets, which will test production cross-section models. Simulations of neutron transport through thick biological shields were made and compared to benchmark data. This is important because biological shielding is necessary for protecting workers, the public and the environment, but it is an expensive element in facility design. The prediction of residual radioactivity and nuclide distributions is needed in determining inventories for facility licensing and for radiation protection purposes. Simulations are then often necessary to obtain residual radioactivity from spallation, fragmentation and fission processes. The results of all of these comparisons can be used to suggest possible improvements to code models.

Simulations carried out by a Michigan State University-led multi-institutional collaboration [8] under DOE award number DE-FG02-ER41313 "Development of a Concept for High Power Beam Dumps and Catchers, and the Pre-separator Area Layout for Fragment Separators for the Rare Isotope Accelerator Project" were based on results solely from the PHITS code system [7]. When this project commenced, there was a lack of other publicly available codes that could transport heavy ions. Iwata *et al.* [9] demonstrated good agreement between their data and simulations using PHITS. Additionally, PHITS can transport ions in regions having magnetic fields, which was a significant requirement for this project.

PHITS [7] is a three dimensional radiation transport code system with the capabilities of transporting heavy ions as well as protons, neutrons, charged pions, electrons, positrons, and gamma rays, *etc.* PHITS has magnetic field capabilities, using a first-order transport-matrix approach for transporting ions in static magnetic dipole and quadrupole fields. The geometry input to PHITS may be either Combinatorial Geometry (CG) or Generalized Geometry (GG).

Thin target secondary neutron production cross sections

Measurements of double-differential (in energy and angle) neutron production cross sections were recently made at the Heavy Ion Medical Accelerator in Chiba (HIMAC) facility of the National Institute of Radiological Sciences (NIRS), Japan. A detailed description of the measurements is given in [10]. In the validation study presented here, data obtained from the 400 MeV/u Xenon beam on a lead target (0.050 cm x 10 cm x 10 cm) and on a lithium target (a cylinder of 5.7-cm diameter and 0.90-cm thick) were compared to simulations using PHITS (version 2.13).

The input file for the simulation was easily adapted from one given in [2] because a description is presented there for simulating the experimental setup used in very similar experiments by Iwata *et al.* [9]. Modifications were made to the source parameters for describing the Xe beam, to the materials descriptions, and to tally modifying factors to account for the Pb or Li target materials.

Comparisons of measured and predicted double differential cross sections for neutron production as functions of neutron energy are shown for the Xe+Li system and for the Xe+Pb system in Figure 1.

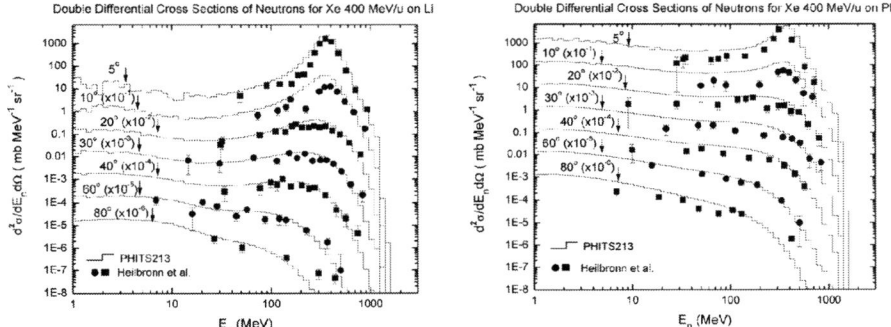

FIGURE 1. Double-differential cross sections for neutron production as a function of neutron energy from a beam of 400 MeV/u Xe on a thin lithium target (left panel) and on a thin lead target (right panel) are shown. The data are shown by filled circles and squares. Simulation results using the PHITS code are shown by the histograms.

With the exception of the $10°$ data set for each target, there is overall very good agreement between measurements and calculations. From the lower energy neutrons, it is apparent that contributions to neutron emission by target-like fragments are adequately described, as evidenced by the predicted and observed differences of nearly two orders of magnitude in cross sections for the production of neutrons having below one-half beam velocity for the Xe+Pb system compared to the Xe+Li system. This can be understood within semi-phenomenological moving source models as discussed in [2].

Production and transport of secondary neutrons through thick biological shields

Biological shielding is an expensive element in a facility design but necessary for protecting workers, the public and the environment from penetrating secondary radiation. It is then highly desirable to use a Monte Carlo simulation code to best estimate the amount of shielding necessary in light of this.

Sasaki et al. [11,12] recently measured neutron fluences as a function of neutron energy behind thick concrete and iron shields at the HIMAC, using a 400 MeV/u carbon beam stopped in a copper target. The thick-target neutron yields for this system as a function of energy and angle were previously measured. Detailed descriptions of those experiments are found in the original publications [11,12]. A synopsis of the experiments is given and the data sets are made available in [2]. In their analyses, Sasaki et al. used the experimentally measured neutron energy spectra from thick target yields as the particle source that was transported through the shields in simulations using the MCNPX [13] code. They then compared their measurements of neutron fluence outside of several thicknesses of concrete shields and of iron shields to predictions from the simulations.

We simulated several of the experiments by Sasaki et al. using the PHITS code. However, in contrast to using an experimentally derived neutron spectrum source, ours was generated by a simulated beam of 400 MeV/u carbon stopping in a 5-cm thick copper target (10 cm x 10 cm). This allowed us to test the neutron production models in PHITS as well as the neutron transport through the shielding. The geometric model used in the PHITS simulations is shown in Figure 2.

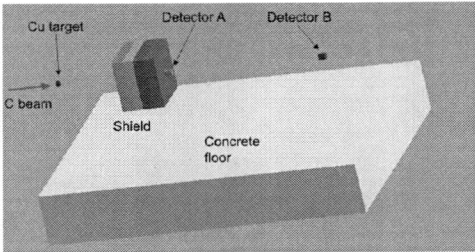

FIGURE 2. The geometry used in the simulations using PHITS is shown. The 400 MeV/u carbon beam stopped in a copper target in front of concrete or iron shields. Neutron detectors placed at position "Detector A" and at position "Detector B" tallied the neutron fluence at these locations.

The shields were 50, 100, 150, and 200 cm of concrete having density 2.2 g/cm^3, and 20, 40, 60, and 80 cm of iron having density 7.87 g/cm^3. The concrete floor was necessarily included in the simulation owing to its close proximity to the target, shields and detectors, relative to other neutron scattering media such as vertical walls and the cave roof. The concrete density was 2.2 g/cm^3 in the concrete shield simulation and 2.25 g/cm^3 in the iron shield simulation. Neutron fluences were calculated at 50 cm ("Detector A") and 200 cm ("Detector B") outside of concrete and iron shields and compared to the experimental values. Our results are shown in Figure 3.

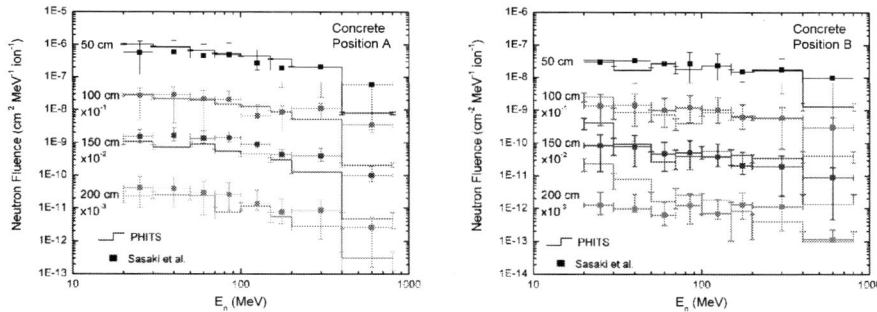

FIGURE 3. The neutron fluences as a function of neutron energy behind 50, 100, 150, and 200 cm thick concrete shields measured (solid symbols) at position "Detector A" (left panel) and at position "Detector B" (right panel) are shown compared to results from simulations using PHITS (solid lines).

For the concrete shielding and at both the Detector A and B positions, the calculated and measured neutron fluences agree very well except at the highest neutron energies. This may in part be due to the under-prediction by PHITS of the highest energy neutrons at forward angles. This is the case for the Xe+Pb system discussed above (see Figure 1) and the 600 MeV/u Ne+Pb system discussed in [2].

Overall, the results of the simulations using PHITS agrees very well with those from the simulations done by Sasaki *et al.* using different codes and sources. Most importantly for shielding design purposes, there is overall satisfactory agreement with experiment.

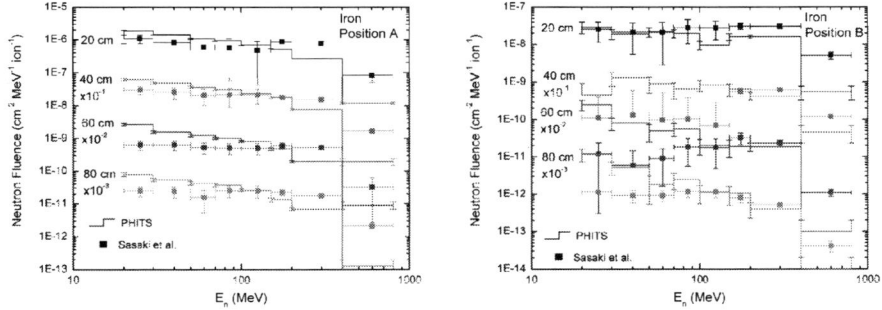

FIGURE 4. The neutron fluences as function of neutron energy behind 20, 40, 60, and 80 cm thick iron shields measured (solid symbols) at position "Detector A" (left panel) and at position "Detector B" (right panel) are shown compared to results from simulations using PHITS (solid lines).

Isotope distributions from projectile fragmentation reactions

Plans for future RIB facilities will likely include consideration of the projectile fragmentation reaction method to produce rare isotopes ever more close to or at demonstrated or predicted neutron and proton drip lines. It is essential to understand

and to model fragment production mechanisms in reactions of heavy-ion primary beams with targets and in addition with beam dump and catcher materials. As part of the validation of the PHITS code as a tool to help design projectile fragmentation target areas, calculated fragment cross sections are compared to experimental data. Recent studies of isotope production cross sections from projectile fragmentation reactions were carried out using ^{40,48}Ca, ^{58,64}Ni beams at 140 MeV/u on beryllium and tantalum targets from the Coupled Cyclotron Facility at the NSCL [3]. These data and other high quality data, obtained [4] using the ^{86}Kr beam at 64 MeV/u at RIKEN, produced nearly 1800 isotope production cross sections. We performed simulations using PHITS to obtain fragmentation production cross sections for the ^{48}Ca + Be, Ta systems, for comparison with the experiments. Figure 5 shows isotopic distributions of the measured and calculated fragmentation reaction production cross sections as a function of neutron excess.

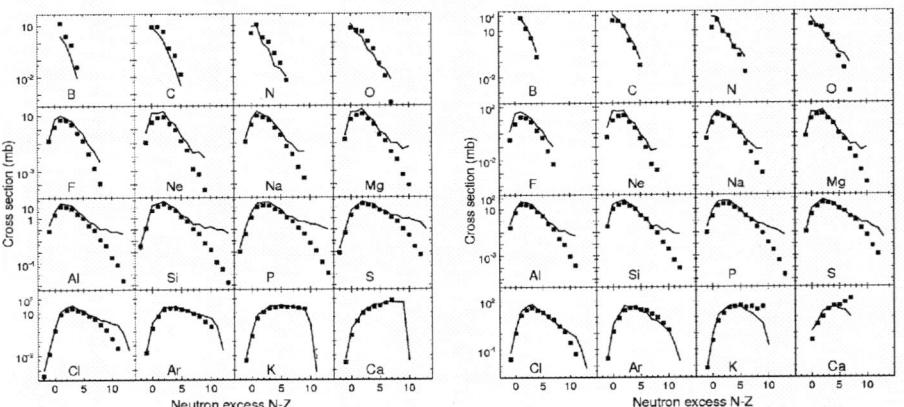

FIGURE 5. Isotopic distributions of measured fragmentation production cross sections are plotted as a function of neutron excess for elements boron through calcium isotopes produced in reactions of 140 MeV/u ^{48}Ca with a beryllium target (left panel) and with a tantalum target (right panel). The data are shown as solid circles. The solid lines connect results of simulations using the code PHITS (version 2.13) for the JQMD time parameter nqtmax = 100 fm/c (see text).

PHITS uses the JQMD model [14] to treat dynamical processes in nucleus-nucleus collisions, *e.g.* direct and non-equilibrium reactions, forming highly excited fragments. The statistical model GEM [15] is then used to describe statistical processes later in the time evolution of the systems, *e.g.* decays of the fragments by fission and evaporation. The time evolution of a system from the dynamical to statistical phase is of significant interest. Within PHITS simulations, the JQMD calculations of dynamical processes are stopped and results transferred to GEM at a time specified by the parameter nqtmax. The default value is 150 fm/c. Beginning with PHITS version 2.13, the parameter nqtmax can be set externally. Simulations of the system ^{48}Ca+Be were carried out using nqtmax = 150, 130 and 100 fm/c. The results are shown in Figure 6 for the sulphur isotopes.

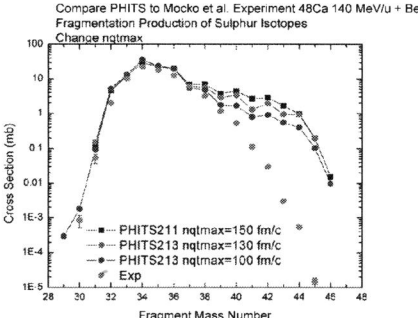

FIGURE 6. Cross sections for sulphur isotopes produced by beam fragmentation in the reaction 140 MeV/u ^{48}Ca on a beryllium target are shown as a function of neutron excess. The solid circles are data and the solid lines are from simulations using the code PHITS (version 2.13) with the JQMD time parameter nqtmax = 150, 130, and 100 fm/c (see text).

There are significant differences for most isotopes when the neutron excess is large. The improvement with decreasing time for the dynamic evolution of fragments having large neutron excesses suggests the hot fragments produced by PHITS are not too excited for significant neutron emission to occur in the statistical phase. Further study of this issue is needed.

Transport of heavy ions in magnetic fields

In simulations of fragmentation pre-separators, design layouts are obtained from design specialists who provide information on magnet function (dipole or quadrupole), lengths, relative positions, dimensions of bores and gaps, and field strengths and directions. This information was placed into the geometry and input parameter sections of PHITS for the pre-separator layout Figure 7.

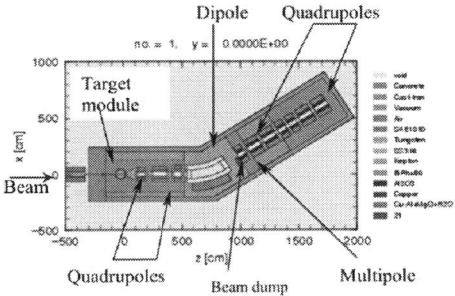

FIGURE 7. The geometry used in PHITS simulations in support of fragmentation pre-separator pre-conceptual designs is shown. The beam enters the pre-separator from the left to strike a target in the Target Module. The unreacted beam and produced fragments are transported through three quadrupoles followed by a dipole. The unreacted beam stops in the beam dump. The desired fragment is transported through the multipole and the six quadrupoles behind it. Acronyms for the materials used in the simulations are shown on the right of the panel.

For a given primary beam and its energy, the optimal target material and thickness for producing a desired fragment, the magnetic rigidity of the fragment is calculated and the magnitudes of the magnetic fields are adjusted using relative design values so that the desired fragment is transported through the pre-separator. There is a check that the primary beam is transported to the correct location at the beam dump. Figure 8 shows an example of a ^{48}Ca beam initially having 500 MeV/u striking a beryllium target to produce a beam of the fragment ^{42}Si. The unreacted beam is indeed transported to the correct location (obtained from the optics design) at the beam dump. The ^{42}Si fragments are transported correctly through the system, as shown in Figure 8.

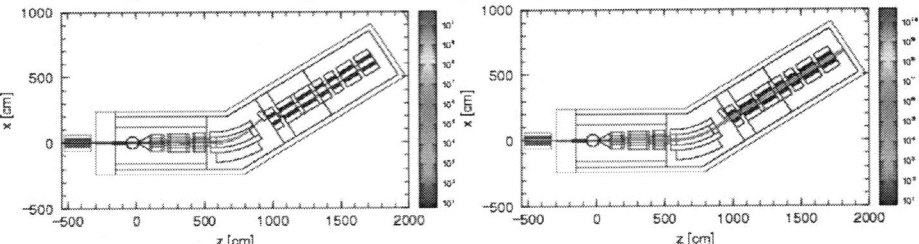

FIGURE 8. In the left panel, the simulation using PHITS, with the geometry shown in Fig. 6, is shown for a 500 MeV/u ^{48}Ca beam transported into the pre-separator target where it interacts with a beryllium target. The unreacted beam is transported to the beam dump. The number of ^{48}Ca ions per cm^2 is shown at the right of the panel. In the right panel, the simulation using PHITS is shown for a 500 MeV/u ^{48}Ca beam interacting with a beryllium target to produce ^{42}Si. The ^{42}Si ions are transported through the pre-separator. The number of ^{42}Si ions per cm^2 is shown at the right of the panel.

One purpose of simulations in the pre-separator area is to obtain fluxes of primary and secondary radiation that induce significant activation of the surrounding structural materials, and produce significant radiation damage (heat deposition, displacements-per-atom, hydrogen and helium gas production, *etc.*) in the target beam dumps and catchers, and surrounding materials. Figure 9 shows an example of the neutron radiation field simulated using PHITS for the system described above. Such results help us make decisions on issues like whether or not highly radiation-tolerant magnets are necessary, are component lifetimes sufficient, what are the needs for active thermal cooling of components and shielding, and how thick should be the biological shielding.

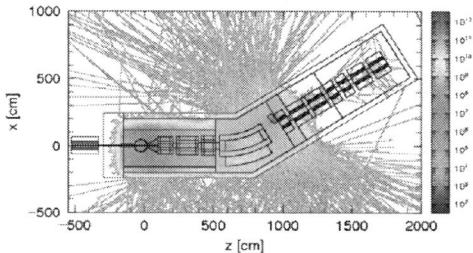

FIGURE 9. A simulation using PHITS of the neutron radiation field in and around a fragmentation pre-separator area is shown for a 500 MeV/u ^{48}Ca beam having 400 kW (~1 x 10^{14} ions/sec) interacting with a beryllium target. The levels of neutron flux are shown at the right, in units of neutrons per cm^2 per second.

CONCLUSIONS

The neutron production in heavy ion reactions and transport of neutrons through concrete and iron shielding as simulated by the PHITS code system agree very well with experiments. The calculated cross sections of isotope production in beam fragmentation also agree very well with experiment except when values of neutron excess are large. The latter suggests further investigation should be made of the evolution of fragments formed within the quantum molecular dynamical framework towards the statistical decay phases. The PHITS code system can also simulate the heavy ion transport and radiation fields after reactions of heavy ion beams with targets and beam dumps in the pre-separator area of a magnetic fragmentation separator system. Overall, our tests validate the PHITS code system show it to be a valuable tool to help design the next generation of radioactive beam facilities.

ACKNOWLEDGMENTS

This material is based on work supported by the U.S. Department of Energy under Grant No. DE-FG02-ER41313 and by the National Science Foundation under Grant No. PHY-01-10253. The author gratefully acknowledges contributions made under the DOE award by Drs. Inseok Baek, Lawrence Heilbronn and Susana Reyes, and contributions made under the NSF award by Drs. Michal Mocko and M. Betty Tsang. The author is also deeply grateful to Dr. Koji Niita for helpful suggestions for using the PHITS code system.

REFERENCES

1. H. Grunder (chair), *ISOL Task Force Report to NSAC* (22 November 1999).
2. Takashi Nakamura and Lawrence Heilbronn, *Handbook on Secondary Particle Production and Transport by High-Energy Heavy Ions*, World Scientific Publishing Co., Singapore, 2005.
3. M. Mocko, M. B. Tsang, L. Andronenko, M. Andronenko, F. Delaunay, M. Famiano, T. Ginter, V. Henzl, D. Henzlova, H. Hua, S. Lukyanov, W. G. Lynch, A. M. Rogers, M. Steiner, A. Stolz, O.

Tarasov, M.-J. van Goethem, G. Verde, W. S. Wallace, and A. Zalessov, *Projectile fragmentation of* ^{40}Ca, ^{48}Ca, ^{58}Ni, *and* ^{64}Ni *at 140 MeV/nucleon*, Phys. Rev. C (December 2006).

4. Michal Mocho, Ph.D. thesis, Michigan State University (September 2006).

5. P. Armbruster, J. Benlliure, M. Bernas, A. Boudard, E. Casarejos, S. Czajkowski, T. Enqvist, S. Leray, P. Napolitani, J. Pereira, F. Rejmund, M.-V. Ricciardi, K.-H. Schmidt, C. Ste´phan, J. Taieb, L. Tassan-Got, and C. Volant, *Measurement of a Complete Set of Nuclides, Cross Sections, and Kinetic Energies in Spallation of 238U 1A GeV with Protons*, Phys. Rev. Lett. **93** 212701 (2004).

6. B. Fernández-Domínguez, P. Armbruster, L. Audouin, J. Benlliure, M. Bernas, A. Boudard, E. Casarejos, S. Czajkowski, J.E. Ducret, T. Enqvist, B. Jurado, R. Legraina, S. Leray, B. Mustapha, J. Pereirad, M. Pravikoff, F. Rejmund, M.V. Ricciardi, K.-H. Schmidt, C. Stéphanc, J. Taiebb, L. Tassan-Got, C. Volant , W. Wlazloa, *Nuclide cross-sections of fission fragments in the reaction* $^{208}Pb + p$ *at 500 A MeV*, Nucl. Phys. A **747** 227-267 (2006).

7. H. Iwase, K. Niita, and T. Nakamura, *Development of General-Purpose Particle and Heavy on Transport Monte Carlo Code*, Journal of Nuclear Science and Tech. **39** 1142-1151 (2002). Hiroshi Iwase, Ph.D. thesis, Tohoku University (23 March 2003).

8. The institutions in this collaboration are Argonne National Laboratory, Lawrence Berkeley National Laboratory, Lawrence Livermore National Laboratory, Michigan State University and Oak Ridge National Laboratory.

9. Y. Iwata, T. Murakami, H. Sato, H. Iwase, T. Nakamura, T. Kurosawa, L. Heilbronn, R.M. Ronningen, K. Ieki, Y. Tozawa, and K. Niita, *Double-differential cross sections for the neutron production from heavy-ion reactions at energies E/A = 290-600 MeV*, Phys. Rev. C **64**, 054609 (2001).

10. L. Heilbronn, Y. Iwata, H. Iwase, T. Murakami, H. Sato, T. Nakamura, R.M. Ronningen, K. Ieki, I. Gudowska, and N. Sobolevsky, submitted to Nucl. Sci. and Engineering (August 2006).

11. M. Sasaki, E. Kim, T. Nunomiya, T. Nakamura, N. Nakao, T. Shibata, Y. Uwamino, S. Ito, A. Fukumura, *Measurements of High-Energy Neutrons Penetrated Through Concrete Shields Using Self-TOF, NE213, and Activation Detectors*, Nucl. Sci. Eng. **141** 140-153 (2002).

12. M. Sasaki, N. Nakao, T. Nunomiya, T. Nakamura, A. Fukumura and M. Takada, *Measurements of high energy neutrons penetrated through iron shields using the Self-TOF detector and an NE213 organic liquid scintillator*, Nucl. Instrum. Meth. B **196** 113-124 (2003).

13. MCNPX, Version 2.4.0, The MCNPX Team, LA-UR-02-5253, Los Alamos National Laboratory (August 2002).

14. Koji Niita, Satoshi Chiba, Toshiki Maruyama, Tomoyuki Maruyama, Hiroshi Takada, Tokio Fukahori, Yasuaki Nakahara, and Akira Iwamoto, *Analysis of the (N,xN') reactions by quantum molecular dynamics plus statistical decay model*, Phys. Rev. C **52** 2620-2635 (1995).

15. S. Furihata, *Statistical analysis of light fragment production from medium energy proton-induced reactions*, Nucl. Instr. Meth. B **171** 251-258 (2000).

The MCNPX Monte Carlo
Radiation Transport Code

Laurie S. Waters, Gregg W. McKinney, Joe W. Durkee,
Michael L. Fensin, John S. Hendricks, Michael R. James,
Russell C. Johns, Denise B. Pelowitz

Los Alamos National Laboratory, MS K575, Los Alamos, New Mexico 87545

Abstract. MCNPX (Monte Carlo N-Particle eXtended) is a general-purpose Monte Carlo radiation transport code with three-dimensional geometry and continuous-energy transport of 34 particles and light ions. It contains flexible source and tally options, interactive graphics, and support for both sequential and multi-processing computer platforms. MCNPX is based on MCNP4c and has been upgraded to most MCNP5 capabilities. MCNP is a highly stable code tracking neutrons, photons and electrons, and using evaluated nuclear data libraries for low-energy interaction probabilities. MCNPX has extended this base to a comprehensive set of particles and light ions, with heavy ion transport in development. Models have been included to calculate interaction probabilities when libraries are not available. Recent additions focus on the time evolution of residual nuclei decay, allowing calculation of transmutation and delayed particle emission. MCNPX is now a code of great dynamic range, and the excellent neutronics capabilities allow new opportunities to simulate devices of interest to experimental particle physics, particularly calorimetry. This paper describes the capabilities of the current MCNPX version 2.6.C, and also discusses ongoing code development.

Keywords: MCNPX, MCNP, Monte Carlo, Radiation Transport, Particle Transport, Neutronics
PACS: 24.10.Lx

INTRODUCTION

MCNPX [1] development began in 1994 with funding from the APT (Accelerator Production of Tritium) project. APT used a 1.0-GeV proton beam on a tungsten spallation target to produce a high neutron flux. The neutrons were moderated in a surrounding lead blanket and captured in He-3 gas to produce tritium. Due to the critical emphasis on neutronics in this project, the primary simulation tool was the LAHET (Los Alamos High Energy Transport) Monte Carlo code system [2]. LAHET is a set of codes allowing transport of 18 particles and includes a high energy package that used the Bertini [3], ISABEL [4,5], and FLUKA [6] models for primary interactions. Dresner evaporation [7], a Fermi Breakup model [8], and both the ORNL [9] and Rutherford-Appleton [10] fission models were included to model residual de-excitation. LAHET was one of the first codes to include a multistage pre-equilibrium exciton model [11] as an intermediate stage between the Intranuclear Cascade (INC) and evaporation phases of a nuclear interaction. Neutrons produced by LAHET below 20 MeV were written to a file for later transport by the MCNP code.

The MCNPX project was initiated to simplify the process of using the LAHET package, to extend the particle base, to enable the addition of new interaction models for

CP896, *Hadronic Shower Simulation Workshop*
edited by M. Albrow and R. Raja

high energy physics, and to provide upgrades to the capabilities of the latest MCNP code. Rather than including low energy neutronics physics in LAHET, it was decided to expand the capabilities of MCNP4c to include more particles, and physics models to calculate interaction probabilities when evaluated data libraries were not available. Initially, the standard LAHET physics packages described above were included; later development added the CEM and LAQGSM models [12], the INCL4 INC model [13], and the ABLA evaporation/fission code [14]. The APT project also sponsored the development of entirely new evaluated data libraries for photonuclear and proton interactions and also expanded the upper energy range of neutron evaluations for 42 isotopes to 150 MeV. APT also funded improvements in the CINDER'90 burn-up code, which was recently incorporated directly into MCNPX and allows simulation of transmutation and delayed particle production.

Since its inception, MCNPX has focused on the needs of the intermediate energy community, here taken to mean incident energies up to a few GeV. This involved a strong focus on accurate INC, evaporation and fission models, and a close coupling with traditional neutronics methods. This has enabled applications for a wide variety of projects, such as spallation target and accelerator shielding design, cosmochemistry, medical physics and Homeland Security applications. The initial addition of certain FLUKA physics raised the energy limits above traditional INC limitations, and the more recent inclusion of LAQGSM has greatly improved code capabilities to model very high energy interactions. The code now ranges from TeV energies to thermal neutrons. The inclusion of reaction time-evolution and the ever expanding set of allowed particles has given MCNPX applicability for experimental particle physics applications, especially in modeling very low energy experimental backgrounds.

THE CODE STRUCTURE

The following sections will describe the code availability and infrastructure, the geometry, and source definition capabilities. Problems in MCNPX are specified with well defined lines in input files, and no user coding is necessary to run a problem.

MCNPX Availability and Basic Structure

MCNPX is available from RSICC, and is accessed by limited number of beta testers, as explained on the official code web site, http://mcnpx.lanl.gov. The RSICC release is bundled with MCNP5, the evaluated data libraries needed to run the code, and several subsidiary libraries needed for the physics models in MCNPX. A few additional codes are included primarily for formatting data libraries and for converting tally information into formats readable by outside analysis packages. Although it is too early to predict a release date, MCNPX and MCNP are in the process of formally merging into one code package. MCNPX also comes with an extensive User's Guide [1], and regular updates on code features are given in release notes which can be examined on the web site. Five-day beginning, intermediate, and advanced classes are held on a regular basis, with the latest schedule available from the web site.

MCNPX is written in Fortran 90, with a few C++ routines primarily involving the graphics capabilities. The code can be downloaded as a binary for PC, LINUX, MAC,

Sun Solaris and IBM AIX platforms. The source code is also available and can be prepared through a special autoconf-generated configure script distributed with the code. This allows the user to set a number of parameters such as the specific compiler, location of libraries, multiprocessing options, and keywords that control specific code features. Actual compiling is done by a MAKE utility, either GNU MAKE, or a version supplied by the system vendor.

MCNPX supports distributed memory multiprocessing for the entire range of all particles. Parallel Virtual Machine (PVM) [15] and Message Passing Interface (MPI) [16] can be used to run the code in parallel. Fault tolerance and load balancing are available, and multiprocessing can be done across a network of heterogeneous platforms. Threading may be used for problems run in the tabular data region only. Many recent MCNPX applications have been done on Beowulf clusters ranging in size from a few to hundreds of nodes.

The code is controlled by a user-written line-formatted input file, with sections describing cell geometry, surface specification, and physics/tallying options. Two interactive GUIs are available for problem setup: MORITZ [17] and VISED [18] which is part of the official RSICC release. CAD geometry input is available through the GUIs and an embedded CAD capability is currently under development.

Geometry and Source Capability

MCNPX uses the MCNP surface-based geometry, in which defined surfaces are combined into geometrical cells. An interactive geometry viewer is included in the code for debugging purposes. Available surfaces include planes, spheres, cylinders, cones, ellipsoid/hyperboloid/ paraboloid, and an elliptical or circular torus. Each surface has a 'sense' defined by the surface normal, and cells are defined by combinations of surfaces, where a point within a cell is defined relative to the various surface senses. Surfaces are combined with Boolean operators: intersection, union and complement. Thus it takes six defined planes to describe a box. Recently this process has been simplified by the addition of macrobody surfaces, which conveniently combine the surfaces needed for common shapes into a single surface description. Ten objects are available, including the box, rectangular parallelepiped, sphere, right circular cylinder, right hexagonal prism, right elliptical cylinder, truncated right angle cone, ellipsoid, wedge, and an arbitrary polyhedron. The macrobody surfaces are used exactly like regular surfaces in the cell description. Due to variable dimensioning in the F90 code, there is in principle no limit to the number of surfaces or cells that may be defined in a problem.

Surfaces can be designated as 'reflecting', whereby any particle hitting the surface will be specularly (mirror) reflected. This aids in setting up a finite boundary problem which mimics an infinite volume. Surfaces can also be designated as 'white', where particles hitting the boundary are reflected with a cosine distribution relative to the surface normal. Periodic boundary conditions may also be applied to pairs of planes to simulate a simple infinite lattice. More complex lattices are described below.

Every cell must be filled with a material at a fixed density. The materials are designated as a combination of elements, either by weight or atomic fraction. The code's "Mix and Match" capability allows the user to specify if interactions will be performed through reference to evaluated nuclear data, directly computed with a model, or some combination of the two, on an isotope by isotope basis. Materials are assumed to be solid, unless a gas phase is specifically identified, and material temperature may also be specified for the free-gas treatment of thermal neutrons. It is also possible to specify conducting materials for electron-transport problems.

Both cells and surfaces can be rotated and translated by applying a transformation description consisting of an offset and/or a rotation matrix. Entire cells may be duplicated elsewhere in the problem, with changes allowed to cell material, material density, orientation, and other values related to variance reduction. The entire physical area of a problem must be specified out to infinity, which defines a 'universe'. Several different 'universes' may be defined in a problem, and cells of finite volume may be 'filled' with different universes. This allows an individual cell to be filled with widely varying geometries, much as a window in a wall shows a view of part of the outside world. Some or all cells in a universe may themselves be filled with other universes. These filled cells may also be replicated into rectangular or hexagonal lattices. The "fully specified fill" capability of the code allows each element of a lattice to be 'filled' with a universe of a different material. Figure 1 shows an example of a CT scan of a head specified with lattice geometry in MCNPX.

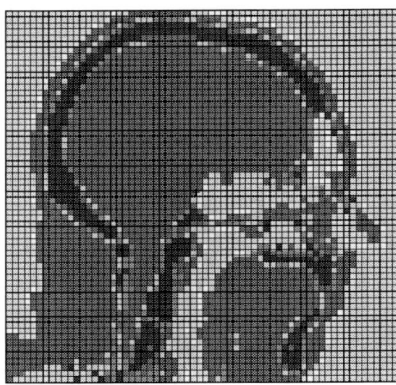

FIGURE 1. CT scan of a head using the fully-specified lattice fill capability of MCNPX.

Sources in MCNPX are designed by identifying the particle type, energy, time, position and direction. Sources can fill a cell, start from a surface, or form a geometry independent of the actual cell geometry. More than one particle type may be specified in a single source. It is often possible to specify one variable as a function of another variable, for example, energy can be a function of particle type. Energies can also be specified with preset functions, such as Maxwell or Watt fission energy spectra. Or, individual energies may be listed along with associated probabilities, either individually or as a histogram. Energy spectra can also be read from an external file, or the results of a calculation may be written to a file and read and sampled for further transport in a different problem. MCNPX can currently automatically generate spontaneous fission neutrons of appropriate multiplicity and energy if the user includes an actinide in the material description. Work is now in progress to allow the user to designate a radioactive material as a source, and the code will read in the decay gammas from a library. If no specifications are made, the default source is an isotropic 14 MeV neutron located at the problem origin.

TRANSPORT AND INTERACTION PHYSICS

MCNPX currently tracks 30 different particles and 4 light ions. Tracking is done to a user-settable lower kinetic energy cutoff, and particles will decay with their standard half-lives. The physics needed to run a simulation will be discussed in three parts. First, we review the MCNPX capabilities for charged particle transport: ionization, multiple scattering and energy straggling. Second, the tabular based physics options are discussed. Lastly, we briefly review the inline model physics used when tabular based options are not available or desired.

Charged Particle Transport

For heavy charged particles, the fundamental Bethe-Bloch formalism has been enhanced to include values and interpolation procedures recommended in ICRU Report 37 [19], bringing the model into closer ICRU compliance. The density effect correction now uses the parameterization of Sternheimer and Peierls [20]. For high-energy protons and other light charged projectiles, the approximate SPAR model [21] has been replaced with a full implementation of the maximum kinetic energy transfer. For intermediate energies, the shell corrections to the stopping power have been adapted from Janni [22]. Finally, a continuous transition in the stopping power between the ranges 1.31 MeV/AMU (Atomic Mass Unit) for the high-energy model, and 5.24 MeV/AMU (the low energy SPAR model) is achieved with a linear interpolation between the two models. Small angle Coulomb scattering uses the Rossi-Griesen algorithm. In the original theory, both angular deflections and small spatial displacements were accounted for. MCNPX does not yet accommodate transverse displacements in charged-particle subsets, therefore we only use the part of the theory that addresses the angular deflections. This is a slight effect and test cases have found negligible effects on the results. Energy straggling uses Vavilov logic, recently improved [23].

Evaluated Data Libraries

Nine classes of evaluated data libraries are available, with the most familiar being continuous-energy neutron, photonuclear, proton, neutron $S(\alpha,\beta)$ thermal data, photoatomic (up to 100 GeV), and electron interaction data (up to 1 GeV). Photoatomic and electron data are atomic in nature, and therefore depend only on Z. All others contain isotopic data. The $S(\alpha,\beta)$ data takes into account scattering in specific molecular structures, and substitutes for very low energy cross sections. These must be evaluated for materials, not isotopes, therefore only commonly used materials have yet been calculated.

Photon interaction libraries include Thompson, Compton, photoelectric and pair production cross sections. The electron libraries contain ionization and bremsstrahlung data. The neutron and photonuclear libraries contain total, elastic, (n,xn), fission, (n,γ) cross sections, and often heating numbers. All library data can be plotted directly in

85

MCNPX, for individual isotopes or for models. Many collections of evaluated libraries, such as ENDF, ENDL and JEF, have been formatted for use in the MCNP-type codes, primarily through the NJOY code. In the future, libraries may become available for very high energy interactions, but presently the community has not yet reached agreement on the physics to sufficient accuracy to warrant this effort.

The traditional process of sampling data libraries sometimes makes individual particle tracking difficult. Upon interaction, all channels are sampled, and the correlation of one interaction to the next is lost. Energy is conserved on average, but not on an individual track basis. Codes like PoliMi [24] are available which restore the correlation.

Physics Models

Most of the physics models included in MCNPX (listed in the Introduction) are discussed in papers in this conference and will not be detailed here. The general sequence of a calculation note starts with an INC model. After a time or energy cutoff (depending on the model), a pre-equilibrium phase is entered. A decision is then made to either fission or de-excite. De-excitation can follow an evaporation or Fermi-breakup model. Below the neutron emission threshold, evaporation proceeds by photon emission. The recent addition of the CINDER'90 transmutation code makes it possible to follow the time evolution of residuals either alone or with interactions in the presence of a neutron flux.

Figure 2 shows the results of Benchmark Problem 7 for this workshop. Energy deposition is calculated for an incident 50-GeV proton beam on a 10-cm-long, 1-cm-radius cylindrical tungsten target. The two MCNPX curves illustrate a problem sometimes encountered by MCNP users, and shows that methodologies which work well at very low energies often do not extrapolate to very high energies. The "Thick Target Bremsstrahlung model" is often used to speed up calculations. Electrons are generated, but their energies are deposited locally. Bremsstrahlung photons produced by the non-transported electrons are fully tracked. This works well at energies where produced photons do not have much energy, but is inappropriate for high-energy physics applications. The upper MCNPX curve shows local electron energy deposition. Full tracking of electrons gives a more consistent answer with other codes.

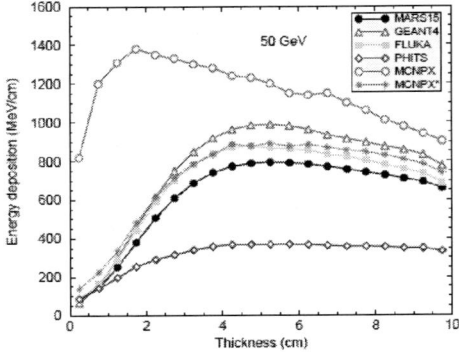

FIGURE 2. Effect of local energy deposition of electron energies..

TALLYING AND ANSWER CONVERGENCE

In the MCNP series of codes, answers are known as 'tallies'. The user specifies what tally is desired for what particles and for what region of the geometry, all with simply constructed entries in the problem input file. With any radiation transport code, three questions must be addressed: what is calculated, what is the accuracy of the answer, and how fast can the calculation be done. MCNPX has well tested predefined tally choices for current, fluence, and energy deposition, eliminating the need for the user to define these quantities from first principles. Rather than rely solely on variance, MCNPX has ten statistical tests that can be used to assess tally convergence. A large variety of variance reduction techniques are available in the code, and the skilled user can substantially reduce the amount of time needed for tally convergence, even with traditionally difficult problems. In discussing tallies and convergence, it is important to note that every particle in MCNPX carries a weight with it at all times. If the weight is 1.0, then every particle is counted as one particle. Weight can change for several reasons, primarily variance reduction. Depending on the technique used, particles can be split or multiplied, with their weights adjusted so that the answer can be proven to be a fair game with the mean value theorem.

MCNPX has four types of tallies, Track Length, Surface, Collision and Next Event estimators. Most of these can be calculated as a function of energy, angle and time. Track length refers to flux/fluence tallies in a volume. Flux is fluence per second, and it is up to the user to normalize the answer per unit time if desired. Flux can easily be weighted by quantities such as number density, cross sections, fluence-to-dose conversion factors, and heating factors in order to calculate physical observables. Many of these factors are contained in the evaluated data libraries. The MCNPX estimator for volume flux in a cell is the sum of all particle weights times their track lengths in the cell, divided by the cell volume.

It is an easy exercise to show that flux across a surface is the sum of particle weights divided by both surface area and the absolute value of the cosine of the particle direction with respect to the surface normal. MCNPX does have the capability to designate angular bins for surface flux, however the normal uses of volume flux do not need such a distinction. The particle will interact no matter what direction it is going. Another example of a Surface tally is current; the simple sum of particle weights crossing a surface, which can be divided into angular bins. A more complex type of Surface tally is the "pulse height" tally, used for calculating the energy deposited in a cell. The energy of particles leaving a volume is subtracted from the energy of the particle entering the volume, all for a single source event. This quantity can be folded in with detector response functions. Recent work in MCNPX has focused on making pulse height tallies work with variance reduction and allowing coincidence and anticoincidence with energy deposition in other detectors.

Collision estimators are primarily used to calculate energy deposition, and proceed in two different ways. If a library is available, heating factors are folded into a flux. In the model regime, separate components of energy deposition (ionization, residual recoil, local energy deposition for non-tracked particles, energies of particles falling below tracking cutoffs) are added together. This differs from the pulse-height tally in that the calculation is done on a particle-by-particle basis, not event-by-event.

Next-Event estimators, which in MCNPX are represented by 'point' and 'ring' detectors, allow the user to estimate fluxes at specific points, or on a ring if the problem has cylindrical symmetry. In evaluated data libraries, angular distributions of scattered particles are tabulated. Thus, the probability that such a particle could head in the direction of interest is known; for a flux calculation, the particle weight is multiplied by this probability. The weight is also divided by the square of the distance to the point and multiplied by an exponential attenuation factor representing the attenuation lengths of intervening materials. Both direct and scattered contributions to the tally are calculated. Arrays of point detectors are used to calculate either pinhole or transmitted radiography images. Point and ring detectors do not currently work with charged particles or in the model region where angular distribution are not tabulated.

Tallies can be viewed with a graphics routine directly in the code, either after completion of a calculation or while the problem proceeds. The 'mesh tally' is also available. This is a method for viewing any of the tally types over a large physical area. A user-defined three-dimensional rectangular, cylindrical, or spherical mesh is laid over the existing geometry, and particles tracked both through the mesh and geometry simultaneously. A quantity such as flux is calculated by summing weighted particle tracks through a mesh cell and dividing by the volume of that cell. Mesh tallies can be viewed with the tally plotter, overlaid on the problem geometry, or reformatted for viewing with an outside plotter.

Five separate quantities are tracked to assess tally convergence. These include the tally mean, relative error, variance of the variance, 'Figure of Merit' (FOM - defined the inverse of the variance multiplied by the time of the calculation), and the slope n in a fit of $1/x^n$ of the largest history scores of the probability density function. The code looks at the value of each quantity and performs certain tests as the problem progresses. For example, the relative error should be less than 5% and decrease as the problem progresses, with decrease rate going as the inverse of the square root of the number of source particles run. The FOM should be constant, with random fluctuations. The user will also have an efficiently running problem if FOM is large. It has been shown that the slope parameter must be greater than 3.0 in order to converge. The statistical check results can be printed for each tally, but it is up to the user to determine if the result is consistent with a converged answer. One common problem is inadequate sampling of the source distribution, which can cause large jumps in these quantities as the problem proceeds; the answer rarely settles down.

Variance reduction options in MCNPX are extensive, and basically include four types: truncation, population control, modified sampling, and partially deterministic methods. Point and ring detectors are examples of partially deterministic methods. Another is the DXTRAN sphere, which is similar to the point detector, except that particles are biased toward a sphere around the point of interest and can be further tracked within the sphere. This allows the calculation of detector responses in small, distant cells. Truncation refers to limiting the problem by stopping the calculation at pre-determined limits, such as a lower energy cutoff in particle tracking, or a time cutoff. Modified sampling methods include options such as biasing source particles in certain angular directions, energies, or times of interest.

Population control relies on particle splitting and Russian roulette to control the number of samples taken in various regions of phase space. One use of this is the 'weight window' methodology. A weight cutoff is specified, and particles above this weight are split, below the weight are Russian Rouletted. A user-defined mesh can be placed over the geometry, and weight window limits written to each cell in the mesh. In 'automated weight window' calculations, the weight windows are automatically adjusted by the code over successive runs to properly converge a particular tally located in a difficult area of the geometry.

SUMMARY

Current developments in MCNPX are largely tuned to the needs of the threat reduction and space science communities. We are adding tallying capabilities that focus on determining the origin of signals – specific reactions, or from particular residual nuclei. The code's ability to calculate delayed particle signals is under aggressive development. Photonuclear physics improvement is a continuing program, while both muon capture and resonant fluorescence physics will be added in the coming year. We are also improving our source capabilities to include automatic generation of gamma lines from specific isotopes as well as individual photon line data from delayed particle emission. Several approaches to correlated particle production are under investigation. Heavy ion transport is close to completion. Above all, MCNPX maintains a high standard of quality assurance and the MCNPX development team welcomes feedback from all users.

REFERENCES

1. D. B. Pelowitz, ed., "MCNPX User's Manual Version 2.5.0," Los Alamos National Laboratory, LA-CP-05-0369 (April 2005).
2. R. E. Prael and H. Lichtenstein, "User Guide to LCS: The LAHET Code System," Los Alamos National Laboratory, LA-UR-89-3014, Revised (September 15, 1989).
3. H. W. Bertini, "Intranuclear-cascade Calculations of the Secondary Nucleon Spectra fro Nucleon-Nucleus Interactions in the Energy Range 340 to 2900 MeV and Comparisons with Experiment," *Phys. Rev.* **188** 1711 (1969).
4. Y. Yariv and Z. Fraenkel, *Phys. Rev. C* **20** 2227 (1979).
5. Y. Yariv and Z. Fraenkel, *Phys. Rev. C* **24** 488 (1981).
6. G. Collazuol, A. Ferrari, A. Guglielmi and P. R. Sala, "Hadronic Models and Experimental Data for the Neutrino Beam Production,", *Nuclear Instruments & Methods* **A449** 609-613 (2000).
7. L. Dresner, "EVAP – A Fortran Program for Calculating the Evaporation of Various Particles from Excited Compound Nuclei", ORNL-TM-196, Oak Ridge National Laboratory (April 1962).
8. D. J. Brenner, R. E. Prael, J. F. Dicello, and M. Zaider, "Improved Calculations of Energy Deposition from Fast Neutrons," in *Proceedings Fourth Symposium on Neutron Dosimetry*, EUR-7448, Munich-Neuherberg (1981).
9. J. Barish et al., "HETFIS High-Energy Nucleon-Meson Transport Code with Fission," ORNL/TM-7882, Oak Ridge national Laboratory (July 1981).
10. F. Atchinson, "Spallation and Fission in Heavy Metal Nuclei under Medium Energy Proton Bombardment," in *Targets for Neutron Beam Spallation Sources*, Jül-Conf-34, Kernforschungsanlage Jülich GmbH (January 1980).
11. R. E. Prael and M. Bozoian, "Adaptation of the Multistage Preequilibrium Model for the Monte Carlo Method (I)," LA-UR-88-3238, Los Alamos National Laboratory (September 1988).

12. S. G. Mashnik, A. J. Sierk, K. K. Gudima, and M. I. Baznat, "CEM03 and LAQGSM03 – New modeling tools for nuclear applications," EPS Euroconference XIX Nuclear Physics Divisional Conference, *Journal of Physics: Conference Series 41*(2006) pp 340-351.

13. A. Boulard, J. Cugnon, S. Leray and C. Volant, "Intranuclear cascade model for a comprehensive description of spallation reaction data," *Phys. Rev. C* **66**, 044615-1 to 044615-28 (2002).

14. A. R. Junghans, M. de Jong, H.-G. Clerk, A. V. Ignatyuk, G. A. Kudyaev, and K.-H Schmidt, "Projectile-fragment yields as a probe for the collective enhancement in nuclear level density," *Nuclear Physics A* **629** 635-655 (1998).

15. CSM, Oak Ridge national Laboratory, "Parallel Virtual Machine (PVM)," http://www.csm.ornl.gov/pvm/pvm_home.html, accessed January 2, 2007.

16. MCS, Argonne National Laboratory, "Message Passing Interface (MPI) Standard," http://www-unix.mcs.anl.gov/mpi/.

17. http://www. whiterockscience.com/moritz.html, accessed January 2, 2007.

18. http://www/mcnpvised.com, accessed January 2, 2007.

19. International Commission on Radiation Units and Measurements (ICRU), "Stopping Powers for Electrons and Positrons," Report 37 (October 1984).

20. R. M. Sternheimer and R. F. Peierls, *Phys. Rev. B3, no. 11* (June 1, 1971), pp 3681.

21. T. W. Armstrong and K. C. Chandler, "Spar, A Fortran Program for Computing Stopping Powers and Ranges for Muons, Charged Pions, and Heavy Ions", ORNL-4869 (May, 1973).

22. J. F. Janni, "Proton Range-Energy Tables, 1keV-10GeV," *Atomic Data and Nuclear Data Tables 27*, (1982) 2/3.

23. R.E. Prael, "Proposed Modification to the Charged Hadron Tracking Algorithm in MCNPX", Los Alamos Research Note X-5-RN (U), LA-UR-00-4027 (August 23, 2000).

24. S.A.Pozzi, E. Padovani, and M. Marseguerra, "MCNP-POLIMI: A Monte Carlo Code for Correlation Measurements," *Nuclear Instruments and Methods in Physics Research A, Vol 513, Issue 3* (2003), pp 550-558.

MCNP/X Transport in the Tabular Regime

H. Grady Hughes

Los Alamos National Laboratory
Group X-3-MCC, MS A143
Los Alamos, NM 87545 USA

Abstract. We review the transport capabilities of the MCNP and MCNPX Monte Carlo codes in the energy regimes in which tabular transport data are available. Giving special attention to neutron tables, we emphasize the measures taken to improve the treatment of a variety of difficult aspects of the transport problem, including unresolved resonances, thermal issues, and the availability of suitable cross sections sets. We also briefly touch on the current situation in regard to photon, electron, and proton transport tables.

Keywords: Monte Carlo; neutron transport; particle transport; nuclear data tables.
PACS: 28.20.Gd

INTRODUCTION

Traditionally, and quite understandably, high-energy particle transport codes have enjoyed less extensive development for the low-energy regime than for high-energy processes. However, many aspects of low-energy transport are important for high-energy facilities and experiments. For example, the low-energy background of an experiment can significantly impact its interpretation. Similarly, the proper characterization of low-energy contributions to calorimeters can be essential, as is discussed elsewhere in this meeting. Serious personnel safety issues in facility shielding design also mandate a full understanding of the transport cascade, including the effects of low-energy particles. Some high-energy facilities (spallation sources) exist entirely for the purpose of producing low-energy secondary particles, such as neutrons for materials science research, production of useful isotopes, etc. Environmental issues, such as air and groundwater activation, are also related to low-energy particle behavior. For these and other reasons, the developers of high-energy transport codes are putting increasing efforts into improving their modeling of transport in the low-energy regime.

In contrast to most high-energy codes, MCNP[1] began life as a low-energy code transporting neutrons and photons (and later electrons) in energy ranges typically below 20 MeV. Important applications included reactor design, criticality safety, radiation shielding and protection, and later electron accelerator design, medical physics, and many other aspects of the low-energy world. Before high-energy transport methods and an extended set of particle types were introduced, first in the MCNPX code[2] and later in the (not yet released) developmental version[3] of MCNP itself, low-energy capabilities benefited from the full attention of the MCNP developers. As a result, the MCNP/X codes are a uniquely valuable resource

CP896, *Hadronic Shower Simulation Workshop*
edited by M. Albrow and R. Raja
© 2007 American Institute of Physics 978-0-7354-0401-4/07/$23.00

especially for high-energy applications in which the low-energy component is important. In this paper we shall review a number of the specific capabilities of the MCNP/X codes in the low-energy regime, with particular attention to those aspects that are associated with use of nuclear data tables.

EVALUATED NUCLEAR DATA TABLES

The essential source of the transport physics for MCNP is the collection of evaluated nuclear data tables that are developed and distributed with the code. MCNP uses data tables in a format based on the Evaluated Nuclear Data File[4] (ENDF), a national (and by now international) standard for format and content established and monitored by the Cross Section Evaluation Working Group (CSEWG). The ENDF tables can be supplemented by other evaluated sets such as the Evaluated Nuclear Data Library[5] (ENDL), leading to a powerful and quite general compilation of data for use in transport calculations. For example, a recent release of cross-section data for MCNP includes 974 "continuous energy" neutron tables covering 250 isotopes among 80 elements ranging from Z=1 to Z=98.

There are a number of different kinds of tables represented in the collection of transport data distributed with the code. We have mentioned "continuous energy" neutron tables, which constitute the most detailed available description of neutron transport. These tables typically address neutron processes from very low energies (on the order of 10^{-5} eV) up to the MeV range. Older tables, driven by applications such as reactor design or radiation shielding, often ended at 20 MeV. More recent ENDF releases, motivated by applications to be addressed by the MCNPX code, continue to 150 MeV. We shall discuss the contents of continuous-energy neutron tables in the next section, and the supplementary $S(\alpha,\beta)$ thermal tables in a later section.

Besides continuous-energy data, MCNP can use multigroup tables familiar from deterministic methods. This kind of data table represents the cross sections as averages over energy groups weighted by assumed energy-flux spectra. The energy groups are generally much larger than the resolution of continuous-energy tables, so that much detail is lost, especially in the neutron resonance region. For example, Figure 1 shows the total cross section for naturally occurring iron over most of the energy range of the ENDF/B-V evaluation, contrasting the detailed presentation of the continuous-energy table (26000.55c) with the relatively crude representation of a typical multigroup table (26000.55m). Further, the dependence on the assumed energy spectrum makes the results quite problem-dependent and requires a great deal of insight in the selection or preparation of the cross sections. For a wide range of Monte Carlo applications, there is now little reason to prefer a multigroup approach. However, there are special circumstances in which multigroup transport offers considerable advantage. For example, in simulations featuring diffuse, extended sources and very localized detectors (tallies), the adjoint method will usually provide a more efficient calculation than the standard forward method. Adjoint Monte Carlo in MCNP is only available as multigroup transport and requires the use of multigroup cross sections. Another popular method requiring multigroup tables is the simulation of coupled electron/photon transport by the Boltzmann-Fokker-Planck[6] (BFP) algorithm. Here the electrons and photon masquerade as neutrons and appropriate multigroup tables can be generated with the CEPXS[7] code developed by Sandia National Laboratory. Another area of recent interest is the use of a deterministic method to pre-calculate phase-space importances or weight-window parameters for variance reduction in a forward Monte Carlo calculation. In this hybrid method, the multigroup cross sections for the deterministic calculation correspond as closely as possible to the continuous-energy cross sections of the eventual Monte Carlo calculation. Finally, multigroup calculations may be needed in the context of validation and verification to compare Monte Carlo results with those of deterministic transport methods, in scoping studies or sensitivity studies, or to simulate isotopes for which continuous-energy cross sections are unavailable.

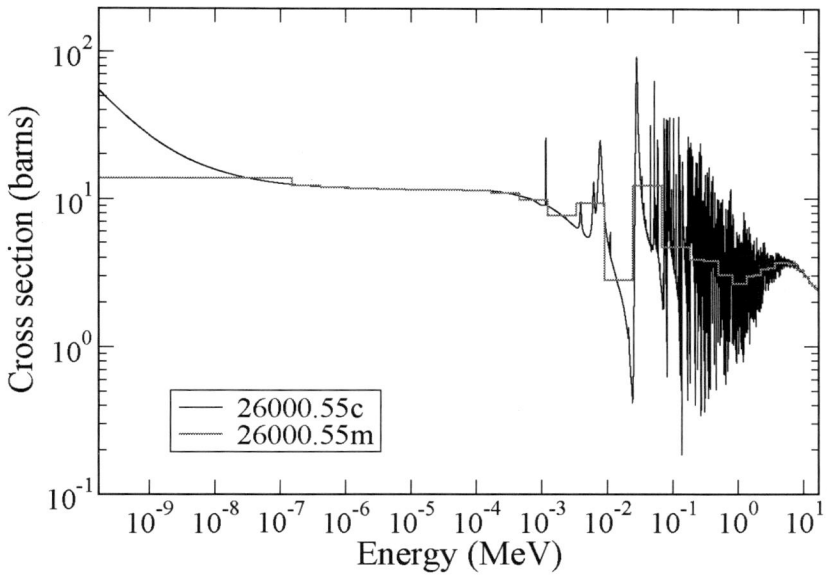

FIGURE 1. Total cross section for naturally-occurring iron from ENDF/B-V (26000.55c) and from an ENDF/B-V-based multigroup tabulation (26000.55m).

The data libraries distributed with MCNP also include an extensive set of neutron dosimetry cross sections. These tables are not full transport tables in the sense of the next section, and cannot be used to model the neutron transport in the specified isotope. Rather they contain energy-dependent cross sections for various specific reactions that neutrons may induce in the particular isotope. Thus these tables are suitable for calculation of reaction rates for trace elements in an experiment by integrating the reaction cross sections weighted by the neutron energy-flux spectrum. This capability is frequently used in neutron dosimetry and neutron activation applications. To generate these data, ENDF sources (ENDF/B-V Dosimetry Tape 531 and Activation Tape 532) are supplemented with ACTL[8], an evaluated neutron activation cross-section library from Lawrence Livermore National Laboratory. The current MCNP distribution includes 463 isotopes of 82 elements from Z=1 to Z=98.

CONTENTS OF CONTINUOUS-ENERGY NEUTRON TABLES

Continuous-energy cross sections are not, of course, continuous in the mathematical sense. They are necessarily tabulations of various data at a discrete set of energies. The word "continuous" is meant to suggest that, unlike multigroup data, the tabulation is for a set of energies sufficiently dense that all of the edges, resonances, and sundry features of the physical evaluation are completely represented to within some prescribed small tolerance. With modern data, that tolerance is typically 1% or better. Many recent tables have been processed with tolerances of 0.1%. In order to provide a complete description of the physical processes, data libraries must incorporate a large amount of information. There must be a transport table available for each isotope present in a simulation. Each of these tables must provide as a function of energy a full accounting of the reactions that a neutron may undergo with the given isotope. For all energies, total, absorption, and elastic cross sections, and average heating numbers are given. For each reaction channel in the evaluation, partial cross sections and reaction-specific Q-values are provided. For every reaction producing secondary particles (other neutron, photons, or other particles types), energy and angular probability distributions for sampling must exist. There may be a large number of reaction channels. For example a recent tabulation for ^{56}Fe (26056.66c) contains individual information for (n,2n), (n,np), and (n,nα) reactions, for excitation to the first 25 excited states, and for excitation to the continuum. For fissionable isotopes, total fission cross sections and average ν (number of fission neutrons) must be present along with fission neutron energy distributions. In many cases, both ν and fission neutron spectra will be tabulated separately for prompt fission and for several modes of delayed fission together with channel probabilities.

In the generation of secondary particles, some sampling distributions are correlated in reaction channel, energy, and angle. For example, elastic scattering is microscopically correct because energy-momentum conservation can be imposed in the sampling process. Some other sampling distribution are coupled in energy and angle as well. However, many distributions for secondary particles are provided as independent energy spectra and angular distributions and therefore lead to uncorrelated sampling of the secondary-particle phase space. In addition, in any reaction generating more than one secondary particle, the various particles from the reaction will be sampled independently. The best-known example of this situation is the fact that neutron-induced photons are generated from the photon-production probability distributions without regard to the selection of the actual neutron reaction. The result of all this lack of correlation is that MCNP, using tabular neutron data, is not well suited to the simulation of coincidence experiments and other application of "microscopically correct" tallies. However, these matters are irrelevant to the macroscopic tallies of MCNP such as particle flux and current, energy spectra, angular distributions, reaction rates, energy deposition, etc. For these standard calculations of integrated quantities, the important issue is the correctness of the transport methods in the average.

UNRESOLVED RESONANCE PROBABILITY TABLES

In Figure 1 we saw an example of a cross-section table containing sufficient data to represent an essentially complete description of the transport process over the entire range of the table. By contrast consider Figure 2, which shows two different tabulations of the total cross section for ^{235}U, specifically an older set from ENDF/B-V (92235.50c) and a newer set from ENDF/B-VI (92235.69c). The abrupt loss of detail (at just above 0.08 keV for the older table and above 2.25 keV for the newer) clearly indicates that the nuclear resonance region is incompletely described in both evaluations. In fact the resonance region for ^{235}U extends up to 25 keV. The meandering curve representing the total cross section above the detailed portion of the resonance region is simply an average of the cross section over energy ranges in which full data are not available. The neglect of the details of the cross section in the unresolved-resonance region would ignore self-shielding effects, which can have a significant impact on the results of the calculation. While the application programmer should be strongly motivated to use the latest cross-section sets, with their increased coverage of the resonance region, the problem of an unresolved region remains. Fortunately, modern evaluations provide an excellent approximation for dealing with this problem. Beginning with MCNP version 4C (and therefore MCNPX) and later versions, the code is able to take advantage[9] of cross-section probability tables derived from statistical information in evaluations such as average level spacings and average resonance widths. When these data are provided in a transport table, a probabilistic approach can be taken in which the cross section encountered by the neutron in the unresolved resonance region is sampled from a probability distribution. With appropriate attention to the transport logic, self-shielding can be well simulated even in the absence of detailed knowledge of the energy-dependence of the cross section. The current data distribution incorporates probability-table data into the cross-section tables of 72 isotopes among 21 elements from Z=40 to Z=98.

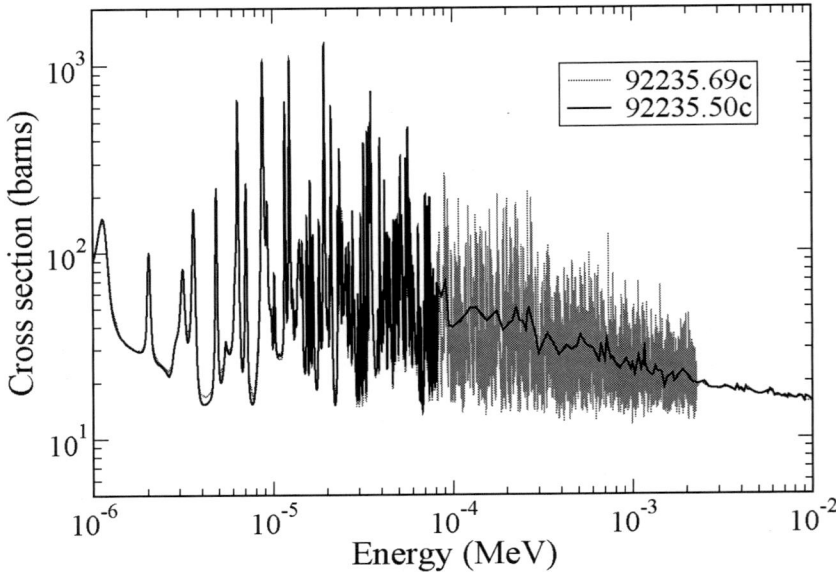

FIGURE 2. Total cross section for ^{235}U from ENDF/B-V and from ENDF/B-VI showing the detailed representation of the resonance region only up to 0.08 keV for the earlier evaluation (92235.50c) and up to 2.25 keV for the later (92235.69c). Neither evaluation provides a complete representation up to the true end of the resonance region (about 25 KeV).

THERMAL ISSUES

Each neutron transport table in the data distribution has been created for a specific temperature. Usually this is room temperature (293.6 K), but for selected isotopes, other temperatures (*e.g.* 20 K, 77 K, 600 K, 3000 K, *etc.*) are provided. For simulations of systems at temperatures other than those provided in the cross-section sets, the results can be affected because of the different thermal velocities of the target nuclei. In MCNP there is a straightforward analytic/stochastic method (the free-gas model) for dealing with this situation, but this method applies only to the elastic scattering channel of the cross section. For problems not dominated by elastic scattering, the temperature dependences can be important. The differences in the resonance region between cross sections evaluated at different temperatures can be quite significant and can have severe effects on the results of the calculation. One really needs cross-section sets evaluated at the correct temperature of the system.

MCNP does not use the ENDF-format files directly. Rather a processing code such as NJOY[10] is used to write the cross-section tables in a format called ACE (A Compact ENDF) with is used directly by the code. It is at this step that a specific temperature is selected. Thus to generate cross sections at a particular temperature not available in the standard distribution, one would expect to have to run NJOY. This has never been an easy matter for the ordinary applications programmer, because NJOY is an extremely complex program with a daunting array of options, and is generally considered something that should be approached only by a specialist in the program. Recently, however, the portion of the logic of NJOY relating to temperature dependence has been isolated in a user-friendly code called DOPPLER[11] and, even more conveniently, has been incorporated into the cross-section librarian called MAKXSF, which is distributed with the MCNP code and data. It is now easy for the user to take an existing cross-section table (or two) and generate a corresponding table at a different temperature. There are two cases. First, for tables in which the resonance region is completely described (or for an application in which the unresolved resonance region is unimportant) one needs only a reference table at a lower temperature than that desired. The MAKXSF code can produce a Doppler-broadened table for the needed temperature. Second, for unresolved-resonance probability tables or for the $S(\alpha,\beta)$ thermal tables to be discussed in the next section, one needs tables evaluated above and below the desired temperature. Then the MAKXSF code can interpolate to produce a table at any temperature between the two reference tables.

THERMAL NEUTRON TRANSPORT TABLES

In the previous section we discussed issues related to thermal motion of target nuclei and associated with Doppler broadening and neutron/nucleus kinematics. There are also more difficult issues affecting low-energy transport, including molecular binding effects, vibrational/rotational levels, lattice spacing, crystal structure, and other solid- and liquid-state properties. The standard ENDF evaluations concentrate on the neutron/nucleus interaction and do not address these matters. Revisiting Figure 1, for example, we see no detail at low energies. In the thermal region, at energies from about 4 eV down to 10^{-5} eV, the cross sections fall into simple $1/v$ form ignoring the complexities of the low-energy regime. In MCNP the best currently-available approach to improving this situation is the use of $S(\alpha,\beta)$ transport tables. These tables are presented as scattering matrices, correlated in scattered energy and angle, and attempting to represent the effects of many of the physical processes that complicate the low-energy neutron transport. When present, they replace the standard cross sections for neutron energies below 4 eV. For applications in which the thermal regime is important, the use of appropriate $S(\alpha,\beta)$ tables can be essential for obtaining accurate results. In the current data distribution there are $S(\alpha,\beta)$ tables available at a variety of relevant temperatures for beryllium metal, beryllium oxide, benzene, ortho and para liquid hydrogen, ortho and para liquid deuterium, graphite, deuterium in heavy water, hydrogen in light water, hydrogen in liquid methane, hydrogen in polyethylene, hydrogen in solid methane, and hydrogen and zirconium in zirconium hydride. Although this collection is a powerful resource for applications that are covered by this list, it is clear that this is a very limited list considering the vast range of applications that can arise in the thermal regime. Obviously this is an area that can benefit greatly from future expansion.

PHOTO-ATOMIC AND ELECTRON TRANSPORT TABLES

MCNP also simulates the coupled photon/electron cascade using methods based on those of the ETRAN[12] codes and of the Integrated TIGER Series[13]. Photo-atomic transport tables provide data describing the four basic photon interaction processes. (1) At photon energies above the threshold ($2mc^2$ for electron mass m), the photon may produce an electron/positron pair. The selection of energies for the two products is based on Bethe-Heitler theory[14], and the angular distribution comes from an adaptation of high-energy theory. Triplet production is not considered separately from pair production. (2) Compton (incoherent) scattering is based on the Klein-Nishina[15] cross section, but is modified by form factors that partially account for the effects of binding energy on the angular distribution[16]. In a recent enhancement, MCNP also takes advantage of data to include the effects of the momentum of bound electrons on the Compton energy distribution. This capability is sometimes confusingly called Compton Doppler broadening, but does not have anything to do with thermal motions of the target atoms, which are not considered for photon or electron transport in MCNP. (3) Thomson (coherent) scattering, arguably the least important of the principal processes, is optionally included in the transport and is also modified by form factors. (4) The photoelectric interaction is the most approximate of the four treatments. MCNP treats only K-shell vacancies and produces at most two fluorescent photons or an Auger electron. The lines that may be selected are restricted to (L3 → K), (L2 → K), (mean M → K), (mean N → K), and (mean upper levels → L). These limitations are somewhat mitigated by the fact that photon/electron transport in MCNP never extends below 1 keV, and is often limited to still higher energies. Nevertheless, the improvement of the treatment of this low-energy regime is a desired future plan for the MCNP/X developers. The upper energy limit for the most recent photo-atomic tables is 100 GeV.

Electron tables, supplemented by analytic models, also support the modeling of the photon/electron cascade. MCNP treats bremsstrahlung as in Version 3 of the Integrated TIGER Series[17], including its effect on stopping powers. Knock-on electrons are sampled from the Møller cross section[18], with angular distributions determined from kinematics. Characteristic X-ray and Auger electron production rely on Kolbenstvedt's cross section[19] and are restricted to the same lines mentioned above for photoelectric events. The electron angular deflection is calculated from the Goudsmit-Saunderson distribution[20], and the electron energy loss straggling is based on Landau's theory[21] with the various extensions described in Ref. 12. The low-energy limit of the MCNP electron tables is 1 keV, as is the case with photo-atomic tables. For historical reasons, the upper energy limit for the electron tables is 1 GeV.

PHOTONUCLEAR TABLES

A recent enhancement of the transport capabilities of MCNP is the addition of code and data to support the simulation of photonuclear reactions at low energies. (Various event-generator models such as the Cascade Exciton Model[22] and Los Alamos Quark-Gluon-String Model[23] deal with photonuclear reactions, but become increasingly uncertain at low energies.) Photonuclear tables contain energy-dependent photon cross sections for nuclear interactions leading to the production of photons, neutrons, protons, deuterons, tritons, helions (^{3}He), and alpha (^{4}He) particles. The tabular data are limited to energies below 150 MeV and address photoabsorption by the excitation of either the giant dipole resonance[24] or a quasi-deuteron nucleon pair[25-27]. Production data including energy and angular distributions are given. In contrast to photo-atomic tables, which are provided for elements and make no distinction among isotopes, the photonuclear tables are isotope-specific. The initial release[28] of photonuclear tables for MCNP/X contained only 13 isotopes: ^{2}H, ^{12}C, ^{16}O, ^{27}Al, ^{28}Si, ^{40}Ca, ^{56}Fe, ^{63}Cu, ^{181}Ta, ^{184}W, ^{206}Pb, ^{207}Pb, and ^{208}Pb. This is an area which will be greatly aided by the eventual availability to MCNP/X of the newly released ENDF/B-VII data[29]. That compilation will include a new sub-library with 163 photonuclear evaluations.

PROTON TRANSPORT TABLES

Proton transport tables were developed at Los Alamos in order to support the Accelerator Production of Tritium program, the same program that was a major driving force behind the original development of the MCNPX code itself. These tables include total, nuclear elastic plus interference, and inelastic cross sections; production cross sections for photons, neutrons, protons, deuterons, tritons, hellions, alphas, and heavy recoils; double-differential (energy and angle) production spectra for neutrons, protons, deuterons, tritons, hellions, and alphas; and angle-integrated emission spectra for photons and heavy recoils. Like the photonuclear tables, proton tables provide data up to 150 MeV. These tables are intended to describe nuclear reactions only: energy loss and angular deflection caused by collisional multiple scattering are to be modeled by other methods. Traditionally in MCNPX, these processes were simulated using the Vavilov[30] model and a Gaussian approximation[31] described by Rossi. Recently we have developed and implemented a new multiple scattering model[32] that takes into account projectile and nuclear target form factors, and provides a coupled sampling of both collisional energy loss and angular deflection.

The initial release[33] of proton transport tables for MCNPX included 42 isotopes ranging from ^{1}H to ^{209}Bi. The release if ENDF/B-VII[29] will expand this list somewhat, adding six more proton tables and introducing five tables for deuteron projectiles, three for tritons, and two for helions. The new projectile types will present a welcome challenge to MCNP/X developers to generalize the code, which currently assumes that only neutrons, photons, electrons, and protons make use of transport tables.

SUMMARY

We have presented a basic introduction to the capabilities of the MCNP and MCNPX transport codes in the low-energy regime, with particular attention to the use of transport data tables. These codes offer powerful and accurate methods for the simulation of particle transport, especially for applications in which the details of all energy ranges are important.

REFERENCES

1. X–5 Monte Carlo Team, *MCNP — A General Monte Carlo N–Particle Transport Code. Version 5, Volume I: Overview and Theory.* Los Alamos National Laboratory report LA–UR–03–1987 (April 24, 2003).
2. H. G. Hughes, K. J. Adams, M. B. Chadwick, J. C. Comly, S. C. Frankle, J. S. Hendricks, R. C. Little, R. E. Prael, L. S. Waters, P. G. Young, "MCNPX — The LAHET/MCNP Code Merger," in Proceedings of the Third Workshop on Simulating Accelerator Radiation Environments (SARE3), May 7-9, 1997, Tsukuba, Japan (KEK Proceedings 97-5, June 1997) pp. 44-51.
3. H. Grady Hughes, Forrest B. Brown, Jeffrey S. Bull, John T. Goorley, Robert C. Little, Lon-Chang Liu, Stepan G. Mashnik, Richard E. Prael, Elizabeth C. Selcow, Arnold J. Sierk, Jeremy E. Sweezy, John D. Zumbro, Nikolai V. Mokhov, Sergei I. Striganov, Konstantin K. Gudima, "MCNP5 for Proton Radiography," in The Monte Carlo Method: Versatility Unbounded in a Dynamic Computing World, proceedings of Monte Carlo 2005, Chattanooga, Tennessee, April 17-21, 2005. On CD-ROM, American Nuclear Society, LaGrange Park, IL (2005).
4. V. McLane, C. L. Dunford, P. F. Rose, ed., "ENDF-102: Data Formats and Procedures for the Evaluated Nuclear Data File, ENDF-6," Brookhaven National Laboratory report BNL-NCS-44945, revised (1995) (available URL: http://www.nndc.bnl.gov/).
5. R. J. Howerton, D. E. Cullen, R. C. Haight, M. H. MacGregor, S. T. Perkins, E. F. Plechaty, "The LLL Evaluated Nuclear Data Library (ENDL): Evaluation Techniques, Reaction Index, and Descriptions of Individual Reactions," Lawrence Livermore Scientific Laboratory report UCRL-50400, Vol. 15, Part A (September 1975).
6. J. E. Morel, L. J. Lorence, Jr., R. P. Kensek, J. A. Halbleib, D. O. Sloan, "A Hybrid Multigroup/Continuous-Energy Monte Carlo Method for Solving the Boltzmann-Fokker-Planck Equation," *Nucl. Sci. Eng.*, 124, p.369-389 (1996).
7. L. J. Lorence, Jr., J. E. Morel, G. D. Valdez, "Physics Guide to CEPXS: A Multigroup Coupled Electron-Photon Cross-Section Generating Code, Version 1.0," Sandia National Laboratory report SAND89-1685 (1989).
8. M. A. Gardner, R. J. Howerton, "ACTL: Evaluated Neutron Activation Cross-Section Library — Evaluation Techniques and Reaction Index," Lawrence Livermore National Laboratory report UCRL-50400, Vol. 18 (October 1978).
9. L. L. Carter, R. C. Little, J. S. Hendricks, R. E. MacFarlane, "New Probability Table Treatment in MCNP for Unresolved Resonances," *1998 ANS Radiation Protection and Shielding Division Topical Conference, April 19-23, 1998, Nashville, TN*, Vol. II, p. 341-347.
10. R. E. MacFarlane, D. W. Muir, "The NJOY Nuclear Data Processing System, Version 91," Los Alamos National Laboratory report LA-12740-M (1994).
11. R. E. MacFarlane, P. Talou, "DOPPLER: A Utility Code for Preparing Customized Temperature-Dependent Data Libraries for the MCNP Monte Carlo Transport Code," Los Alamos National Laboratory internal report (Oct. 3, 2003).
12. Stephen M. Seltzer, "An Overview of ETRAN Monte Carlo Methods," in *Monte Carlo Transport of Electrons and Photons*, edited by Theodore M. Jenkins, Walter R. Nelson, and Alessandro Rindi, p. 153 (Plenum Press, New York, 1988).
13. J. Halbleib, "Structure and Operation of the ITS Code System," *ibid*, p. 249.
14. H. A. Bethe, J. Ashkin, "Passage of Radiations through Matter," in *Experimental Nuclear Physics, Vol. I*, edited by E. Segre (John Wiley, New York, 1953) p. 166.
15. O. Klein, Y. Nishina, "Über die Streuung von Strahlung durch freie Elektronen nach der neuen relativistischen Quantendynamik von Dirac," *Z. Phys.* 52 (1929) 853.
16. J. F. Williamson, F. C. Diebel, R. L. Morin, "The Significance of Electron-Binding Corrections in Monte Carlo Photon Transport Calculations," *Phys. Med. Biol.* 29 (1984) 1063.
17. S. M. Seltzer, "Cross Sections for Bremsstrahlung Production and Electron-Impact Ionization," in *Monte Carlo Transport of Electrons and Photons*, edited by Theodore M. Jenkins, Walter R. Nelson, and Alessandro Rindi, p.81.

18. C. Møller, "Zur Theorie des Durchgang schneller Elektronen durch Materie," *Ann. Physik*. **14** (1932) 568.
19. H. Kolbenstvedt, "Simple Theory for K-Ionization by Relativistic Electrons," *J. Appl. Phys*. **38** (1967) 4785.
20. S. Goudsmit, J. L. Saunderson, "Multiple Scattering of Electrons," *Phys. Rev*. **57** (1940) 24.
21. L. Landau, "On the Energy Loss of Fast Particles by Ionization," *J. Phys. (USSR)* **8** (1944) 201.
22. Gudima, K. K., Mashnik, S. G., and Toneev, V. D. *Cascade-exciton model of nuclear reactions*, Nucl. Phys. **A401**, 329 (1983).
23. Gudima, K. K., Mashnik, S. G., and Sierk, A. J. *User Manual for the Code LAQGSM*, Los Alamos National Laboratory report LA-UR-01-6804 (2001).
24. A. Bohr, B. R. Mottelson, *Nuclear Structure*, 2nd Edition (World Scientific: Singapore 1998).
25. J. S. Levinger, "Neutron Production by Complete Absorption of High-Energy Photons," *Nucleonics* **6** #5 (1950) 64.
26. J. S. Levinger, "The High-Energy Nuclear Photoeffect," *Phys. Rev*. **84** #1 (1951) 43.
27. M. B. Chadwick, P. Oblozinsky, P. E. Hodgson, G. Reffo, "Pauli-Blocking in the Quasideuteron Model of Photoabsorption, *Phys. Rev. C* **44** #2 (1991) 814.
28. M. C. White, "Release of the LA150U Photonuclear Data Library," Los Alamos National Laboratory memorandum X-5:MCW-00-87(U), July 26, 2000, revised March 21, 2001.
29. M. B. Chadwick, P. Oblozinsky, M. Herman, *et al.*, "ENDF/B-VII.0: Next Generation Evaluated Nuclear Data Library for Nuclear Science and Technology," *Nuclear Data Sheets* **107** #12 (2006) 2931.
30. P. V. Vavilov, "Ionization Losses of High-Energy Heavy Particles," *Soviet Physics JETP* **5** (4) (1957) 749.
31. B. Rossi, *High-Energy Particles*, Prentice-Hall, Incorporated, New York 1952.
32. S. Striganov "On Theory and Simulation of Multiple Coulomb Scattering of Heavy Particles," in *Proceedings of the ICRS 10/RPS 2004*, Madeira, Portugal, 9-14 May 2004.
33. M. B. Chadwick, P. G. Young, S. Chiba, S. Frankle, G. M. Hale, H. G. Hughes, A. J. Koning, R. C. Little, R. E. MacFarlane, R. E. Prael, L. S. Waters, "Cross Sections Evaluations to 150 MeV for Accelerator-Driven Systems Implementation in MCNPX," *Nucl. Sci. Eng*. **131** (1999) 293.

High energy hadron production Monte Carlos

J.Ranft

Siegen University, Germany

Abstract. We discuss here Quantum molecular dynamics models (QMD) and Dual Parton Models (DPM and QGSM). We compare RHIC data to DPM–models and we present a (Cosmic ray oriented) model comparison.

Keywords: <Monte Carlo models, Inclusive hadron production, Dual Parton model, Quantum Molecular Dynamics model>
PACS: <12.40.Nn, 13.85.Ni, 13.85.Tp>

QUANTUM MOLECULAR DYNAMICS MODELS (QMD)

The emphasis of this contribution is to high energies, therefore, QMD non–relativistic models ($E < 2AGeV$) are not treated here.

The first relativistic QMD model was RQMD [1]. This model is no longer supported since the year 2000 ,RQMD is used in FLUKA for A–A collisions below 5AGeV.

A second relativistic model is UrQMD [2], it is used in CORSIKA Cosmic Ray cascade code below 80AGeV.

Let me mention some efforts within the FLUKA collaboration which will not be treated here: F.Cerutti et al. add (approximate) energy conservation , evaporation, and residual nuclei to RQMD [3]. M.V.Garzelli et al. construct a low energy QMD for A–A collisions in FLUKA [4]. There are efforts in Milano and Houston to construct a fully relativistic model for A–A collisions to be inserted into FLUKA.

A relativistic QMD model is a Lorentz invariant cascade (molecular dynamics) with nucleons of both nuclei and all produced hadrons as participants. Properties of such models are: (i)A formation zone cascade of all produced hadrons. (ii) Elementary interactions used in the models include: (1) h + h $\longrightarrow$ resonance; resonance + resonance and resonance decay (similar to HADRIN in FLUKA), (2) high eneny: h + h $\longrightarrow$ hadronic chain; 2 hadronic chains and Lund like chain fragmentation, (3) chain fusion (called formation of color ropes) in RQMD, (4) empirical parametrization of all cross sections, (5) pQCD description of hard collisions (UrQMD).

The kinematics of the UrQMD model

The model is fully described in [2]. It is based on the covariant propagation of all hadrons on classical trajectories in combination with stochastic binary scatterings, color string formation and resonance decay. It includes the Monte Carlo solution of a set of coupled partial integro-differential equations. Each nucleon is represented by a coherent

CP896, *Hadronic Shower Simulation Workshop*
edited by M. Albrow and R. Raja

state ($\hbar, c = 1$)

$$\phi_i(\vec{x}, \vec{q}_i, \vec{p}_i, t) = \left(\frac{2}{L\pi}\right)^{3/4} \exp\left\{-\frac{2}{L}(\vec{x} - \vec{q}_i(t))^2 + i\vec{p}_i(t)\vec{x}\right\} \tag{1}$$

which is characterized by 6 time-dependent parameters, $\vec{q}_i$ and $\vec{p}_i$. L, (the extension of the wave packet in coordinate space) is fixed. The total n-body wave function is a product of coherent states (1) $\Phi = \prod_i \phi_i(\vec{x}, \vec{q}_i, \vec{p}_i, t)$. The Hamiltonian H of the system contains a kinetic term and mutual interactions V_{ij} ($H = \sum_i T_i + \frac{1}{2}\sum_{ij} V_{ij}$). This yields an Euler-Lagrange equation for each parameter.

$$\dot{\vec{p}}_i = -\frac{\partial \langle H \rangle}{\partial \vec{q}_i} \quad \text{and} \quad \dot{\vec{q}}_i = \frac{\partial \langle H \rangle}{\partial \vec{p}_i}. \tag{2}$$

These are the time evolution equations which are solved numerically. The UrQMD Hamiltonian contains: $E_{kin}, E_{jk}^{Sk2}, E_{jkl}^{Sk3}, E_{jk}^{Yukawa}, E_{jk}^{Coulomb}, E_{jk}^{Pauli}$.

Please note: As one sees from the Hamiltonian UrQMD is not really a Lorentz invariant molecular dynamics. Therefore, the results of the model might depend strongly on the reference frame in which the calculation is done. To minimize this frame dependence the authors [2] use a frame–independent definition of the cross sections (via using the impact parameters in the two–particle rest frames). They give as example for this minimization in S–S collisions: The multiplicities and collision numbers vary only by less than 3% between the lab and CM frames.

Let us mention, that RQMD has a manifestly Lorentz invariant eq. of motion. Using 4–vectors for positions and momenta, each particle carries its own time. The 2N additional degrees of freedom are fixed by 2N constraints: The N mass shell constraints: $H_i = p_i^2 - m_i^2 - V_i = 0$, ($V_i$: quasi potential) and (N-1) time fixations. The 2Nth constraint: A relation of times of particles to the evolution parameter τ.

Projectile or target nucleus are modeled according to a Fermi-gas ansatz. The centroids of the Gaussians are randomly distributed within a sphere with the radius $R(A)$,

$$R(A) = r_0 \left(\frac{1}{2}\left[A + \left(A^{\frac{1}{3}} - 1\right)^3\right]\right)^{\frac{1}{3}} \quad r_0 = \left(\frac{3}{4\pi\rho_0}\right)^{\frac{1}{3}}. \tag{3}$$

ρ_0 is the nuclear matter ground state density. If the phase-space density at the location of each nucleon is too high (i.e. the area of the nucleus is already occupied), then the location of that nucleon is rejected and a new location is randomly chosen. The initial momenta of the nucleons are randomly chosen between 0 and the local Thomas-Fermi-momentum: $p_F^{max} = \hbar c \left(3\pi^2\rho\right)^{\frac{1}{3}}$, with ρ being the corresponding local proton- or neutron-density. One disadvantage of this type of initialization: the initialized nuclei are not in their ground-state with respect to the Hamiltonian used for the propagation. The parameters of the Hamiltonian are tuned to the equation of state of infinite nuclear matter and to properties of finite nuclei (such as their binding energy and their root mean square radius). One can use a so-called Pauli-potential in the Hamiltonian: This has the advantage that the initialized nuclei remain stable whereas in UrQMD with

103

the conventional initialization and propagation without the Pauli-potential the nuclei start evaporating single nucleons after approximately 20 - 30 fm/c. One drawback of this potential: the kinetic momenta of the nucleons are not anymore equivalent to their canonic momenta, i.e. the nucleons carry the correct Fermi-momentum, but their velocity is zero. The impact parameter of a collision is sampled according to the quadratic measure $(dW \sim b\,db)$. At a given impact parameter the centers of projectile and target are placed along the collision axis in such a manner that a distance between surfaces of the projectile and the target is equal to 3 fm. The momenta of the nucleons are transformed into the system where the projectile and target have equal velocities directed in opposite directions of the axis. After that the time propagation starts. During the calculation each particle is checked at the beginning of each time step whether it will collide within that time step.

The relativistic QMD models are compared to a large sample of data in the publications of the authors. Up to energies of about $\sqrt{s} = 200$ GeV usually a good agreement is found. The experimental collaborations at RHIC compare their data quite often to RQMD results, One example: The PHOBOS Collaboration in their white paper [5] find a good agreement to dN_{ch}/dy from RQMD for Au–Au collisions to their data, the reason for this is the chain fusion build from the beginning into RQMD.

The main problems of the relativistic QMD models, which make them difficult to apply for shower simulations are (i) problems with the energy conservation, (ii) the missing evaporation and residual nuclei and (iii) the partly excessive computer running times.

DPM AND QGSM MODELS

The Dual Parton Model DPM and the Quark Gluon String Model QGSM are two models, which are largely equivalent in their construction, only with some characteristic differences. The DPM is due to Capella, Tran Than Van and collaborators [6]. The most detailed Monte Carlo versions of the DPM are PHOJET [7, 8] for h–h and γ–h collisions and DPMJET for h–A, A–A and γ–A collisions [9].

The QGSM is due to Kaidalov, Ter-Martyrosian and collaborators [10]. The Monte Carlo version QGSJET for h–h, h–A and A–A collisions is due to Kalmykov, Ostapchenko and collaborators [11].

The construction of the PHOJET multichain model

We restrict us in this contribution to describe the PHOJET model, which is used directly for h–h collisions and for all elementary Glauber collisions in h–A and A–A collisions in DPMJET . There are no essential differences in the formulations of the Glauber model between different Monte Carlo models.

The (soft) Born cross section of the supercritical pomeron has the form $\sigma_s = g^2 s^{\alpha(0)-1}$. The supercritical pomeron has $\alpha(0) > 1.$, therefore it clearly violates unitarity. According to the Froissart bound the cross section asymptotically should not rise faster than $(\log s)^2$.

If we start to construct the full model, which is unitarized, we should introduce some more input Born cross sections. Very important is the hard cross section, which we calculate according to the QCD improved parton model:

$$\sigma^{\text{hard}}(s, p_\perp^{\text{cutoff}}) = \int dx_1 dx_2 d\hat{t} \sum_{i,j,k,l} \frac{1}{1+\delta_{k,l}}$$
$$f_{a,i}(x_1, Q^2) f_{b,j}(x_2, Q^2) \frac{d\sigma^{\text{QCD}}_{i,j \to k,l}(\hat{s}, \hat{t})}{d\hat{t}} \Theta(p_\perp - p_\perp^{\text{cutoff}}), \qquad (4)$$

where $f_{a,i}(x_1, Q^2)$ is the distribution of the parton i in a.

One of the most important difference between PHOJET/DPMJET and QGSJET is in the $p_\perp$ cutoff $p_\perp^{\text{cutoff}}$: PHOJET/DPMJET use $p_\perp^{\text{cutoff}}$ rising with energy. QGSJET uses $p_\perp^{\text{cutoff}}$ constant, independent of the energy.

We introduce furthermore the cross sections for high–mass single and double diffraction σ_D and for high–mass central diffraction σ_C according to the standard expressions. The amplitudes corresponding to the one-pomeron exchange are unitarized applying an eikonal formalism. In impact parameter representation, the eikonalized scattering amplitude has the structure

$$a(s, B) = \frac{i}{2} \left(\frac{e^2}{f_{q\bar{q}}^2} \right)^2 \left(1 - e^{-\chi(s,B)} \right) \qquad (5)$$

with the eikonal function

$$\chi(s, B) = \chi_S(s, B) + \chi_H(s, B) + \chi_D(s, B) + \chi_C(s, B). \qquad (6)$$

Here, $\chi_i(s, B)$ denotes the contributions from the different Born graphs: (S) soft part of the pomeron and reggeon, (H) hard part of the pomeron (D) triple- and loop-pomeron, (C) double-pomeron graphs.

The eikonals $\chi_i(s, B)$ are defined as follows

$$\chi_i(s, B) = \frac{\sigma_i(s)}{8\pi b_i} \exp[-\frac{B^2}{4b_i}]. \qquad (7)$$

The free parameters are fixed by a global fit to proton-proton cross sections and elastic slope parameters. Once the free parameters are determined, the probabilities for the different final state configurations are calculated from the discontinuity of the elastic scattering amplitude (optical theorem).

The total discontinuity can be expressed as a sum of graphs with k_c soft pomeron cuts, l_c hard pomeron cuts, m_c triple- or loop-pomeron cuts, and n_c double-pomeron cuts by applying the Abramovski-Gribov-Kancheli cutting rules. In impact parameter space one gets for the inelastic cross section

$$\sigma(k_c, l_c, m_c, n_c, s, B) = \frac{(2\chi_S)^{k_c}}{k_c!} \frac{(2\chi_H)^{l_c}}{l_c!} \frac{(2\chi_D)^{m_c}}{m_c!} \frac{(2\chi_C)^{n_c}}{n_c!} \exp[-2\chi(s, B)] \qquad (8)$$

with

$$\int d^2B \sum_{k_c+l_c+m_c+n_c=1}^{\infty} \sigma(k_c,l_c,m_c,n_c,s,B) \approx \sigma_{\text{tot}} \qquad (9)$$

where σ_{tot} denotes the total cross section

In the Monte Carlo realization of the model, the different final state configurations are sampled from Eq. (8). For pomeron cuts involving a hard scattering, the complete parton kinematics and flavors/colours are sampled according to the Parton Model. For pomeron cuts without hard large momentum transfer, the partonic interpretation of the Dual Parton Model is used: mesons are split into a quark-antiquark pair whereas baryons are approximated by a quark-diquark pair. The longitudinal momentum fractions of the partons are given by Regge asymptotics . We give it here for an event with n_s soft and $n_h (n_h \geq 1)$ hard cut pomerons, sea–quarks are used at the chain ends if we have more than one soft pomeron.

$$\rho(x_1,...,x_{2n_s},...,x_{2n_s+2+n_h}) \sim \frac{1}{\sqrt{x_1}} \left(\prod_{i=3}^{2n_s+2} \frac{1}{x_i} \right) x_2^{1.5}$$

$$\prod_{i=2n_s+3}^{2n_s+2+n_h} g(x_i,Q_i) \delta(1 - \sum_{i=1}^{2n_s+2+n_h} x_i). \qquad (10)$$

The distributions $g(x_i,Q_i)$ are the distribution functions of the partons engaged in the hard scattering. The momentum fractions of the constituents at the ends of the different chains are sampled from this exclusive parton distribution,

After all this we have all chains defined and PHOJET/DPMJET continues with hadronizing all multiple chains using the Lund code JETSET (PYTHIA).

Now we are able to compare the multichain model PHOJET with particle production data. There are many comparisons to data published in the PHOJET, DPMJET and QGSJET literature. Here we present only two examples: the average multiplicity of all kinds of secondary particles in p–p collisions as function of the energy in Fig.1 and the rapidity distribution of charged hadrons in central S-S and S–Ag collisions in Fig.2 at SPS energies.

Next we present comparisons of the DPM–models (DPMJET–III) with RHIC data We first present some comparisons, where DPMJET-III is used in its pre–RHIC form. In Fig.3 we compare rapidity distributions of charged hadrons in p–p and d–Au collisions according to DPMJET with RHIC data from the PHOBOS collaboration. In Fig.4 we compare the transverse momentum distribution measured at RHIC by the PHENIX collaboration with PHOJET calculations, we find the hard collisions very well represented in PHOJET.

For other comparisons DPMJET needs some modifications to get agreement with the RHIC data. One of the most important modification is the **Percolation of hadronic strings in** DPMJET–III

Using the original DPMJET–III with enhanced baryon stopping and a centrality of 0 to 5 % the DPMJET multiplicities are larger than the ones measured in Au–Au collisions at RHIC. A new mechanism needed to reduce N_{ch} and $dN_{ch}/d\eta|_{\eta=0}$ in situations with a produced very dense hadronic system. We consider only the percolation and fusion of

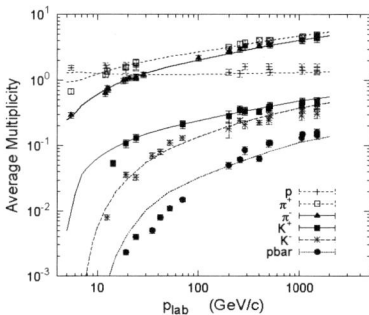

FIGURE 1. Average particle multiplicity proton-proton interactions. PHOJET results (curves) are compared to experimental data (symbols).

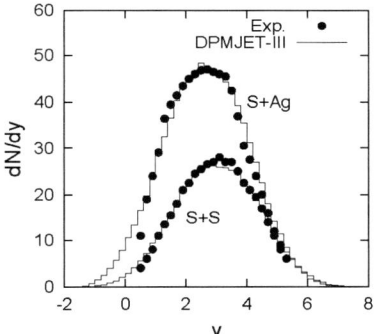

FIGURE 2. Rapidity distributions of negative hadrons in central nuclear collisions at 200 GeV/nucleon.

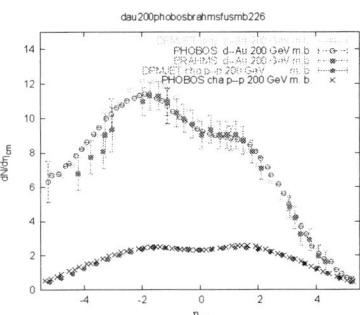

FIGURE 3. Pseudorapidity distribution of charged hadrons produced in minimum bias $\sqrt{s} = 200$ GeV d–Au and p–p collisions. The results of DPMJET are compared to experimental data from the BRAHMS–Collaboration and the PHOBOS–Collaboration. At some pseudorapidity values the systematic PHOBOS–errors are given.

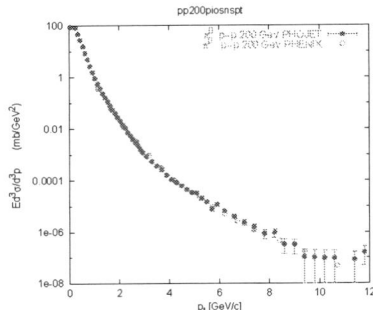

FIGURE 4. Transverse momentum distribution as measured in p–p collisions at $\sqrt{s}$ = 200 GeV by the PHENIX collaboration at RHIC compared to the calculation by PHOJET

soft chains (the transverse momenta of both chain ends are below a cut–off $p_\perp^{fusion} =$ 2 GeV/c). The condition of percolation is, that the chains overlap in transverse space. We calculate the transverse distance of the chains L and K R_{L-K} and allow fusion of the chains for $R_{L-K} \leq R^{fusion} = 0.75$ fm. The chains in DPMJET are fragmented using the Lund code. Only the fragmentation of color triplet–antitriplet chains is available in JETSET, however fusing two arbitrary chains could result in chains with other colors. Therefore, we select only chains for fusion, which again result in triplet–antitriplet chains. Examples are:

(i) A $q_1 - \bar{q}_2$ plus a $q_3 - \bar{q}_4$ chain become a $q_1 q_3 - \bar{q}_2 \bar{q}_4$ chain.
(ii) A $q_1 - q_2 q_3$ plus a $q_4 - \bar{q}_2$ chain become a $q_1 q_4 - q_3$ chain.
(iii) A $q_3 - q_1 q_2$ plus a $q_4 - \bar{q}_1$ plus a $\bar{q}_3 - q_5$ chain become a $q_4 - q_2 q_5$ chain.
(iv) A $q_4 - \bar{q}_1$ plus a $q_5 - \bar{q}_3$ plus a $\bar{q}_5 - q_1$ chain become a $q_4 - \bar{q}_3$ chain.

The expected results of these transformations are a decrease of the number of chains. Even when the fused chains have a higher energy than the original chains, the result will be a decrease of the hadron multiplicity $N_{hadrons}$. In reaction (i) we observe new diquark and anti–diquark chain ends. In the fragmentation of these chains we expect baryon–antibaryon production anywhere in the rapidity region of the collision. Therefore, (i) helps to shift the antibaryon to baryon ratio of the model into the direction as observed in the RHIC experiments.

In Fig. 5 we compare the pseudorapidity distributions in Au–Au collisions at 200*A GeV as measured by the PHOBOS collaboration at RHIC for different centralities with the DPMJET–III results obtained with the model including chain percolation and fusion.

Further RHIC related improvements (not treated here because of the limited space) in DPMJET include:

(i) Modified $p_\perp$ distributions in PYTHIA: For the fragmentation of soft chains the Gaussian transverse momentum distributions in PYTHIA have to be replaced by expo-nential ones.

(ii) Collision scaling in h–A collisions: To obtain collision scaling in DPMJET we have to change the sampling of hard chains.

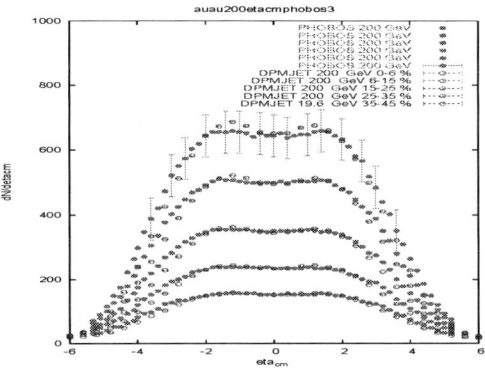

FIGURE 5. Pseudorapidity distributions of charged hadrons in Au–Au collisions at $\sqrt{(s)} = 200$ GeV for centralities 0–5 % up to 40–50 %. The points with rather small error bars are from the DPMJET–III Monte Carlo with chain fusion as described in the text. The data points are from the PHOBOS Collaboration.

(iii) Anomalous baryon stopping: New diagrams lead to more baryons in central region.

(iv) Modified diquark fragmentation: Find missing diagram in diquark fragmentation, to get Antihyperon to Hyperon ratios into agreement with experiment.

MODEL COMPARISONS

This section is based on a talk of D.Heck, Karlsruhe: Comparison of models in the CORSIKA Cosmic Ray cascade code at the VIHKOS CORSIKA School 2005, Lauterbad, Germany, May 31 – June 5, 2005 and Dieter Heck, private communication.

High energy models in CORSIKA used for $E_{lab} \geq$ **80 GeV** include: DPMJET 2.55 (J.Ranft,Phys.Rev.D51 (1995) 64), NEXUS 2/3 (J.Drescher et al., Phys.Rep.350 (2001) 93), QGSJET 01/II/III (S.Ostapchenko, Nucl.Phys.B(Proc.Suppl.)2005), SIBYLL 2.1 (R.Engel et al.,Proc.26th ICRC 1(1999)415) .

Low energy models in CORSIKA used for $E_{lab} \leq$ **80 GeV** include: FLUKA 2003 (only hadron production model) (A.Fasso et al.,Proc.Monte Carlo 2000 (2001)955), GHEISHA 2002 (H.Fesefeld, PITHA–85/02 Aachen (1985)), UrQMD 1.3 (S.A.Bass et al.,Prog.Part.Nucl.Phys.41 (1998)225).

In Fig.6 we compare the average charged multiplicity in p–p collisions as function of the energy between the high energy models. Only the three QGSJET models differ strongly from the rest, at high energies the multiplicity increases much stronger than in the other models. This is the result of the energy independent $p_\perp$ cut–off. In Fig.7 we compare the X_{max} of proton and iron induced vertical showers according to the high energy models with X_{max} data. Such a comparison is hoped to determine finally the composition of the highest energy particles in the cosmic radiation. Even, when the models differ considerably like in Fig.6 in their properties, we find here only modest differences in the X_{max} predictions.

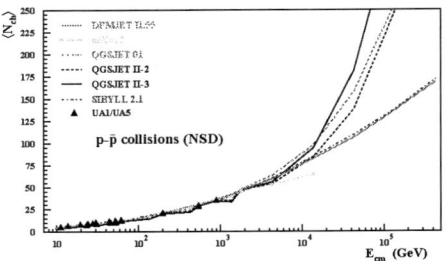

FIGURE 6. Table 1 Charged particle average multiplicity in p–$\bar{p}$ collisions.

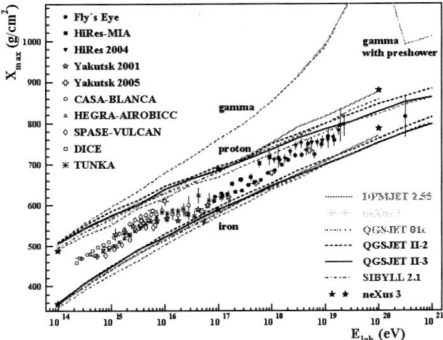

FIGURE 7. Penetration depth X_{max} for gamma, proton and iron induced vertical showers as function of the energy.

In Table 1 we compare the CPU times of the high energy and low energy models. We find only NEXUS and UrQMD to need much more running time than the other models.

CPU–times (sec)	for DEC–Alpha 1000XP		
low energy model	100000 p-Air coll 10 GeV	high energy model	10000 p-Air coll 1 PeV
FLUKA	181	DPMJET 2.55	271
GHEISHA 2002	108	NEXUS 2	3145
UrQMD 1.3	12200	QGSJET II	693
		SIBYLL 2.1	186

SUMMARY AND CONCLUSIONS

Code comparisons

* Within 10 years of CORSIKA code comparisons: models have much improved.

* Accelerator physics oriented code comparisons could help in a similar way.
* include evaporation particles and residual nuclei.
* compare also hadron calorimeter performance, produced and residual radioactivity.

Relativistic QMD models

* Impressive performance for nucleus–nucleus collisions up to RHIC energies.
* Missing: exact energy conservation, excited residual nuclei and evaporation, residual nuclei. (Patches to include this into FLUKA). Computer running times of these models excessively long.
* Construct improved relativistic model which includes all properties needed for cascades at accelerators, this could become a genuine alternative to DPM, QGSM models.

DPM, QGSM models

* Impressive performance for hadron–hadron, hadron–nucleus, nucleus–nucleus, photon–hadron and photon–nucleus collisions up to present collider energies.
* Improvements through CORSIKA code comparisons.
* Acceptable agreement of all models up to Auger Cosmic Ray energies. Includes also predictions for all cross sections.
* These are the models which include best evaporation and residual nuclei needed for accelerator applications.

ACKNOWLEDGMENTS

I thank very much Dr. Dieter Heck from Karlsruhe for the permission to present material from his code comparison in this talk.

REFERENCES

1. H. Sorge, H. Stöcker, and W. Greiner, *Ann. Phys. (NY)* **192**, 266 (1989).
2. S.A.Bass et al., *Prog. Part. Nucl. Phys.* **41**, 225 (1998).
3. G.Battistoni et al. (F.Cerutti) , *Braz. J. Phys.* **34**, 897 (2004).
4. M.V.Garzelli et al. , *J. Phys. Conf. Ser.* **41**, 519 (2006).
5. B. B.Back et al, PHOBOS Collaboration (2002), nuxl–ex/0201005.
6. A. Capella, U. Sukhatme, C. I. Tan, and J. Tran Thanh Van, *Phys. Rep.* **236**, 225 (1994).
7. R.Engel, *Z. Phys.* **C66**, 203 (1995).
8. R.Engel and J.Ranft, *Phys. Rev.* **D54**, 4244 (1996).
9. S. Roesler, R. Engel, and J. Ranft, *hep–ph/0012252, Proc. of Monte Carlo 2000, Lisboa, Oct.2000,Springer,p.1033* , 1033 (2000).
10. A. B. Kaidalov, and K. A. Ter-Martirosyan, *Phys. Lett.* **B117**, 247 (1982).
11. N.N. Kalmykov et al., *Physics of Atomic Nuclei* **58**, 1728 (1995).

Hadronic Shower Code Inter-Comparison and Verification[1]

N.V. Mokhov and S.I. Striganov[2]

Fermilab, Batavia, IL 60510, U.S.A.

Abstract. To evaluate the quality of general purpose particle interaction and transport codes widely used in the high-energy physics community, express benchmarking is conducted. Seven tasks, important for high-energy physics applications, are chosen. For this first shot, they are limited to particle production on thin and thick targets and energy deposition in targets and calorimetric setups. Five code groups were asked to perform calculations in the identical conditions and provide results to the authors of this report. Summary of the code inter-comparison and verification against available experimental data is presented in this paper. Agreement is quite reasonable in many cases, but quite serious problems were revealed in the others.

Keywords: Hadrons, cascades, simulation, particle production, energy deposition
PACS: 13.85.-t, 24.10.Lx

INTRODUCTION

Predictive power and reliability of hadron-nucleus event generators and general purpose particle interaction and transport codes are of a great importance in numerous detector, accelerator, shielding and cosmic ray applications. All code development groups do perform their code verifications and usually well document them. At the same time, several instances were discussed in the community over last few years, with a puzzling disagreement between simulations performed by the code users and data, and between different versions of the same code. Therefore, it was decided to conduct an express code benchmarking, limiting it for this first shot by the energy range and two values important primarily for neutrino experiments and calorimetry as well as for accelerator and shielding applications: particle production on thin and thick targets and energy deposition in targets and calorimetric setups.

Certainly, we realized that there are many other cases to consider. Some members of the Workshop Organizing Committee have proposed several important calorimetric tasks. Many things – such as neutronics, shielding, low energies, nuclide production, etc. – were intentionally left aside as not directly related to the workshop and rather time-consuming to be performed by the meeting. We have limited the list to the tasks which can easily be simulated by the code teams on a very short notice.

[1] Work supported by Fermi Research Alliance, LLC, under contract No. DE-AC02-07CH11359 with the U.S. Department of Energy.
[2] On behalf of Benchmarking Working Group: M. Baznat, P. Folger, K. Gudima, H.-S. Lee, N. Matsuda, N. Mokhov, N. Nakao, K. Niita, P. Sala, T. Sanami, S. Striganov, L. Waters, D. Wright.

CP896, *Hadronic Shower Simulation Workshop*
edited by M. Albrow and R. Raja

TASKS FOR INTER-COMPARISON

Seven tasks were chosen to cover the workshop primary goals and be simple in modeling by all the codes involved, reliable and well documented on the experimental side. Five codes used worldwide in the above applications and discussed in detail at this workshop are involved in this analysis: FLUKA [1], GEANT4 [2], MARS [3], MCNPX [4], and PHITS [5]. In addition, two stand-alone event generators – LAQGSM03 [6] and DPMJET-III [7] – were involved in benchmarking for a particle production on thin targets. The principal developers of the codes were asked to submit results of their simulations to the authors of the report. This assured one that the latest versions of the codes were used and that the calculations are performed in the most "optimal" way.

The following tasks were proposed to be calculated by each of the code group:

1. HARP experiment (2006). 12.9 GeV/c $p + Al \rightarrow \pi^+ + X$: 1D and 2D inclusive production cross-sections.
2. NA49 experiment (2006). 158 GeV/c $p + A \rightarrow \pi^+, \pi^- + X$: 1D and 2D inclusive production cross-sections, for (a) proton and (b) carbon targets.
3. IHEP experiment (1980). 67 GeV/c protons on a thick aluminum target: double differential yields for $p, \overline{p}, \pi^+, \pi^-, K^+, K^-$.
4. PAL experiment (2004). Neutron spectra at several angles from a thick target irradiated by a 2-GeV electron beam.
5. KEK experiment (2004). 12 GeV protons on a thick target: energy deposition distribution in a surrounding cylindrical absorber.
6. CDHS-measured longitudinal profiles of hadronic showers for 10, 20, 50 and 100 GeV pion beams on an iron-scintillator calorimeter: longitudinal and lateral energy deposition profiles.
7. Energy deposition longitudinal profiles in a 10-cm thick tungsten target for proton beam energies of 1, 20 and 50 GeV.

Five groups sent their results to S. Striganov at Fermilab by the beginning of the Workshop. Table 1 lists the codes and the contributor names. The QGSP physics list was used in GEANT4. MCNPX-2.40 was used for Tasks 4 and 5, while the newest version MCNPX-2.6b03 was used for Task 7. Stand-alone event generator results were additionally provided for Task 1 (LAQGSM03, K. Gudima) and Task 2 (LAQGSM03, K. Gudima and DPMJET-III, M. Baznat).

TABLE 1. Summary of main contributors.

Task	FLUKA-2005	GEANT4-8.1	MARS15	MCNPX	PHITS-2.13
1	P. Sala	-	S. Striganov	-	N. Matsuda
2	P. Sala	P. Folger	S. Striganov	-	N. Matsuda
3	-	D. Wright	S. Striganov	-	N. Matsuda
4	-	-	T. Sanami	T. Sanami	-
5	-	D. Wright	N. Mokhov	N. Matsuda	N. Matsuda
6	-	-	N. Nakao	-	N. Matsuda
7	P. Sala	D. Wright	N. Mokhov	L. Waters	N. Matsuda

TASK 1

Recently, the HARP collaboration has published new data on double differential cross sections of a positive pion production in proton-aluminum interactions at 12.9 GeV/c [8]. A comparison of FLUKA-2005, LAQGSM03, MARS15 and PHITS results with the HARP data is presented in Fig. 1. The codes agree with the data at p>2 GeV/c, while at lower momenta and small angles, LAQGSM and MARS tend to overestimate experiment by up to 15%.

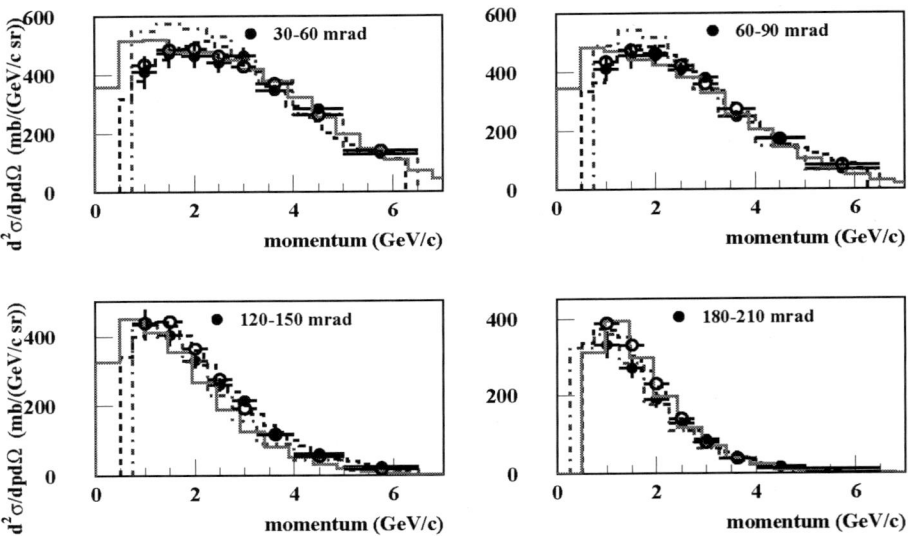

FIGURE 1. Comparison of HARP data (filled circles) with simulations: PHITS - opaque circles, LAQGSM - solid, FLUKA - dashed, MARS - dot-dashed lines.

TASK 2

The NA49 collaboration has measured transverse momentum distributions at fixed Feynman x_F values as well as integrated x_F and rapidity distributions for charged pion production in proton-proton and proton-carbon collisions at 158 GeV/c [9, 10].

Proton target

Results for comparison with data [9] are available from the DPMJET-III, LAQGSM03, MARS15, and PHITS groups. Integrated rapidity distributions are compared in Fig. 2. MARS and PHITS agree very well with measurements, DPMJET overestimates high-momentum and mid-rapidity positive pion yields, LAQGSM somewhat overestimates fast and underestimates central negative pion yields. The invariant cross sections as a function of transverse momentum at two x_F are presented in Fig. 3 together with the three code results. A comparison of simulations with data for the other six x_F values can be found in Ref. [11]. MARS results agree quite well with data while both LAQGSM and DPMJET underestimate central pion production and overestimate it at high transverse momenta.

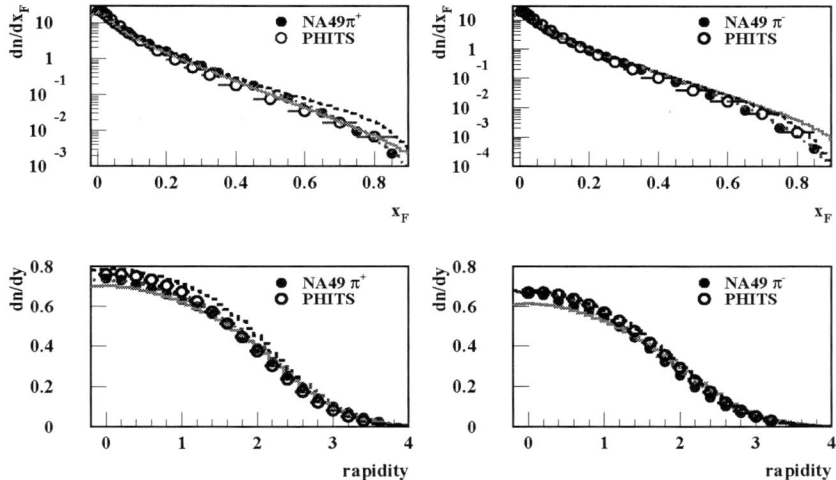

FIGURE 2. Comparison of NA49 pp-data (filled circles) with simulations: PHITS - open circles, LAQGSM - solid, DPMJET - dashed, MARS - dot-dashed lines.

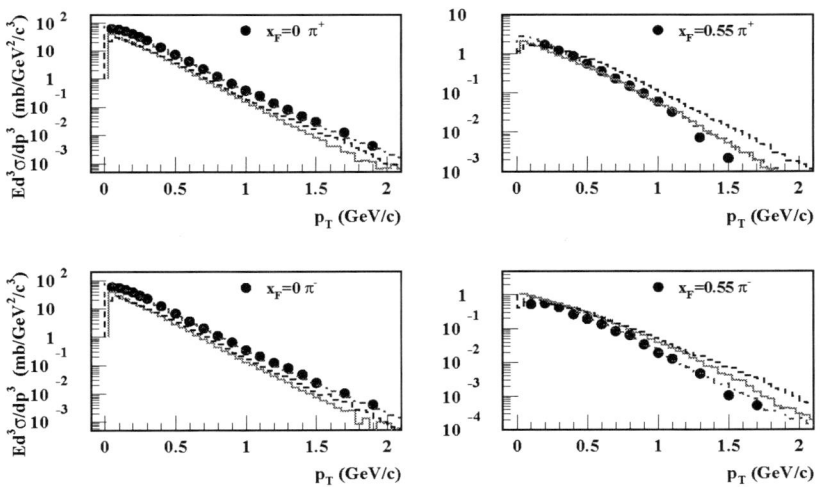

FIGURE 3. Comparison of NA49 pp-data (filled circles) with simulations: LAQGSM - solid, DPMJET - dashed, MARS - dot-dashed lines.

Carbon target

Figs. 4-5 show comparisons of results from six codes with data [10] for charged pion production in proton-carbon collisions. All the codes agree quite well with data for x_F distributions (there are no measurements at $x_F > 0.5$ for carbon target where DPMJET and LAQGSM overestimate data for a proton target). The codes reproduce well measured rapidity distributions except for LAQGSM at mid-rapidities. Transverse momentum distributions in central ($x_F = -0.1$) and fragmentation ($x_F = 0.5$) regions are compared in Fig. 5. Additional comparisons for intermediate values of x_F can be found in [11]. MARS and FLUKA are very close to the data. LAQGSM and DMPJET underestimate high-$p_\perp$ central pion production. LAQGSM overestimates a negative pion yield at low-$p_\perp$ for $x_F = 0.5$. GEANT4 agrees well with data for positive pions but overestimates a negative pion yield in the fragmentation region.

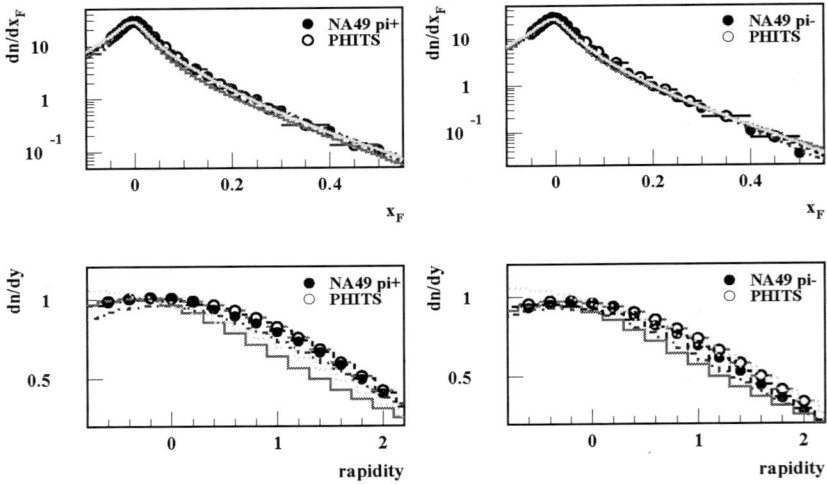

FIGURE 4. Comparison of NA49 pC-data (filled circles) with simulations: PHITS - open circles, LAQGSM - solid, DPMJET - dashed, MARS - dot-dashed, FLUKA - dotted histograms.

TASK 3

Double-differential cross sections of charged pions and kaons, protons and antiprotons in 67 GeV/c proton interactions with Al and Al_2O_3 thick targets were measured for neutrino experiments at IHEP [12]. Ratios of calculated results for particle yields from an aluminum target (L=60 cm, R=3 cm) to the data are presented in Fig. 6 for p, π^+, π^- and two angles of 5 and 25 mrad. Results for other particles and angles can be found in Ref. [11]. MARS15 agrees with data within a factor of two. The disagreement of both GEANT4 and PHITS results with data is quite substantial except for positive pions at 5 mrad.

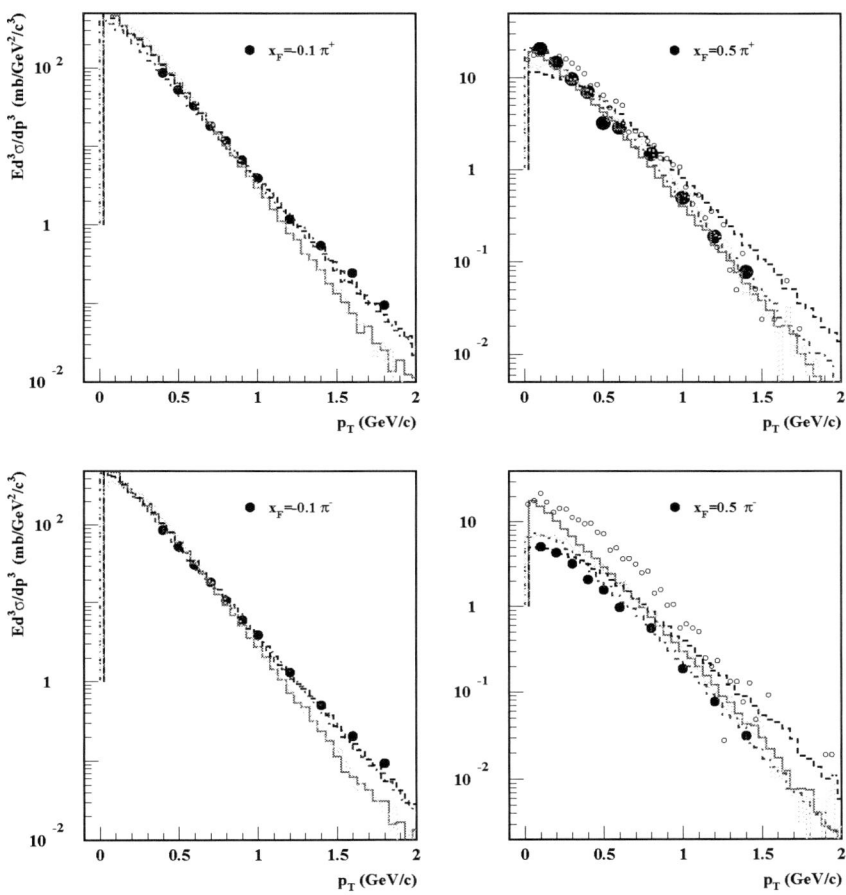

FIGURE 5. Comparison of NA49 pC-data (filled circles) with simulations: G4 - open symbols, LAQGSM - solid, DPMJET - dashed, MARS - dot-dashed, FLUKA - dotted histograms.

TASK 4

Double-differential neutron yields from thick targets irradiated by a 2-GeV electron beam were measured in the PAL experiment [13]. The copper target was 14-cm long with a radius of 2.5 cm. MARS15 and MCNPX-calculated results are presented in Fig. 7 together with this data. MARS results nicely agree with data at $9 \leq E \leq 40$ MeV for all angles, and underestimate the data at higher energies by up to a factor of two. MCNPX results practically coincide with the MARS's ones at $E > 25$ MeV but are lower by about a factor of two at $E < 15$ MeV.

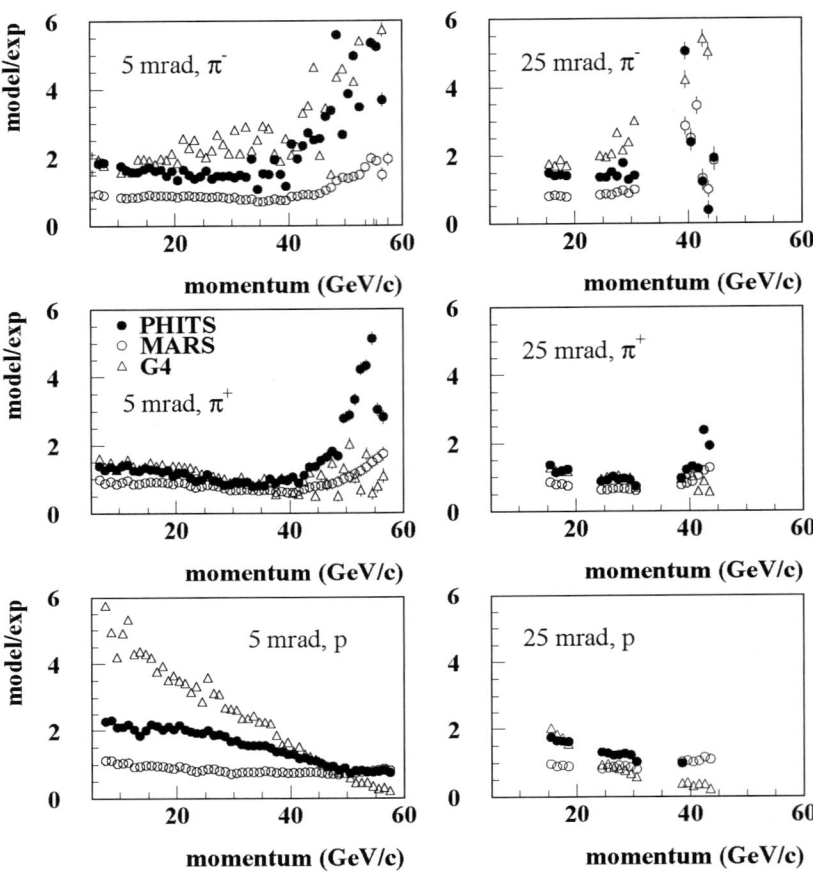

FIGURE 6. Ratio of calculated and measured cross sections of particle production in proton collisions with a thick aluminum target at 67 GeV/c.

TASK 5

Energy deposition in a cylindrical copper absorber (L=24 cm, R_{in}=6.5 cm, R_{out}=8.5 cm) around a copper target (L=3 cm, R=1.5 cm) irradiated by a 12-GeV proton beam was measured at KEK [14]. Experimental data versus the target longitudinal position with respect to the absorber center are shown in Fig. 8 along with four sets of calculation results. GEANT4 predictions here are almost identical for different physics lists, therefore only QGSP simulations are shown. MARS15, MCNPX and PHITS are very close to the measurements, except for the middle point where MARS15 underestimates data by 5%. GEANT4 underestimates data by about 10% for two points but it is closer to data than MCNPX and PHITS for the last point.

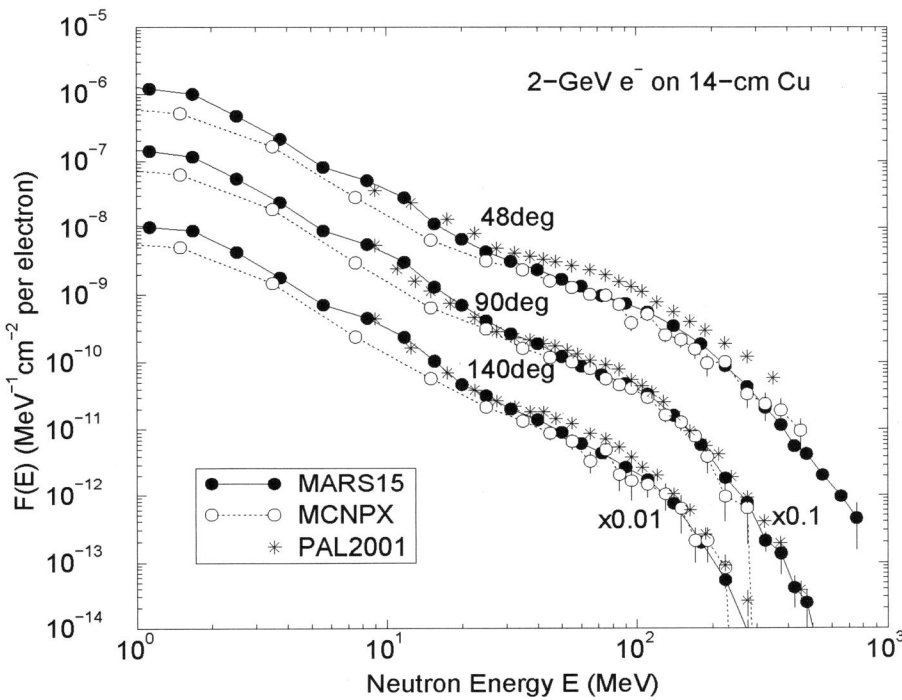

FIGURE 7. Neutron spectra calculated with MARS15 and MCNPX at three angles for a 2-GeV electrons on a thick copper target *vs* data [13].

TASK 6

The CERN-Dortmund-Heidelberg-Saclay-Warsaw (CDHS) collaboration measured longitudinal profiles of hadronic showers in iron-scintillator calorimeters for 10 to 100 GeV pion beams at the CERN SPS [15]. The ratios of lateral integrated energy deposition calculated with MARS15 and PHITS to this data are presented in Fig. 9. MARS15 results agree with the experiment within about 15%. PHITS does not transport photons with energy E>1 GeV just ignoring them [16]. This explains a growing with a primary energy underestimation by PHITS. Note that the 10-GeV PHITS's results behave quite differently compared to other energies.

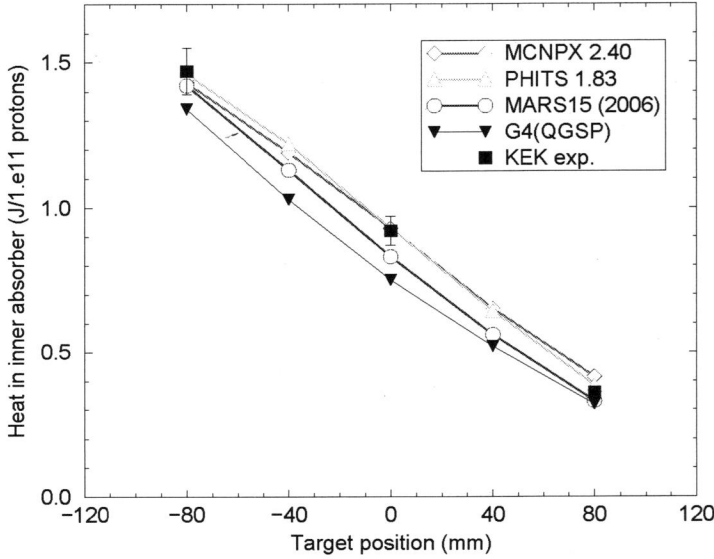

FIGURE 8. Energy deposition in a cylindrical absorber surrounding a copper target irradiated by a 12-GeV proton beam.

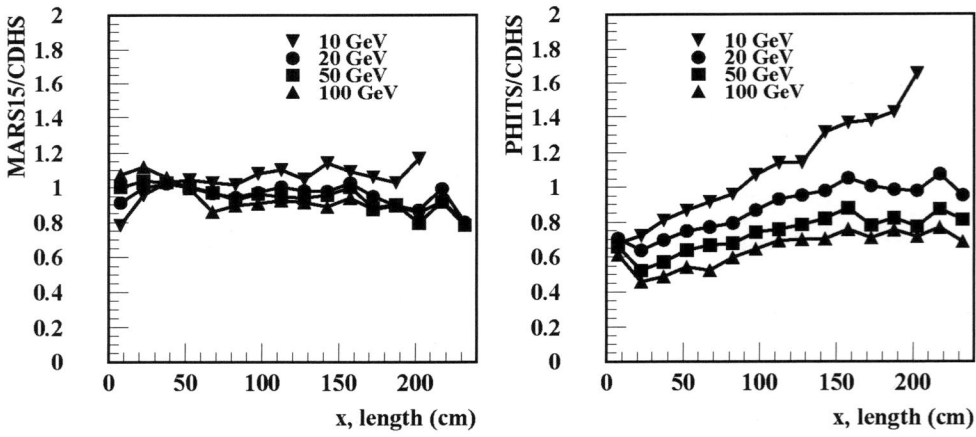

FIGURE 9. Ratio of calculated with MARS15 and PHITS to measured by CDHS laterally integrated energy deposition in the iron-scintillator calorimeter for four energies of a pion beam.

TASK 7

Longitudinal profiles of energy deposition in a cylindrical tungsten target (L=10 cm, R=1 cm) irradiated by pencil proton beams of 1, 20 and 50 GeV were calculated by all the participated code groups. Fig. 10 shows results of this inter-comparison for proton energies of 1 and 50 GeV. The 20-GeV results are very similar in shape to the 50-GeV case and therefore are not shown here. Before the workshop, MCNPX version 2.6b03 was used in the mode where electrons are not tracked; their energies are deposited locally when created, which results in a pileup near the front of the shower, but greatly decreased tracking times. Corresponding energy deposition results at 20 and 50 GeV for this task were substantially off compared to the other codes. Therefore, MCNPX was recently re-run for 50 GeV in a mode where all particles are fully tracked. As one can see from Fig. 10, all the codes predict rather similar (within 10%) energy deposition for the 1-GeV protons, with some pileup at the beginning for MCNPX. At high energies, FLUKA, GEANT4, MARS15, and MCNPX agree again within 10%, while PHITS shows a substantial flaw. The reason is that PHITS does not simulate showers induced by high-energy photons from $\pi^0 \rightarrow 2\gamma$ decays. Their energy at E>1 GeV is not included into energy deposition that explains a substantial underestimation in PHITS predictions at high primary energies.

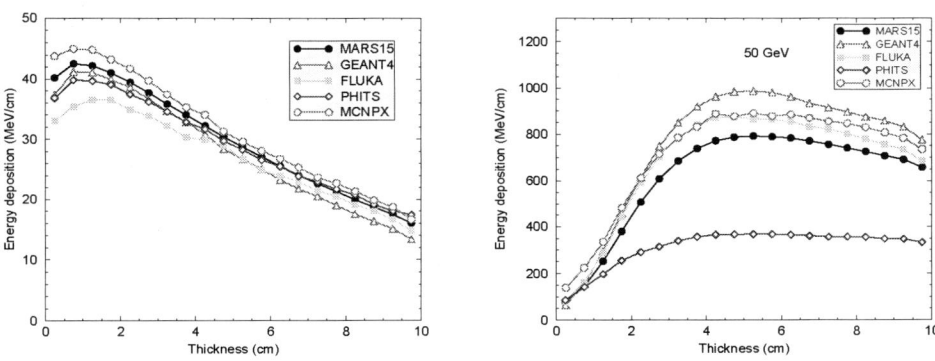

FIGURE 10. Energy deposition in a cylindrical tungsten target irradiated by 1 and 50 GeV protons as calculated with five codes.

CONCLUSION

Seven tasks, important for high-energy physics, accelerator, shielding, and space applications, were studied for the inter-comparison and verification against experimental data by the five major hadronic shower simulation code groups. As a result of this benchmarking, it was found that agreement between the codes and data is quite reasonable in many cases, but quite serious problems were revealed in the others. Obviously, much more verification is needed, especially for calorimetry-specific applications.

REFERENCES

1. A. Fasso et al, *CERN-2005-10, SLAC-R-773* (2005); A. Fasso et al, *CHEP 2003, hep-ph/0306267* (2003); http://www.fluka.org.
2. J. Allison et al, *IEEE Trans. Nucl. Sci.* **530**, 270–278 (2006); http://geant4.web.cern.ch
3. N.V. Mokhov, *Fermilab-FN-628* (1995); N.V. Mokhov et al, *Rad. Prot. Dosimetry*, **116**, 99 (2005); N.V. Mokhov et al, *Proc. Int. Conf. on Nucl. Data Sci. Technology*, **AIP Conf. Proc. 769**, 1618 (2005); http://www-ap.fnal.gov/MARS/.
4. J.S. Hendricks et al, *LA-UR-05-2675* (2005); http://mcnpx.lanl.gov.
5. H. Iwase et al, *J. Nucl. Sci. Technol.* **39**, 1142 (2002).
6. K. K. Gudima et al, *LA-UR-01-6804* (2001).
7. http://sroesler.home.cern.ch/sroesler/dpmjet3.html
8. M. G. Catanesi et al, *Nucl. Phys.* **B732**, 1–45 (2006).
9. C. Alt et al, *Eur. Phys. Jour.* **C45**, 343–381 (2006).
10. C. Alt et al, *hep-ex/06106028*.
11. *http://conferences.fnal.gov/hss06/*
12. N. I. Bozhko et al, *Sov. Jour. Nucl. Phys* **31**, 644 (1980).
13. H.-S. Lee et al, *Rad. Prot. Dosimetry*, **116**, 653–657 (2005).
14. H. Matsuda, private communication.
15. E. Huges, *Proc. 1 Int. Conf. on Calor. in HEP*, 525, FANL, Batavia (1990).
16. K. Niita, private communication.

Toward Meaningful Simulations of Hadronic Showers

Richard Wigmans

Department of Physics, Texas Tech University, Lubbock TX79409-1051, USA

Abstract. The physics processes that are crucial for the description of hadronic shower development in calorimeters are π^0 production, the release of protons in nuclear reactions and (in calorimeters with hydrogenous active material) elastic scattering of soft neutrons. In this paper, I discuss how we know that these elements are crucial, and I describe experimental data that are sensitive to a correct implementation of these elements in simulation codes. Therefore, these data should serve as benchmarks for (generic) validation of these codes. I also illustrate the practical importance of reliable shower simulations with some recent real-life examples.

Keywords: Calorimetry, calibration, shower sampling
PACS: 29.40Ka, 29.40Mc, 29.40Vj

INTRODUCTION

In the past 25 years, our knowledge and understanding of hadronic shower development, and the application of that in calorimetry, has considerably improved. Issues such as compensation, which were very mysterious in the early 1980s, are now fully understood. We also understand in great detail the major factors that dominate and limit the performance of hadron calorimeters. However, **none** of this is the result of the simulation codes that are commonly being used. It is **exclusively** the result of experimental R&D efforts, in which new ideas were tried out and gradually improved.

Of course, simulations are as good as the physics on which they are based, and that is precisely where the problem lies. The physics which is essential to hadron calorimetry is not well implemented in any of the codes that are commonly used. What is this physics? There are two areas that are absolutely crucial. The first is the generation of π^0s in the absorption process. The π^0s develop electromagnetic showers and typical calorimeters respond differently to these particles than to the non-em shower components. The level of π^0 production, the energy dependence of this level, and (very important) event-to-event fluctuations in the π^0 fraction all translate directly into measurable calorimeter performance characteristics, such as e/π signal ratios, response non-linearity and response function (line shape).

The second area that is absolutely crucial for understanding hadron calorimetry are the nuclear reactions. Just as in the case of em showers, most of the hadronic energy is deposited by very soft particles. In the case of em showers, a large fraction (up to $\sim 50\%$) of the energy is deposited by sub-MeV electrons, in the case of hadron showers an even larger fraction of the (non-π^0) signal comes from MeV-type protons. These protons are produced either in nuclear spallation or transmutation reactions, or in elastic neutron scattering (in the case of hydrogenous active material).

CP896, *Hadronic Shower Simulation Workshop*
edited by M. Albrow and R. Raja

If one wants to really test the validity of hadronic Monte Carlo simulation programs, it is thus essential to use experimental data that are particularly sensitive to a correct implementation of the mentioned physics aspects. In this paper, I describe such data. Meaningful simulations, *i.e.* simulations that may be expected to have some predictive value for the performance of new types of hadron calorimeters (such as the ones proposed for the ILC), should first and foremost reproduce these particular data.

I start, however, by describing some recent examples that illustrate the crucial importance of reliable simulations for optimizing, calibrating and operating calorimeters in modern experiments.

THE PRACTICAL IMPORTANCE OF RELIABLE SIMULATIONS

The first example concerns the AMS experiment [1], which is designed to measure high-energy cosmic rays outside the Earth's atmosphere. The em calorimeter of this experiment consists of 18 Pb layers (each $\sim 1X_0$ thick), interleaved with plastic scintillating fibers. Each layer of this detector is read out separately, but because of its limited depth ($17X_0$), high-energy electron and photon showers are not fully contained.

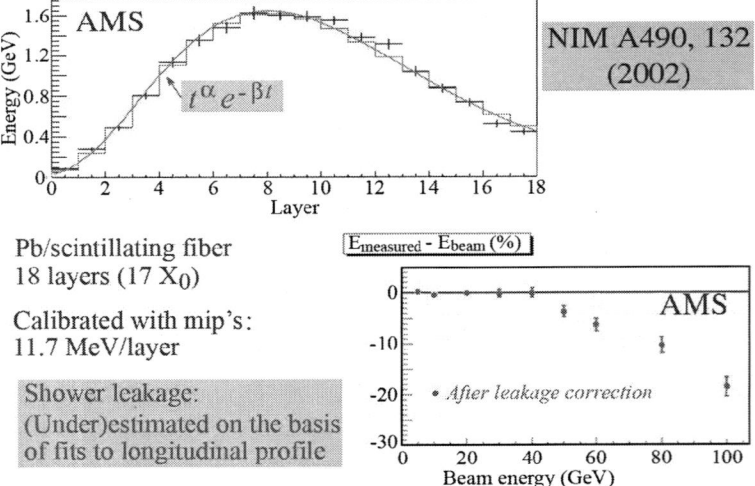

FIGURE 1. Average signals for 100 GeV electrons in the 18 longitudinal sections of the AMS lead/scintillating fiber calorimeter (top). Average difference between the measured energy and the beam energy, after leakage corrections based on extrapolation of the fitted shower profile (bottom).

This is illustrated in Figure 1, which shows the average signals from 100 GeV electron showers developing in this structure. These signals were translated into energy deposits based on a calibration with minimum-ionizing particles (mips). The measured data were fitted to a Γ-function and at high energies, where the showers were not fully contained, the average leakage was estimated by extrapolating this fit to infinity. As shown in the bottom diagram of Figure 1, this procedure systematically underestimated this leakage fraction, more so as the energy (and thus the leakage) increased. The reason for this

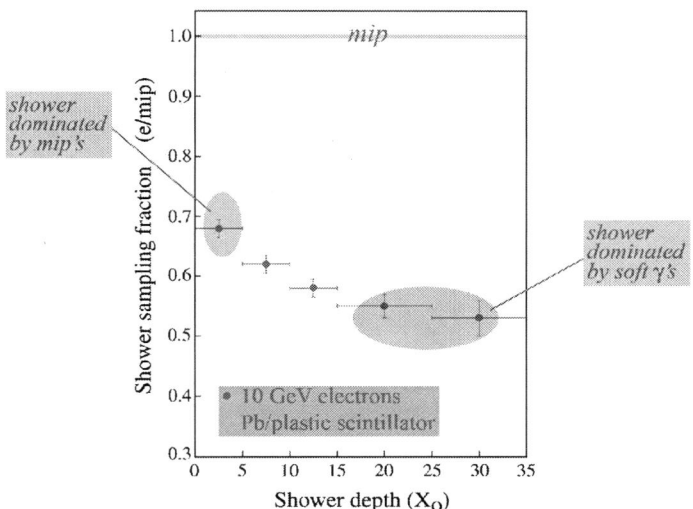

FIGURE 2. The sampling fraction of 10 GeV electron showers in a Pb/plastic-scintillator calorimeter, as a function of depth. The ratio of deposited energy and signal is normalized to that for mips [2].

phenomenon is illustrated in Figure 2, which shows that the sampling fraction of em showers in this type of sampling calorimeter decreases by as much as 25-30% over the volume in which the absorption takes place. In the early stage of its development, the shower still resembles a collection of mips, but especially beyond the shower maximum, the energy is predominantly deposited by soft (< 1 MeV)γs. The latter are much less efficiently sampled than mips in this type of structure, where dominant processes such as photo-electric effect and Compton scattering strongly favor the high-Z absorber material [2]. If one thus uses mips for calibrating this device, *i.e.* for establishing the relationship between the measured signal and the corresponding deposited energy, it is clear that the latter is systematically underestimated, and that leakage estimates based on this procedure are systematically too low.

The fact that the em shower sampling fraction decreases as the shower develops may have very important practical consequences. Apart from the *systematic mismeasurement of energy* described above, these include:

- Electromagnetic signal *non-linearity* [3], and
- *Differences in response* to showers induced by electrons, photons and π^0s [4].

These issues are especially relevant in longitudinally segmented calorimeters, where one has to decide which calibration constants to assign to the different segments. This *intercalibration* is more often than not done incorrectly, resulting in the consequences mentioned above.

A recent example of an experiment that has to deal with this intercalibration issue is ATLAS, whose Pb/LAr ECAL consists of three longitudinal segments. Figure 3 shows how the sampling fraction evolves as a function of depth, for electrons of three differ-

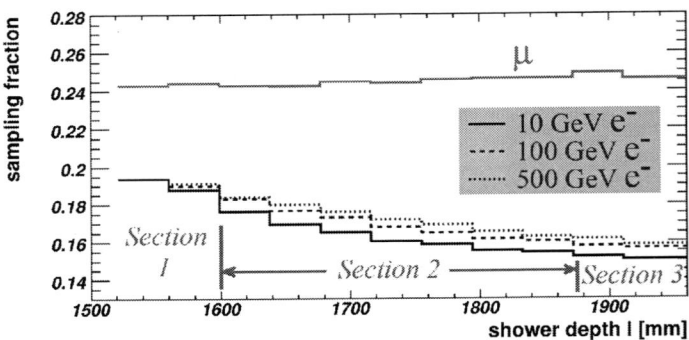

FIGURE 3. Energy sharing between the 3 longitudinal segments of the ATLAS em calorimeter, for electron showers of different energies [5].

ent energies. Both Figures 2 and 3 concern results of detailed Monte Carlo simulations of em shower development. Such simulations have played an absolutely crucial role in ATLAS' solution of the intercalibration problem, for which they developed a very sophisticated procedure, based on a variety of energy-dependent parameters. This procedure was tested in great detail with Monte Carlo events and yielded both excellent signal linearity and good energy resolution [5]. Typically, in more empirical approaches to this problem, only one of these performance characteristics is pursued in isolation, and the results are far from optimal (or even correct) for other aspects [2, 6].

The examples discussed in this section thus illustrate that Monte Carlo simulations are an invaluable practical tool, not only for calibrating and operating a calorimeter system, but also and more importantly for acquiring a detailed understanding of the functioning of these largely non-trivial detectors. This is of course as true, if not more, for hadronic calorimeters as for electromagnetic ones.

CRUCIAL ELEMENTS OF HADRONIC SHOWER SIMULATIONS

The absorption of high-energy particles in dense matter is a very complicated process. However, for the purpose of calorimetry, only very few aspects of this process are of critical importance. In the examples discussed above, what mattered was the fact that the soft-γ shower component is sampled with a different (smaller) efficiency than the mip component. There is no need to simulate in detail what happens at energies below 100 keV to get this crucial aspect right.

Also for hadron calorimetry, only a few aspects of the shower development process are of crucial importance. It is important to get these aspects right in the simulations. For example, a correct application of the laws of conservation of energy and baryon number is essential. On the other hand, hadronic reactions with a relative probability < 1% can be safely ignored in the simulations, except for some very specific applications, *e.g.*, studies of muon production in hadronic shower development. In the following, we discuss the crucial aspects of hadronic shower simulations. Only *experimental* data are shown. They have been chosen with the sole purpose to illustrate this cruciality.

So which aspects of hadronic shower development are crucial for understanding the signals from hadron calorimeters and how do we know this?

What physics is crucial and how do we know?

Hadronic showers consist of two distinctly different components:

1. An *electromagnetic* component; π^0s and ηs generated in the absorption process decay into γs which develop em showers.
2. A *non-electromagnetic* component, which combines essentially everything else that takes place in the absorption process.

For the purpose of calorimetry, the main difference between these two components lies in the fact that some fraction of the energy contained in the non-em component does *not* contribute to the calorimeter signals. The numerous nuclear reactions that take place in the absorption process are responsible for the overwhelming majority of this so-called *invisible energy*. The nuclear binding energy of protons, neutrons and heavier aggregates released in these reactions has to be supplied by the shower particles that induce the reactions, and does not contribute to the signals. Other, relatively minute contributions to the invisible energy come from neutrinos and muons that escape the calorimeter.

The invisible energy may represent a considerable fraction of the total shower energy. If we define the *response* as the average calorimeter signal per unit of energy, and the responses of the em and non-em shower components are e and h, respectively, then the e/h ratio quantifies the importance of the invisible energy effects for a given calorimeter. For example, in some crystal calorimeters (*e.g.,* $PbWO_4$) $e/h \approx 2$, which means that, on average, about 50% of the non-em energy is invisible.

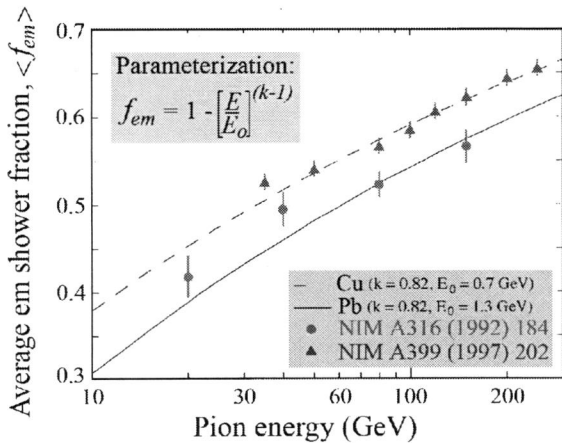

FIGURE 4. The average em fraction of showers initiated by pions in lead and copper-based calorimeters, measured as a function of the pion energy [7, 8].

127

First, we concentrate on the em shower component. Let f_{em} be the fraction of the total shower energy contained in this component. Among the characteristics of the em component that have profound implications for the performance of hadron calorimeters, we mention:

1. The fact that the average value of the em shower fraction, $\langle f_{em} \rangle$, increases with energy. This effect, which is illustrated in Figure 4, is directly responsible for the hadronic signal non-linearity exhibited by all non-compensating hadronic calorimeters (*i.e.* calorimeters with $e/h \neq 1$). The figure shows that the em shower component represents typically of the order of half the total energy, and that $\langle f_{em} \rangle$ not only depends on the energy, but also on the type of absorber material. The curves in this figure represent Groom's parameterization [9].

2. Event-to-event fluctuations in f_{em} are large and non-Poissonian. Figure 5 shows a measurement of these fluctuations for showers initiated by 150 GeV π^- in a lead-based hadron calorimeter. The observed asymmetry in this distribution is a conse-

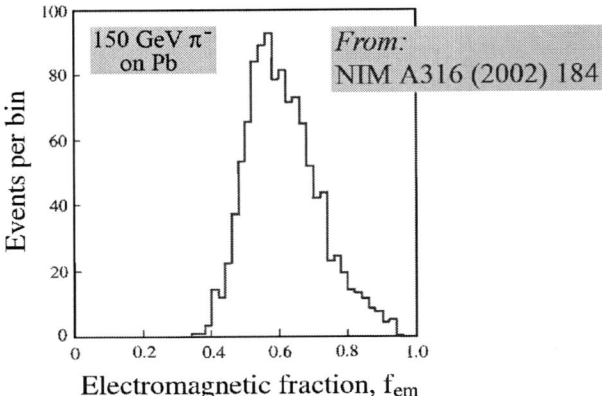

FIGURE 5. Event-to-event fluctuations in the em shower fraction [7].

quence of the fact that π^0 production is a "one-way street", *i.e.* energetic charged pions produced in the interactions may transfer (part of) their energy into π^0s in subsequent reactions, but the reverse process does not take place because of the extremely short π^0 lifetime. This effect is responsible for the asymmetric response function (line shape) observed for pions in non-compensating calorimeters (see, for example, Figure 14).

Another consequence of the non-Poissonian nature of the fluctuations in f_{em} is the fact that the width of distributions such as the one shown in Figure 5 does not scale with the square root of the shower energy. Measurements of the energy dependence of the fractional width (σ_{rms}/mean) are shown in Figure 6, for pion-induced showers in a copper-based calorimeter. For comparison, the $(\sqrt{E})^{-1}$ behavior expected for fluctuations governed by Poisson statistics (such as sampling fluctuations) is shown as well. This deviation from $E^{-1/2}$ scaling is responsible for the energy dependence of calorimeter performance characteristics such as the energy resolution

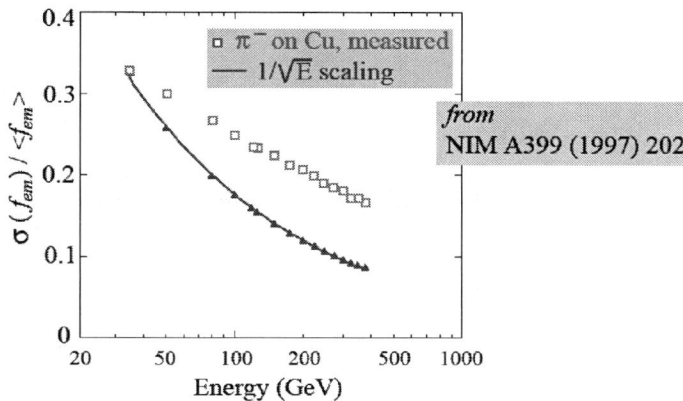

FIGURE 6. Energy dependence of the fractional width of the f_{em} distribution. Shown are the results of measurements [8] and the expected dependence for fluctuations governed by Poisson statistics.

and the position resolution in non-compensating calorimeters (see, for example, Figure 14).

3. There are substantial differences between the generation of π^0s in proton and pion-induced showers. These lead to differences in the calorimeter response, the response function and the energy resolution. These differences derive from the fact that the requirement of baryon number conservation tends to produce a leading baryon in the primary collision of proton-induced showers, whereas pions may produce a leading π^0. As we saw above, the latter effect is responsible for the asymmetric response function. Since π^0s are not produced as leading particles in proton-induced showers, the response function tends to be much more symmetric in that case. This is illustrated in Figure 7.

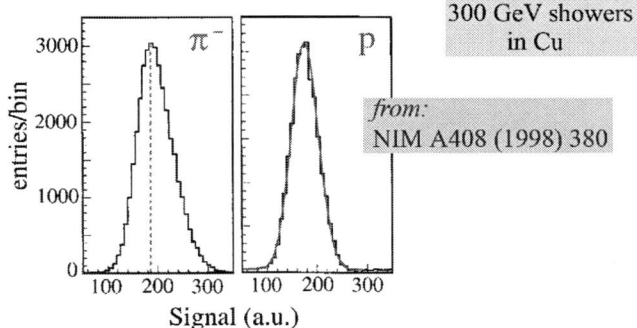

FIGURE 7. Calorimeter response functions for 300 GeV pions and protons [10].

The absence of leading π^0s in proton showers also reduces the *average* em shower fraction ($\langle f_{em} \rangle$), compared to pion showers. This effect is illustrated in Figure 8a. On the other hand, the fractional width of the $\langle f_{em} \rangle$ distribution is also smaller for

129

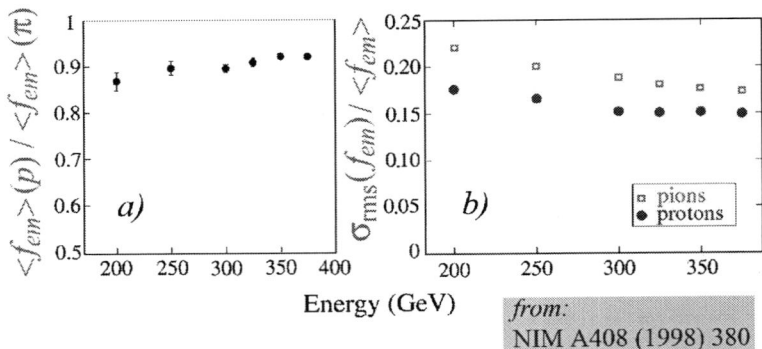

FIGURE 8. The average em shower fraction $\langle f_{em} \rangle$ (a) and the fractional width of the $\langle f_{em} \rangle$ distribution (b) as a function of energy, for proton and pion-induced showers in a copper-based calorimeter [10].

proton showers. The larger average em shower fraction in pion showers thus goes hand in hand with larger event-to-event fluctuations in this fraction (Figure 8b).

4. The final characteristic of π^0 production in hadron showers with profound consequences for hadron calorimetry concerns the fact that the resulting em shower component is distributed over the entire calorimeter volume in which the shower develops, and **by no means** limited to the electromagnetic calorimeter section (typically the first nuclear interaction length). This effect is illustrated in Figure 9, which

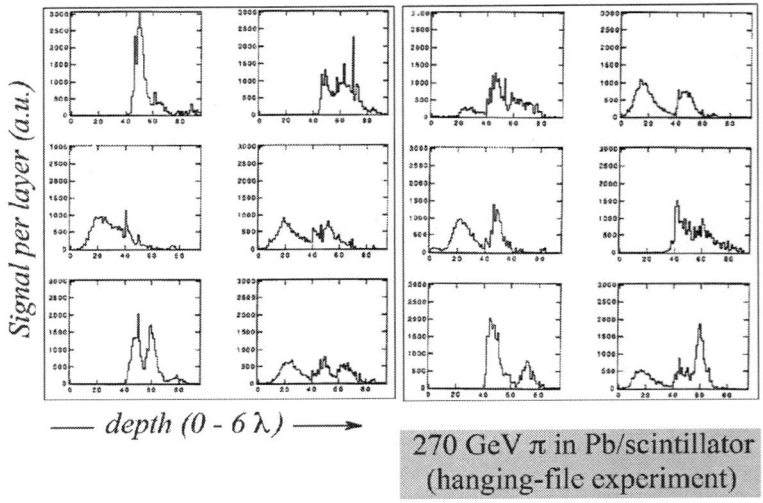

FIGURE 9. Longitudinal energy deposit profiles for 270 GeV π^- showers in lead [11].

shows longitudinal energy deposit profiles for 270 GeV pion showers developing in a $6\lambda_{int}$ deep lead-based calorimeter. The em components are characterized by the peaks in these profiles, a consequence of the large λ_{int}/X_0 ratio of lead. This figure illustrates that, unlike for em showers, there is no such thing as a "characteristic hadronic shower profile" and that, therefore, the average hadronic shower profile is not a very meaningful validation tool for simulations. Because of the abundant production of em shower components beyond $1\lambda_{int}$, the validity of some calibration calibration schemes applied to calorimeter systems consisting of separate em and hadronic compartments is also questionable [12].

Next, we turn our attention to the *non-electromagnetic* shower component. The most crucial aspect of this component is that it is dominated by the products of nuclear reactions, **not** by pions, kaons and other mips. Since nuclear breakup determines many aspects of the performance of hadron calorimeters, it is thus crucial to simulate the processes in which these protons, neutrons and other nuclear fragments are being produced as accurately as possible.

In the case of em showers, a large fraction (up to $\sim 50\%$) of the energy is deposited by sub-MeV electrons [2], in the case of hadron showers an even larger fraction (80% or more) of the (non-em) signal comes from MeV-type protons. How do we know that?

- Because of the small hadronic signals (*i.e.* large e/h values) of calorimeters that are blind to such protons. For example, calorimeters based on the detection of Čerenkov light only produce signals for relativistic shower particles. In quartz-fiber calorimeters (refractive index 1.46), the threshold of $\beta > 0.69$ corresponds to 350 MeV protons and thus eliminates almost all nuclear reaction products. The

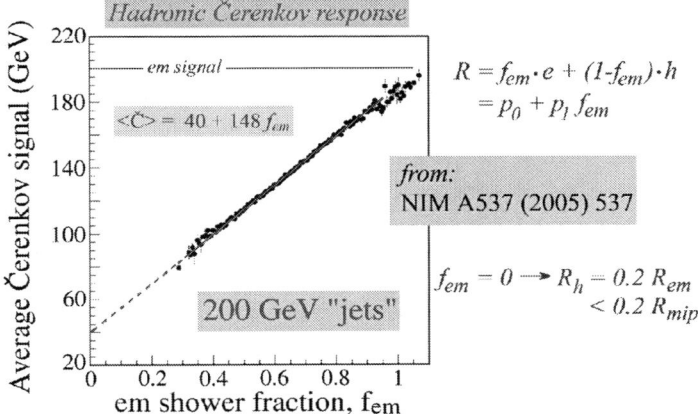

FIGURE 10. Hadronic Čerenkov signal for 200 GeV showers, as a function of f_{em} [13].

DREAM calorimeter measures not only the Čerenkov signal, but also f_{em}, event by event. The hadronic Čerenkov response is a linear function of this variable, as illustrated in Figure 10. When extrapolated to $f_{em} = 0$ (*i.e.* the shower is 100% non-electromagnetic), the response is found to be only 20% of the em response. And since $e/mip < 1$ (see Figure 2), we have to conclude that mips account for less than 20% of the energy contained in the non-em shower component. This type of calorimeter is thus, for all practical purposes, only sensitive to the em shower component (the threshold for shower electrons producing a Čerenkov signal in it is ~ 200 keV).

- Detailed measurements by the ZEUS Collaboration revealed that there was almost no correlation between the signals produced in subsequent sampling layers of their calorimeter (Figure 11). In order to measure the contribution of sampling fluctu-

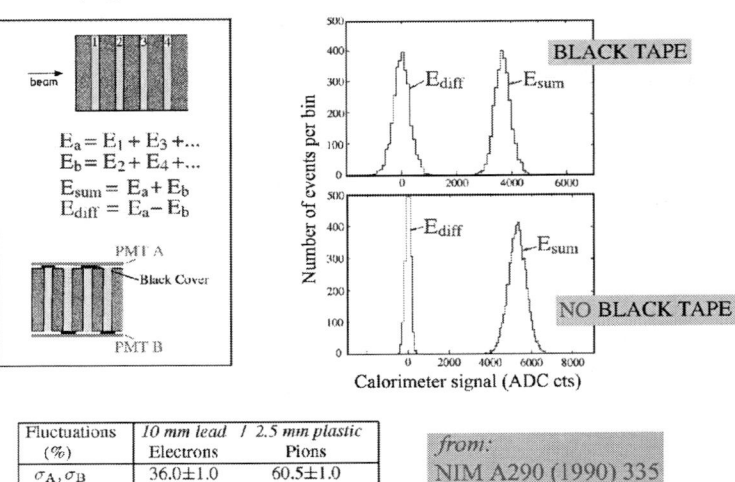

| Fluctuations | 10 mm lead / 2.5 mm plastic | |
(%)	Electrons	Pions
σ_A, σ_B	36.0±1.0	60.5±1.0
σ_{sum}	24.5±1.0	43.5±1.0
σ_{diff}	25.8±1.0	42.3±1.0

from:
NIM A290 (1990) 335

FIGURE 11. Measurement of the contribution of sampling fluctuations to the energy resolution of a Pb/scintillator calorimeter [14]. See text for details.

ations to the energy resolution of their calorimeter, they covered the scintillator plates, which are read out by WLS plates coupled to photomultiplier tubes, alternatingly on one side with black tape, as illustrated in the figure. In this way, the calorimeter is split into two independent calorimeters, whose signals may be compared and combined. Minimum ionizing particles lost, on average, only 12 MeV in each absorber layer and, therefore, a mip generated in the shower development would typically produce a signal in many subsequent sampling layers. If mips formed a dominant contribution to the shower signals, then the distribution of the *differences* between the two calorimeter signals should be expected to be narrower than the distribution of the *sums* of these two signals, because of the effects of cancellation of fluctuations in the energy carried by the mip component. This is illustrated by the results observed when the black tape was removed, and all fluctuations except those in photoelectron statistics cancelled in the distribution of signal differences. The fact that no significant differences between the widths of the distributions of sums and differences was observed in the "black-tape geometry" means that the particles that generated the signals in plates 1,3 were not the same particles that were responsible for the signals in plates 2,4.

These results can be understood from the fact that protons produced in nuclear reactions dominate the non-em signal component. The stopping power of the absorber layers is ~ 100 MeV for protons, which therefore rarely produce a signal in two consecutive sampling layers.

What is the source of the protons discussed here? They are produced either in nuclear spallation or in nuclear transmutation reactions. The latter are often induced by neutrons, which are abundantly produced in any process involving nuclei. Spallation protons carry typically ~ 100 MeV, protons from reactions such as (n, p) carry energies similar to the particles that initiated the reaction, typically $5 - 10$ MeV.

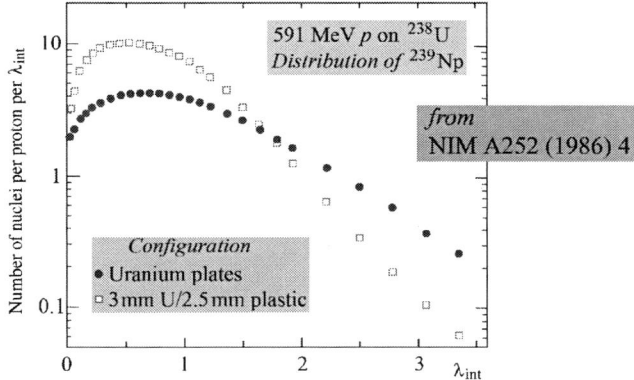

FIGURE 12. Longitudinal distribution of ^{239}Np produced in the absorption of 591 MeV protons in a stack of uranium plates [15].

Among the nuclear reaction products, neutrons outnumber protons by a considerable factor, typically an order of magnitude. The characteristics of neutron production in hadronic shower development have been measured by studying the distributions of radioactive nuclides produced in neutron-induced reactions [15]. An example is shown in Figure 12, where the amount of ^{239}Np produced in the absorption of 591 MeV protons in a stack of uranium plates is plotted as a function of depth. This nuclide is produced by the capture of *thermalized* neutrons in ^{238}U and the subsequent β decay of ^{239}U. From this type of experiments, it was found that in high-Z materials the neutron production may be as high as ~ 40 per GeV deposited energy.

Figure 12 shows one other very interesting aspect, namely the enormous effects of inserting sheets of plastic in between the uranium plates. As a result, the total amount of ^{239}Np was considerable increased and also concentrated closer to the front face of the absorber stack. Since this nuclide is the result of thermal neutron capture, the plastic apparently plays a very important role in the thermalization process. Indeed, in this structure, the (MeV type) neutrons produced in nuclear reactions lose more than 90% of their kinetic energy through elastic neutron-proton scattering in the plastic.

This phenomenon forms the basis of *compensating* calorimetry, where these recoil protons are an essential element of the non-em calorimeter signals. By reducing the sampling fraction of the calorimeter structure, the relative contribution of these recoil protons to the non-em signals can be boosted, to the point where the non-em and em calorimeter responses become equal ($e/h = 1$). This effect is beautifully illustrated in Figure 13, which shows the results of tests of two lead/plastic-scintillator calorimeters

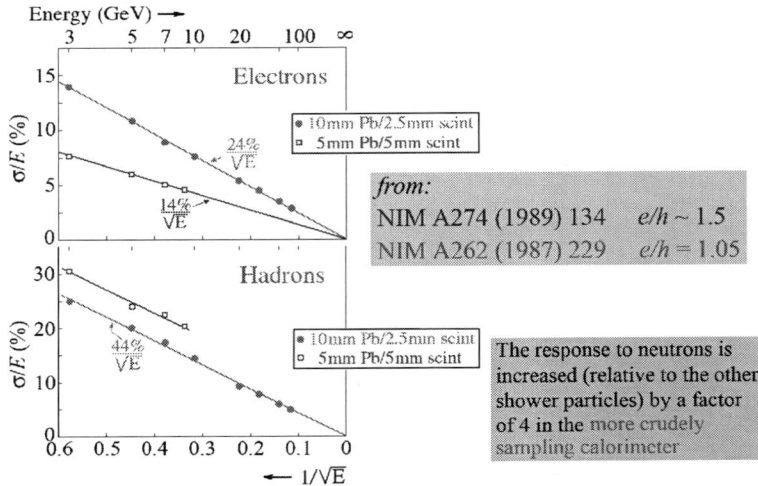

FIGURE 13. Energy resolution for electrons and pions as a function of energy for two lead/plastic-scintillator calorimeters, which differ by a factor 4 in sampling fraction [16, 17].

with different sampling fractions [16, 17]. The *fine-sampling* structure consisted of 5 mm thick lead plates, read out by 5 mm thick scintillator plates, in the *crude-sampling* structure the thickness of the lead plates was doubled (10 mm), while the thickness of the scintillator plates was halved (2.5 mm). As a result, the sampling fraction (for mips) was reduced by a factor of 4, while the (recoil p) signals from neutrons produced in the hadronic absorption process was barely changed. Therefore, the relative contribution of these neutrons to the calorimeter signals was boosted by a factor of ~ 4 in the crude-sampling device. And while the energy resolution for electrons deteriorated considerably as a result of the increased sampling fluctuations (from $0.14E^{-1/2}$ to $0.24E^{-1/2}$), the resolution for hadrons became considerably *better*, especially at high energies, because of the elimination of the (non-compensation)effects that led to deviations from $E^{-1/2}$ scaling. The change in the sampling fraction of the calorimeter, and the resulting increase in the contribution of (recoil protons from) neutrons to the signals decreased the e/h value of the calorimeter from 1.5 to 1.05.

(Generic) code validation

If one wants to really test the validity of hadronic Monte Carlo simulation programs for the purpose of hadron calorimetry, it is essential to use experimental data that are particularly sensitive to the effects discussed above:

- Peculiarities of π^0 production in the absorption process
- Signal contributions from nuclear reactions
- Neutron transport (in calorimeters with hydrogenous active material)

In the previous sections, I have discussed such experimental data. They are summarized in Figures 4 - 8 (π^0 production), 10 (signal contributions from nuclear reactions) and 12,13 (neutron signals). Meaningful simulations, *i.e.* simulations that may be expected to have some predictive value for the performance of new types of hadron calorimeters (such as the ones proposed for the ILC), should first and foremost reproduce these particular data.

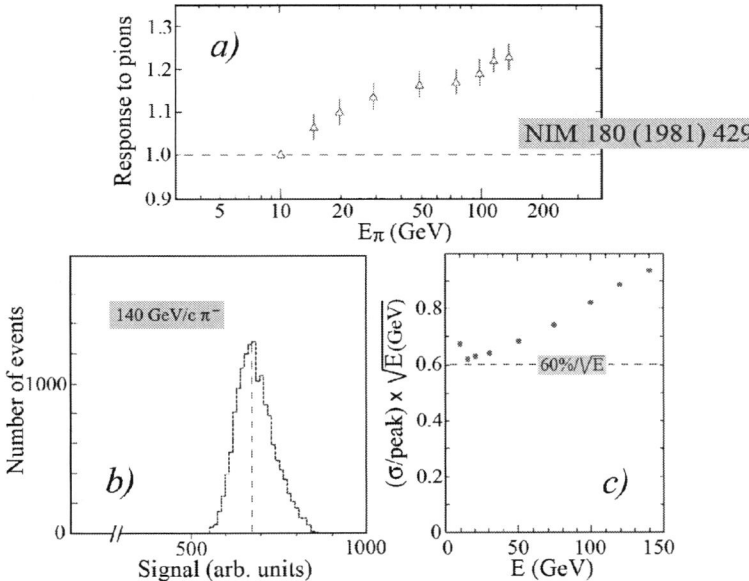

FIGURE 14. Effects of the peculiarities of π^0 production in hadronic shower development on the signal linearity (*a*), response function (*b*) and energy resolution (*c*) for pions in a non-compensating calorimeter [18].

Of course, the described effects are also observed in other types of calorimeters. For example, Figure 14 shows the signal non-linearity, the response function and the energy resolution for pion detection in a calorimeter with a $e/h = 1.5$ value [18]. These characteristics all reflect the peculiarities associated with π^0 production in the hadronic shower development discussed above. However, the closer e/h gets to 1.0, the less pronounced these effects become. Therefore, it is best to test the validity of simulations for the case in which these effects are maximized, *i.e.* in Čerenkov calorimeters.

One should also be very careful in using data from longitudinally *segmented* calorimeters for this generic validation. Usually, the authors of papers in which these data are described have used weighting factors to make the results look better[1], and this introduces an element of arbitrariness in the comparison.

---

[1] This is actually an illusion, but that is besides the point of this discussion.

CONCLUSIONS

The performance of hadron calorimeters depends sensitively on some very specific aspects of the shower development process: π^0 production, the nuclear component and (in calorimeters with hydrogenous active material) neutron transport. Several sets of experimental data, discussed in this paper, document this sensitivity in precise, quantitative ways. Unfortunately, none of these data sets were considered in the context of the "grand validation" undertaken for the occasion of this Workshop[2].

Reliable shower simulations are absolutely essential for designing, optimizing, calibrating, operating and acquiring the necessary understanding of the calorimeter one is working with. Examples drawn from recent experience with electromagnetic calorimeters underscore that point. Unfortunately, the reliability of available hadronic codes leaves very much to be desired.

REFERENCES

1. AMS Collaboration, Nucl. Instr. and Meth. **A490** (2002) 132.
2. R. Wigmans, *Calorimetry, Energy Measurement in Particle Physics*, International Series of Monographs on Physics, Vol. 107, Oxford University Press (2000).
3. T. Åkeson *et al.* , Nucl. Instr. and Meth. **A262** (1987) 243.
4. R. Wigmans and M.T. Zeyrek, Nucl. Instr. and Meth. **A485** (2002) 385.
5. M. Aharrouche *et al.* (ATLAS Collaboration), *Energy Linearity and Resolution of the ATLAS Electromagnetic Barrel Calorimeter in an Electron Test-Beam*, Nucl. Instr. and Meth. (in press, 2006).
6. R. Wigmans, *On the Calibration of Longitudinally Segmented Calorimeters*, Proc. of the 12th International Conference on Calorimetry in High Energy Physics (CALOR06), Chicago, June 2006.
7. D. Acosta *et al.* , Nucl. Instr. and Meth. **A316** (1992) 184.
8. N. Akchurin *et al.* , Nucl. Instr. and Meth. **A399** (1997) 202.
9. T.A. Gabriel *et al.* , Nucl. Instr. and Meth. **A338** (1994) 336; *see also* D. Groom, contribution to these Proceedings.
10. N. Akchurin *et al.* , Nucl. Instr. and Meth. **A408** (1998) 380.
11. D. Green, *Proc. 4th Int. Conf. on Calorimetry in High Energy Physics*, La Biodola, Italy, 1994, eds. A. Menzione and A. Scribano, (Singapore: World Scientific), p. 1.
12. M. Albrow *et al.* , Nucl. Instr. and Meth. **A487** (2002) 381.
13. N. Akchurin *et al.* , Nucl. Instr. and Meth. **A537** (2005) 537.
14. G. Drews *et al.* , Nucl. Instr. and Meth. **A290** (1990) 335.
15. C. Leroy, Y. Sirois and R. Wigmans, Nucl. Instr. and Meth. **A252** (1986) 4.
16. E. Bernardi *et al.* . Nucl. Instr. and Meth. **A262** (1987) 229.
17. G. d'Agostini *et al.* , Nucl. Instr. and Meth. **A274** (1989) 134.
18. H. Abramowicz *et al.* , Nucl. Instr. and Meth. **180** (1981) 429.

---

[2] During the workshop, it was agreed that this would be an important part of the follow-up activities

A simplistic view of hadron calorimetry

Donald E. Groom

50R6008, Lawrence Berkeley National Laboratory, Berkeley CA 94720
deg@lbl.gov

Abstract. All too often we rely on Monte Carlo simulations without worrying too much about basic physics. It is possible to start with a very simple calorimeter (a big cylinder) and learn the functional form of π/e by an induction argument. Monte Carlo simulations provide sanity checks and constants. A power-law functional form describes test beam results surprisingly well. The prediction that calorimeters respond differently to protons and pions of the same energy was unexpected. The effect was later demonstrated by the CMS forward calorimeter group, using the most noncompensating calorimeter ever built. Calorimeter resolution is dominated by fluctuations in π^0 production and the energy deposit by neutrons. The DREAM collaboration has recently used a dual readout calorimeter to eliminate the first of these. Ultimate resolution depends on measuring neutrons on an event-by-event basis as well.

Keywords: Hadron calorimetry, hadron cascades, sampling calorimetry
PACS: 02.70.Uu, 29.40.Ka, 29.40.Mc, 29.40Vj, 34.50.Bw

Introduction

After more than three decades of creative thought, development, and testing, hadron calorimeters have become highly evolved and sophisticated backbones of high-energy physics detectors. Their design concept remains the domain of experienced physicists who often rely on modern hadronic cascade simulation codes, but the actual design and construction has moved to expert engineering design teams. It is easy to lose sight of the underlying simplicity of the processes taking place inside the calorimeter. The object of this paper (colloquium) is to shed considerable baggage and to go back to the elemental physics situation.

The "calorimeter" we consider is nothing more than a cylinder which is long enough and big enough to completely contain the cascade produced by a, usually a pion or proton, arriving along the axis. It can be homogenous or sampling; if the latter, the sampling is sufficiently fine that the material can be regarded as uniform. There is no front em compartment and no rear catcher; these can be added after the conceptual framework is in place.

I do rely on Monte Carlo calculations made by other people—for the most part, the creators of HETC (CALOR), MARS, and FLUKA. The only thing of interest here is the π^0 fraction; this is a robust feature of all of the codes and has changed little over a decade or two of code improvement.

The objects are to understand the energy dependence of the π/e ratio, the difference between pion and proton induced cascades, and, finally, the exciting new results from a dual-readout calorimeter. The content of this paper (colloquium) derives mainly from the 1994 paper by Gabriel *et al.*[1] (Paper I) and a paper presently in process with Nucl.

CP896, *Hadronic Shower Simulation Workshop*
edited by M. Albrow and R. Raja
2007 American Institute of Physics 978-0-7354-0401-4/07/$23.00

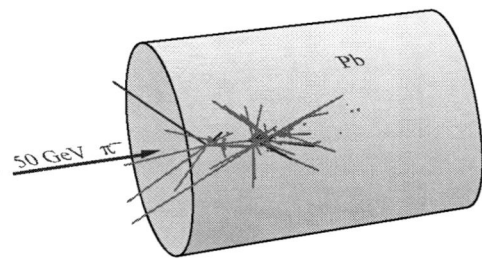

FIGURE 1. The "calorimeter" used in this discussion. The cylinder (usually lead or iron in the simulations) has a big enough radius and length to totally contain the hadronic cascade except for front-surface albedo losses, and the projectile, usually a pion but sometimes a proton, is incident along the axis.

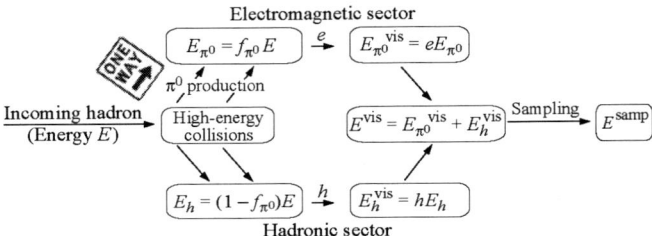

FIGURE 2. Energy flow in a hadronic cascade. A fraction f_{π^0} (with energy-dependent mean $f_{\pi^0}^0$) is transferred to the electromagnetic sector through π^0 production in repeated hadronic inelastic collisions. The π^0 and hadronic energy deposits after the division are separately stochastic, and so must be treated as parallel statistical processes. Each produces a potentially detectable signal, whose sum E^{vis} is sampled.

Instrum. and Meth.[2] (Paper II).

Electron and hadron energy deposit

My model of energy flow in a hadronic cascade is shown in Fig. 2. A hadron with energy E generates a cascade in which there are repeated hadronic collisions. In each of these π^0's are produced, which immediately decay to photons. A fraction f_{π^0} (with mean $f_{\pi^0}^{0}$[1]) of the energy is irrevocably removed from the hadronic part of the cascade in this way, and is instead deposited in electromagnetic (em) showers. It is converted to a potentially observable signal with efficiency e, which is in general different, usually larger, that the hadronic detection efficiency h. Gammas from nuclear excitation are considered as part of the hadronic signal, since the energy they carry scales as the hadronic fraction.

---

[1] A superscript 0 is used to denote the mean of a stochastic variable.

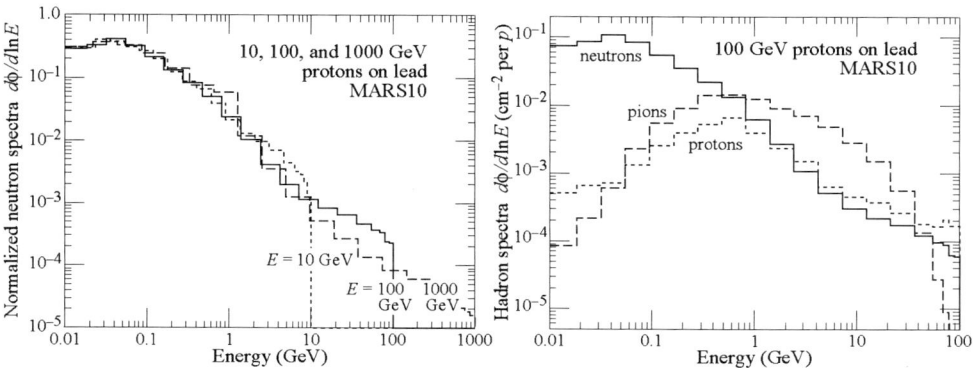

FIGURE 3. (a) MARS10 simulations of the neutron spectra in a lead beam stop for incident proton energies of 10, 100, and 1000 GeV. They are normalized for relative agreement at low energies to emphasize the shape similarity below the beam energy cutoff. (b) MARS10 simulation of the neutron, proton, and pion spectra for 100 GeV protons on lead. At low energies the charged particles are removed by ionization loss. Above about 1 GeV protons and neutrons have similar spectra. The relative numbers depend upon the material.

The "universal spectrum"

In either case, the cascade or shower grows exponentially until the lowest-energy particles stop or are absorbed. Except in the case of Cherenkov light, the ionization by charged particles in the sensitive part of the calorimeter constitutes the observable signal. In the em case much of the ionization is by electrons below the critical energy, of order 10 MeV (13.8 MeV in Fe and 6.0 MeV in U). In the case of hadronic deposit, it is by ionization losses by pions, protons, neutron interaction products, Compton recoil electrons, spallation fragments, *etc.* The low-energy particles responsible for most of the deposit have some interesting features:

- The spectra have no memory of the initial hadron or even of the early stages of the cascade. A proton, pion, or kaon generates the same spectrum of a given species and the same ratios of the spectra of different species. For example, the relative low-energy particle spectra are the same for an incident pion or proton.
- There is also no memory of the energy of the incident hadron.

MAR10 simulations illustrate these points[1]. The neutron spectra shown in Fig. 3(a) are normalized for relative agreement. Over two orders of magnitude in the incident proton energy, the shapes of the neutron spectra below the lowest incident proton energy (or slightly lower) agree. The flux is 700 times higher at the peak than at 10 GeV; contributions to the total flux by high-energy particles have little affect. The neutron, pion, and proton fluxes shown in Fig. 3(b) have ratios dependent on the calorimeter environment but independent of the nature of the incident hadron.

The universality of the low-energy spectra in a given calorimeter environment has important consequences for calorimetry, among these the possibility of defining a constant

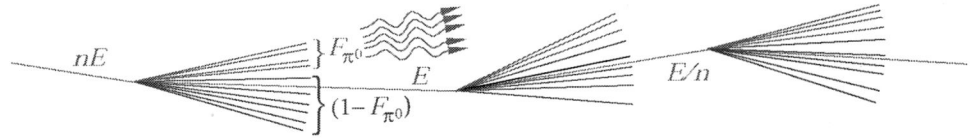

FIGURE 4. Cartoon of a hadronic cascade. It is assumed that in each generation the average energy of cascade particles decreases by a factor n and than an average fraction F_{π^0} of the energy leaves the hadronic sector via π^0 production.

hadron detection efficiency h.

Can we understand the π/e ratio from basic principles?

The mean hadronic fraction in a cascade

Real hadronic cascades are lumpy affairs, often with a few separated regions of high energy deposit. A low-energy cascade of this sort is shown in Fig. 1. Dramatic examples of this were shown by Richard Wigmans in his talk at this Symposium. But in the spirit of my approach, we instead imagine the cartoon cascade shown in Fig. 4.

In this Figure, a hadron with energy nE interacts with the production of π^0's and secondary hadrons. If the multiplicity is n, then an average hadron after the interaction has energy E. In this *single collision* an average energy fraction of F_{π^0} goes into π^0's. (This is not to be confused with f_{π^0}, indicated in Fig. 2 as the total fraction going into π^0's in the cascade.)

Now let $A(E)$ be some measure of the level of hadronic activity induced by a hadron with energy E. There are many candidates: radioactivation, the length of all tracks made by particles with energies above some cutoff, the number of stars with energies above some cutoff, ionization energy deposit, the total energy of nuclear gamma rays, or, in the case of interest here, hadronic activity. So we can write

$$A(nE) = \sum_{\text{secondaries} \neq \pi^0} A(E_i) , \tag{1}$$

since except for the ionization of a stray track or two the activity induced by $A(nE)$ has to be the same as that induced by the hadronic daughters.

So far this is incontestable; nothing has been said about the energy distribution of the 2nd generation hadrons. But now I replace the summation by n times the activity induced by the *average* hadron. A fraction F_{π^0} has been carried out of the hadronic sector by π^0's, so

$$A(nE) \approx (1 - F_{\pi^0})nA(E) . \tag{2}$$

This is immediately recognizable as the recursion equation for a power law, so we can write

$$A(E) = KE^m . \tag{3}$$

Substitution into Eqn. 2 yields

$$1 - m = \frac{\ln(1/(1 - F_{\pi^0}))}{\ln n} \,. \tag{4}$$

From an isotopic spin argument we might expect F_{π^0} to be about 1/3; the Monte Carlo's indicate that it is closer to 1/4. The multiplicity n might be 6 or 7; it has a $\ln E$ energy dependence but since we take the log again, the denominator doesn't vary much with energy. These rather hand-waving arguments say that m is in the range 0.82–0.87. It is ultimately an experimental number. From the construction, we must not expect the power-law *approximation* to work very well below about 10 GeV, although it seems to work down to about 5 GeV. It is asymptotically zero; at very high energies the cascade is nearly entirely electromagnetic.

If we say that $A(E)$ is the mean hadronic fraction f_h^0, then

$$f_h^0 \approx (E/E_0)^{m-1} \,. \tag{5}$$

Here E_0 is introduced as the scale factor for dimensional reasons, but it can be understood as roughly the threshold for π-p inelastic collisions, or about 1 GeV.

The π/e response ratio

A calorimeter usually has a linear response eE to electrons, where e is the efficiency for converting the em energy deposit to a visible signal. In a hadronic cascade a mean fraction $f_{\pi^0}^0$ is observed with efficiency e, and a mean fraction f_h^0 ($= 1 - f_{\pi^0}^0$) with efficiency h, so that the mean signal is $E(e f_{\pi^0}^0 + h f_h^0)$. After a little manipulation, the ratio of pion to electron response is found to be

$$\begin{aligned} \pi/e &= 1 - (1 - h/e) f_h^0 \\ &\approx 1 - (1 - h/e)(E/E_0)^{m-1} \equiv 1 - a E^{m-1} \,. \end{aligned} \tag{6}$$

In the second line we have explicitly used the power-law dependence given in Eqn. 5. It is important to note that only $a = (1 - h/e)/E_0^{m-1}$ can be determined by measuring the energy dependence of π/e. Any claim about the value of h/e rests on an assumed value for E_0. But since E_0 is raised to a small power, a is insensitive to its exact value, and $E_0 \approx 1$ GeV is often assumed.

Fits to a wide variety of test-beam data have been made. A subset is shown in Fig. 5.

Inclusion of nuclear gamma rays

Substantial em energy can be deposited by nuclear gamma rays. Most of this energy comes from slow neutron capture followed by nuclear deexcitation, which occurs on a time scale of μs rather than ns. A typical acceptance gate is open for 100 ns, and so some of the energy contributes to the visible signal. It is convenient to define this contribution

141

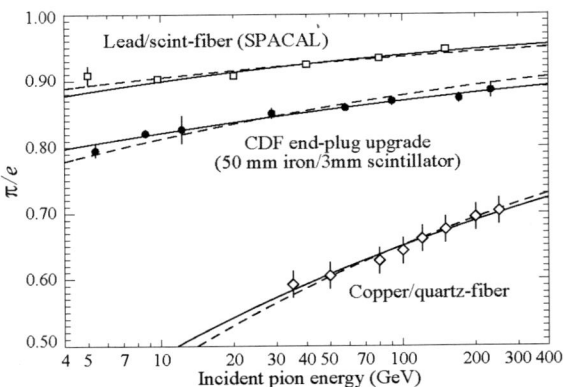

FIGURE 5. Fits to test-beam results for a lead/scintillator-fiber)[4], for the CDF upgrade end-cap hadron calorimeter (50 mm iron/3 mm scintillator sheets)[5] and for a copper/quartz-fiber test calorimeter[6].

as $f_h f_\gamma$, so that the total em energy contribution to the visible signal is $E(f_{\pi^0} + f_h f_\gamma)$, with mean $E(f_{\pi^0}^0 + f_h^0 f_\gamma)$. We can repeat the algebra leading to Eqn. 6 to find

$$\begin{aligned} \pi/e &= & 1 - (1 - h'/e)(1 - f_\gamma)f_h^0 \\ &\approx & 1 - (1 - h'/e)(1 - f_\gamma)(E/E_0)^{m-1} \equiv 1 - aE^{m-1}. \end{aligned} \quad (7)$$

The most important conclusion from Eqn. 7 is that the power law is retained, with the same experimental parameter a, even though part of the em contribution tracks with f_h. One can use either Eqn. 6 or Eqn. 7. The only difference is that the hadron detection efficiency h' does not contain contributions from nuclear gamma rays.

The p/π response ratio

In a hadronic collision a leading particle (highest energy secondary) tends to have the same quark number as the incident particle. If it is a pion, there is a high probability that the leading particle is a π^0. If this is the case, a large fraction of the incident energy is removed to the electromagnetic sector; if not, a leading $\pi^\pm$ can dump a significant fraction of the energy into the em sector on the next collision. A smaller energy fraction (with mean $f_{\pi^-}^0$) is available for the hadronic sector than would be the case without leading particle effects. If the incident hadron is a proton, the leading particle is usually a proton or neutron, and the net effect is a larger mean hadronic fraction than in the pion case: $f_p^0 > f_{\pi^-}^0$.

The "universal spectrum" idea tells us that the cascade's low-energy composition is the same in either case, not only as a function of energy but for either projectile. The ratio $f_p^0/f_{\pi^-}^0$ should therefore be the same at any incident energy, even though both $f_{\pi^-}^0$ and f_p^0 are functions of energy.

142

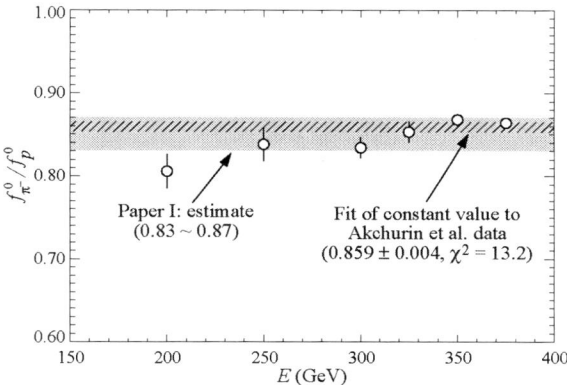

FIGURE 6. The mean hadronic fraction ratio $f_{\pi^-}^0/f_p^0$ as calculated from the copper/quartz-fiber calorimeter data of Ref. [3]. The gray band is the range expected from Paper I; the crosshatched band is a constant value fitted to the data. The PDG scale factor is 1.6.

But since they are different at a given energy, there should be a measurable π/p ratio different from unity in a noncompensating calorimeter. We can rewrite Eqn. 6 for the incident pion and incident proton cases, and rearrange to get

$$f_{\pi^-}^0/f_p^0 = \frac{1-\pi/e}{1-p/e} \approx (E_{0\pi^-}/E_{0p})^{1-m} . \tag{8}$$

The factor $(1-h/e)$ cancels in the reduction. If the power-law approximation is invoked, as in the second line of Eqn. 8, the ratio is just the ratio of the scale energies to the $m-1$ power. It is clear from the equation that the statistical sensitivity is maximal for small h/e, when π/e and p/e are as small as possible. The opportunity to make the measurement was provided by the CMS quartz-fiber forward calorimeter test module (QFCAL), whose Cherenkov readout had very low sensitivity to hadrons. It had the lowest h/e of any known calorimeter, about 1/5. A plot of $f_{\pi^-}^0/f_p^0$ as a function of energy, derived from results presented in the paper, is shown in Fig. 6[3]. Since the constancy of the ratio follows from simple physical arguments, the apparent energy dependence in the Figure is not understood.

Dual readout calorimeters

The resolution of hadron calorimeters is worse that that of em calorimeters, mostly because of fluctuations in the π^0 content (in the case of noncompensating calorimeters) and fluctuations in neutron content. (Many of the neutrons which are produced endothermically do not "return" their energy before the acceptance gate of the calorimeter has closed.) The "Holy Grail" in the design of hadron calorimeters is to measure, correct for, and thus remove the effects of both of these fluctuations[7]. There have been many proposals concerning how to do this, but the most significant advance has been by the

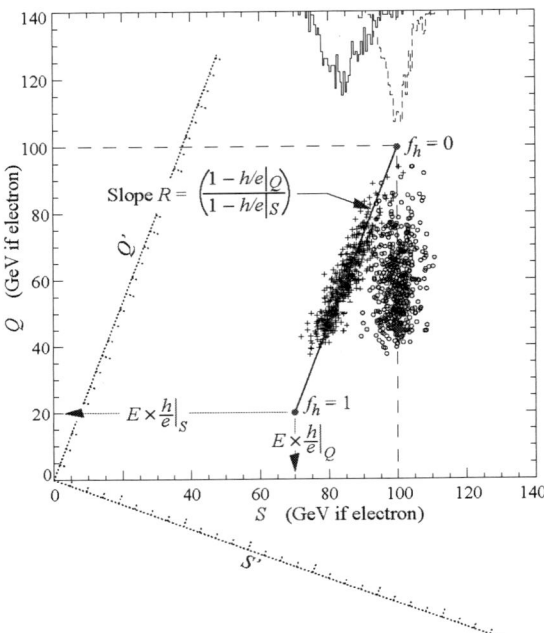

FIGURE 7. A toy model showing energy correction for 100 GeV pions in an idealized DREAM calorimeter, where Q is the response in the quartz-fiber readout and S is the response in the scintillator-fiber readout. The observed "events" are shown by the +'s, and the corrected events by the o's. Rotating to a frame in which the Q' axis is parallel to the event locus provides an equivalent reduction.

DREAM collaboration[8]. Their test calorimeter consisted of copper tubes stuffed with 7 fibers each—3 of scintillator, and 4 of either quartz or acrylic nonscintillating fibers. Each kind had its own readout. The scintillators were sensitive to both photons and hadrons, while the nonscintillating fibers were sensitive only to Cherenkov light generated mostly by electrons. The "compensation contrast" was as large as possible, with $h/e \approx 0.7$ for the scintillators and $h/e \approx 0.2$ for the Cherenkov radiators. The Cherenkov output (Q) vs the scintillator output (S) is shown by the +'s in Fig. 6 for a set of 100 GeV toy events. They scatter around the diagonal locus in the figure, whose position depends upon h/e in each channel. The marginal distribution in S is shown by the solid histogram at the top, with $\sigma = 5.3\%$. Given the locus (either by a fit to the data or by measuring the energy dependence of π/e for each channel), the energy can be corrected for em content. With the definition $R = (1 - h/e|_Q)/(1 - h/e|_S)$, the corrected energy is

$$E_{\text{corr}} = \frac{RS - Q}{R - 1} . \tag{9}$$

The circles in Fig. 7 show the same events after reduction via Eq. (9), and the dashed histogram shows the marginal distribution. The mean is 100.1 GeV, the fractional standard deviation is 3.4%, and there is no evident skewness. Complete compensation has been achieved by using the simultaneous readouts.

An alternative approach is to move to a rotated coordinate system (S', Q') where the event locus is perpendicular to the S' axis. The marginal distribution on the S' axis has minimal width. After correction for foreshortening effects, this procedure yields a result which is algebraically identical to Eqn. 9.

The future

The remaining problem is to correct for fluctuations in the number of neutrons in each event. Ways to do this are under active investigation, particularly by the NLC "Fourth concept" group[9]. Neutron signatures include the lateral distribution of the energy deposition, ionization produced by *n-p* elastically scattered protons in hydrogenous scintillator, and the highly ionizing spallation products. The possibility of observing the fast, blue, directional, polarized Cherenkov light in an inorganic scintillator is the object of present test beam work, using the scintillator $PbWO_4$[10]. (The slow component has only a 50 ns decay time, $\lambda_{max} = 560$ nm (yellow), but the scintillation efficiency is only 0.1% that of NaI.) This scheme might achieve the em separation of DREAM with a single detector. It remains to do something like sandwiching $PbWO_4$ with thin organic scintillator sheets to resolve the neutron component via *n-p* elastically scattered protons. For each event the $PbWO_4$ scintillation signal, the $PbWO_4$ Cherenkov signal, and the organic scintillator signal would be represented as a point in a data cube analogous to the two-dimensional "data square" shown in Fig. 7. It will be interesting to see if a correction formula as simple as Eqn. 9 can be found.

These are very promising directions, and it seems likely that the hadron calorimeter of the future will achieve something close to the energy resolution of em calorimeters.

Appendix: The uselessness of dE/dx

"The expression dE/dx should be abandoned; it is never relevant to the signals in a particle-by-particle analysis." —Hans Bichsel[11]

Calorimeters are often calibrated with muon beams. Here, as in other situations, it is sometimes assumed that the Bethe-Bloch dE/dx[2] (including density-effect corrections) is being measured. One plots a histogram of the signals at a given muon energy, calculates the average, and reduces it to the "*mip*," the signal that would be produced at the minimum of the Bethe-Bloch dE/dx function at $\beta\gamma \approx 3$–4. The measurement is typically based on several hundred events, so the average is biased by events in the Landau tail and background events. Sometimes the more stable peak of the distribution, the "most probable" energy deposit, is used instead. In either case, particularly the former, the result for high-energy muons must also be corrected for radiative effects.

---

[2] This is the *mean dE/dx*; we follow convention by ignoring the fact that it is negative.

To understand the situation, it is convenient to write[12]:

$$\frac{dE}{dx} = \begin{pmatrix} \text{Contribution from} \\ \text{low-energy electron} \\ \text{scattering} \end{pmatrix} + \begin{pmatrix} \text{Contribution from} \\ \text{high-energy elec-} \\ \text{tron scattering,} \\ T_{cut} < T_{max} \end{pmatrix} + \begin{pmatrix} \text{Contribution} \\ \text{from } \delta \text{ rays,} \\ T_{cut} \leq T < T_{max} \end{pmatrix} \quad (10)$$

dE/dx is derived (nearly) this way[13, 14], except that the last two terms are usually combined. (In his very thorough treatment, Fano[14] also introduces a third term in the region where the low-energy an high-energy terms meet.) The division into dE/dx contributions from low-energy electron scattering and high-energy electron scattering is made on the basis of the approximations used. For example, in calculating the high-energy electron contribution, atomic binding energy can be neglected.

Semiclassically, the low-energy region corresponds to distant collisions in which the atom is excited or frees an electron with energy somewhat larger than its binding energy. We might imagine the incident particle's electric field as extending outward from the smallest impact parameter consistent with the upper energy cutoff. As the incident particle becomes more and more relativistic, the field flattens and extends. This produces a $\ln \beta \gamma$ additive contribution. But this extension is limited by polarization of the medium, the "density effect." Sternheimer developed a rather accurate theory for predicting the effect in any material[15]. Asymptotically,

$$\delta/2 \rightarrow \ln(\hbar \omega_p / I) + \ln \beta \gamma - 1/2 . \quad (11)$$

where the plasma energy $\hbar \omega_p$ and effective ionization potential I are not of interest here. The important points are that (a) the density-effect term goes with the low-energy contribution, not elsewhere, and (b) this $\ln \beta \gamma$ cancels the $\ln \beta \gamma$ in the low-energy contribution, as we might expect on physical grounds. The first term's contribution to dE/dx thus approaches a constant as the energy increases. Both of these points are widely misunderstood.

Together, the sum of the first two terms in Eqn. 10 is the "restricted energy loss"[16], which means that dE/dx is calculated with the prohibition of single energy transfers in excess of some arbitrary $T_{cut} \leq T < T_{max}$. T_{max} is the kinematic limit on kinetic energy transfer to one electron. The middle (second) term depends only on β as well, and so also quickly approaches a constant. Examples of restricted energy loss are shown in Fig. 8.

It is easy to integrate $T \, d^2N/dx \, dT$[16] over the interval $T_{cut} < T < T_{max}$ to obtain the last term in Eqn. 10. Here N is the number of δ rays. The result is exactly the difference between the full dE/dx and the restricted dE/dx. The integral is proportional to $\ln T_{max}$, which automatically introduces a $\ln \beta \gamma$ dependence. *The asymptotic logarithmic rise of dE/dx comes entirely from the δ rays.* An integration to find dN/dx shows that the number of δ rays above T_{cut} in a reasonable detector is very small, even for T_{cut} as low as the minimum energy loss in the detector. For example, a 500 MeV pion traversing a 300 μm thick silicon detector produces a δ ray with energy greater than "1 *mip*" in only one event out of 20. The increase in the number and length of δ-ray tracks along the trajectory of a particle in a bubble chamber was once used to estimate the particle's energy.

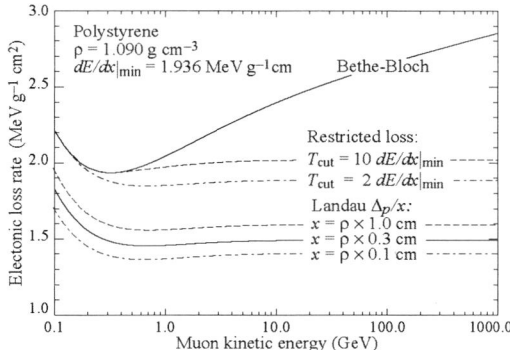

FIGURE 8. Bethe-Bloch dE/dx, two examples of restricted energy-loss rate, and the Landau most probable energy deposit per unit thickness in polystyrene scintillator, in which $dE/dx|_{min} = 1.936\ \text{MeV g}^{-1}\text{cm}^2$. The incident particles are muons. The calculations of Δ_p/x were by Hans Bichsel.

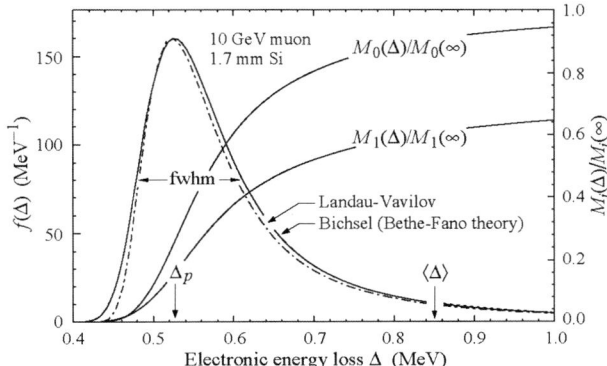

FIGURE 9. Bichsel's calculations of the electronic energy deposit distribution for a 10 GeV muon traversing a 1.7 mm silicon detector (which has roughly the stopping power of a 3-mm thick scintillator)[11, 17, 18, 20]. The Landau-Vavilov function (dot-dashed) uses a Rutherford cross section without atomic binding corrections but with a maximum kinetic energy transfer limit T_{max}. The solid curve was calculated using Bethe-Fano theory. $M_0(\Delta)$ and $M_1(\Delta)$ are the cumulative 0th and 1st moments of $f(\Delta)$, respectively. The fwhm of the Landau-Vavilov function is about 4ξ for detectors of moderate thickness.

For detectors of moderate thickness x (*e.g.*, the scintillator tiles or LAr cells used in calorimeters),[3] the energy-loss probability distribution $f(\Delta; \beta\gamma, x)$ is adequately described by the Landau (or Landau-Vavilov-Bichsel) distribution[19, 20, 21]. A special

---

[3] $G \lesssim 0.05$–0.1, where G is given by Rossi [[13], Eq. 2.7.0]. It is Vavilov's κ[19].

case is shown in Fig. 9. The most probable energy loss is given by

$$\Delta_p = \xi \left[\ln \frac{2m_e c^2 \beta^2 \gamma^2}{I} + \ln \frac{\xi}{I} + j - \beta^2 - \delta(\beta\gamma) \right]$$

$$\xrightarrow[\beta\gamma \gtrsim 100]{} \xi \left[\ln \frac{2m_e c^2 \xi}{(\hbar\omega_p)^2} + j \right] . \tag{12}$$

where $\xi = 0.153537 \langle Z/A \rangle (x/\beta^2)$ MeV for a detector whose thickness x is in g cm^{-2}, and $j = 0.200$.[4] While dE/dx is independent of thickness, Δ_p/x scales as $a\ln x + b$. The density correction $\delta(\beta\gamma)$ was not included in Landau's or Vavilov's work, but it was later included by Bichsel[20]. For scintillator (polystyrene), the mean excitation energy is 68.7 eV and the plasma energy $\hbar\omega_p$ is 21.8 eV.

For the case shown in Fig. 9, the mean energy loss is about 50% greater than the most probable loss. With increasing energy it moves to the right because of occasional δ-ray production. With T_{max} in the GeV region, the mean is strongly influenced by contributions to the tail hundreds or thousands of times the range of the plot. It makes sense to determine the most probable energy loss, not the ephemeral and undependable average.

Radiative contributions to dE/dx rise almost linearly with energy, becoming as important as ionization losses at some "muon critical energy" $E_{\mu c}$: 1183 GeV in plastic scintillator, 347 GeV in iron and 141 GeV in lead[12, 23]. The contributions are significant well below the critical energy. Monte Carlo calculations by Striganov and collaborators[24] indicate that, while a high-energy shoulder appears on the energy-loss distribution, the most probable energy loss increases only slightly in "thin" absorbers, e.g., for 1000 GeV muons incident on 100 g cm^{-2} of iron. They regard radiative effects as "important" when the most probable height of the normalized energy-loss distributions are lowered by $\gtrsim$ 10% when radiative effects are included. This can be the case under some circumstances.

Acknowledgments

I am indebted to a large fraction of my calorimetry and radiation physics friends for profitable discussions in the course of formulating these ideas, but particularly so to Richard Wigmans and my collaborators on Paper I: Tony Gabriel, P. K. Job, Nikolai Mokhov, and Graham Stevenson. Conversations with and input from Nural Akchurin, Alberto Fassó, Alfredo Farrari, and John Hauptman have been especially welcome and useful. The dE/dx appendix is the result of long conversations with Hans Bichsel.

This work was supported by the U. S. Department of Energy under Contract No. DE-AC02-05CH11231.

---

[4] Rossi[13], Talman[22], and others give somewhat different values for j. The most probable loss is not sensitive to its value.

148

REFERENCES

1. T. A. Gabriel, D. E. Groom, P. K. Job, N. V. Mokhov, and G.R. Stevenson, Nucl. Insrum. and Meth. Phys. Res., Sect. A **338** (1994) 336–347.
2. D. E. Groom, Nucl. Insrum. and Meth. Phys. Res., Sect. A (in process) (2006); arXiv:physics/0605164 v3.
3. N. Akchurin *et al.*, Nucl. Insrum. and Meth. Phys. Res., Sect. A **408** (1998) 380.
4. D. Acosta *et al.*, Nucl. Insrum. and Meth. Phys. Res., Sect. A **308** (1991) 481.
5. J. B. Liu, "Testbeam results for the CDF endplug hadron calorimeter," Proc. VII Inter. Conf. on Calorimetry in High Energy Physics, Tucson, Arizona, 9–14 November 1997, ed. E. Cheu, T. Embry, J. Rutherfoord, R. Wigmans, World Scientific, (1998) 237–240.
6. N. Akchurin *et al.*, Nucl. Insrum. and Meth. Phys. Res., Sect. A **399** (1997) 202.
7. R. Wigmans, "Quartz Fibers and the Prospects for Hadron Calorimetry at the 1% Resolution Level," Proc. VII Inter. Conf. on Calorimetry in High Energy Physics, Tucson, Arizona, 9–14 November 1997, ed. E. Cheu, T. Embry, J. Rutherfoord, R. Wigmans, World Scientific, (1998) 182–193.
8. N. Akchurin *et al.*, Nucl. Insrum. and Meth. Phys. Res., Sect. A **537** (2005) 537
9. P. Le Du *et al.*, "Detector outline document for the Fourth Concept Detector at the International Linear Collider," http://physics.uoregon.edu/~lc/wwstudy/concepts/ (May 2006).
10. R. Wigmans, private communication (2006).
11. H. Bichsel, Nuc. Insrum. and Meth. Phys. Res., Sect. A 562 (2006) 154–197.
12. D. E. Groom, N. V. Mokhov, and S. I. Striganov, Atomic and Nuclear Data Tables **78** (2001) 183.
13. B. Rossi, *High-Energy Particles*, (Prentice-Hall, Inc., Englewood Cliffs, NJ, 1952).
14. U. Fano, Ann. Rev. Nucl. Sci. **13** (1963) 1.
15. R. M. Sternheimer, Phys. Rev. **88**, 851 (1952); Phys. Rev. **93**, 351 (1953); Phys. Rev. **103**, 511 (1956); Phys. Rev. **145**, 247 (1966); Phys. Rev. **164**, 349 (1967).
16. W.-M. Yao *et al.*, "The Review of Particle Physics," J. Phys. G **33** (2006) 1.
17. H. Bichsel, private communications (2006).
18. H. Bichsel, Ch. 87 in the Atomic, Molecular and Optical Physics Handbook, G. W. F. Drake, editor (Am. Inst. Phys. Press, Woodbury NY, 1996).
19. P. V. Vavilov, Sov. J. Phys. JETP **5** (1957) 749.
20. H. Bichsel, Rev. Mod. Phys. **60** (1988) 663.
21. L. Landau, J. Phys. VIII, (1944) 201; P. V. Vavilov, Sov. J. Phys. JETP **5** (1957) 749.
22. R. Talman, Nucl. Insrum. and Meth. **159** (1979) 189.
23. http://pdg.lbl.gov/AtomicNuclearProperties
24. N. V. Mokhov, S I. Striganov, A. V. Uzunian, On Fluctuations of Energy Losses of Ultrarelativistic Muons, (in Russian), IFVE-80-56, (Serpukhov, IHEP) (Apr 1980), 12 pp.

Hadron Interaction Simulation in Atmospheric Neutrino Flux Calculations

G.D. Barr

Department of Physics, University of Oxford,
Denys Wilkinson Building, Keble Road, Oxford, UK, OX1 3RH

Abstract. The aspects of hadron simulation affecting the calculation of unoscillated atmospheric neutrino fluxes is discussed. The method of simulating the atmospheric cascades is described. An evaluation of the errors on the atmospheric neutrino fluxes and various experimentally convenient ratios is presented. A new measurement of hadron production with the NA49 experiment at CERN is briefly reported.

Keywords: hadronic interaction simulation, cosmic rays, atmospheric neutrinos, hadronic interaction measurement

PACS: 13.60.Le, 95.85.Ry

INTRODUCTION

Hadron shower simulation is a crucial part of the simulation of cosmic ray induced showers in the atmosphere. One of the uses of cosmic ray shower simulation is the determination of the neutrino flux in underground detectors. A comparison of these predicted fluxes with experiment has been used extensively for the measurement [1, 2, 3, 4, 5, 6] of the oscillatory properties of neutrinos. Several groups have computed atmospheric neutrino fluxes and use a variety of different hadron production simulation codes, many of which have been discussed at this symposium.

This paper will briefly discuss the components of an atmospheric Monte-Carlo calculation. Meurer [7] covers aspects of extensive air shower (EAS) simulation in a separate paper at this symposium. These two problems are similar. The main differences are (1) atmospheric neutrino codes cover interactions occurring over the entire globe and therefore must run fast while EAS simulations cover only the sky above the detector and (2) EAS simulations carefully propagate the electromagnetic part of the shower which is ignored by atmospheric neutrino codes because it doesn't produce neutrinos.

This paper provides an introduction to a recently published [8] attempt to evaluate uncertainties associated with hadron production on the flux measurements. To reduce the effects of hadron production uncertainties in neutrino oscillation measurements, the experimental measurements are made by looking at ratios of fluxes. The challenge here is to evaluate the uncertainties on the simulated flux ratios themselves to see that the cancellation of hadron simulation effects actually occurs. The ratios which are commonly used are the type-ratios ν_μ/ν_e, $\nu_\mu/\bar{\nu}_\mu$, $\nu_e/\bar{\nu}_e$ and the directional ratios up/down and up/horizontal of both muon and electron type neutrinos. Finally, we briefly report a recent measurement of hadron production, from the NA49 experiment at incident proton momentum 158 GeV/c [9], which has been done in connection with this study.

CP896, *Hadronic Shower Simulation Workshop*
edited by M. Albrow and R. Raja

SIMULATION OF SHOWERS TO COMPUTE THE ATMOSPHERIC NEUTRINO FLUX

Modern calculations [10, 11, 12, 13, 14, 15, 16] are performed using full three-dimensional propagation of showers generated at any point on the globe and traveling in any direction. This is in contrast to earlier calculations, e.g. [17, 18] which used a one-dimensional approximation in which all secondaries were generated collinear with the primary to save computational time. The one-dimensional approach was found to neglect a geometrical effect which was most apparent near the horizon for low energy neutrino fluxes [19, 20].

The calculation proceeds by injecting a cosmic ray (which are approximately 70% free protons and 30% fully ionised heavier nuclei) with random direction and position at a distance of 80 km altitude. Two things then happen to this particle. Firstly, the cosmic ray is tracked backwards in the Earth's magnetic field to determine whether the particular region of phase space selected is in the shadow of the Earth (any which are are dropped from the calculation). This process and assumptions are described in more detail in [16]. Secondly, the cosmic ray is propagated forward in the fashion similar to most other Monte-Carlo programs to produce a cascade of secondary particles. The effect of energy loss in the atmosphere, the variation of atmospheric density with height and the bending in the Earth's magnetic field are taken into account.

The main feature of these calculations is the wide range of primary energies which is involved in producing even a limited energy range of neutrinos. Contained events in underground detectors are produced by neutrinos with energies around 1 GeV and most of these are produced from cosmic rays between 2 and 200 GeV. Underground detectors can also detect events which are partially contained or appear as a through going muon (which must be traveling upward as downward muons are usually produced directly in a cosmic ray shower and are not from neutrinos). The energies involved in producing a significant amount of neutrino flux involve primaries of at least 2 orders of magnitude in energy higher. Monte-Carlo simulations therefore extend into energies of hadron interactions which are higher than the highest interactions which have been seen in accelerators to date.

UNCERTAINTY DETERMINATION

The work reported here is aimed at understanding and reducing the level of uncertainties in the neutrino fluxes. By far the largest uncertainty is in the generation of hadronic showers. For example, the effects of a badly modeled atmospheric profile causes less than 1% changes in absolute fluxes [21] because the showers are generally fully contained within the atmosphere.

The method is described in detail in [8] and is a straightforward and general method for evaluating uncertainties. One of the main goals is to avoid using the difference between Monte-Carlo simulation outputs as an estimator of the error.

For atmospheric neutrinos, the particular information that is needed on hadron interactions is rather large. As indicated above, the primary energies which are important extend from the pion production threshold to well beyond the highest energies available by accelerators. The secondary particle phase space of interest is essentially the entire phase space in both x_F and p_T. The shape of the differential cross section in x_F is important. The shape of the differential cross section in p_T is not so important as cosmic ray showers are incident and produce neutrinos at a wide range of angles and an error here is lost in smearing. It is however important to be sure that the x_F shape distributions represent the distributions when integrated over the full range of p_T. Air is composed mainly of nitrogen and oxygen which are both nuclei containing equal numbers of protons and neutrons. Correction is needed (e.g. in the π^+/π^- ratio) if accelerator data is used from targets which have a different ratio of neutrons to protons e.g. beryllium or aluminium.

The evaluation of the uncertainties proceeds in three steps

1. The level of uncertainty in percent is assigned to each region of phase space (as a function of primary energy, secondary energy and meson type π^+, π^-, K^+ and K^-). This has been done by Robbins [21] who has taken account of the experimental errors and the amount of extrapolation and interpolation which is required (e.g. in some regions, measurements are available which cover only a fraction of the p_T range) [22].

2. The question of how the uncertainties in one region are correlated with the uncertainties in another region is considered. Clearly if there is a shift in the hadron production related to one experiment (e.g. the actual hadron production is one-sigma away from the central value measured by an experiment) and the result has been used for extrapolation into other regions of the phase space, then the effect in these different regions is correlated.

3. Finally, the Monte-Carlo is run with the shifts computed above applied to find the effect on the neutrino flux. The shift in ratios of neutrino fluxes are also computed to investigate by how much the uncertainty is suppressed by cancellation in the ratio. A shift is applied by weighting according to the first interaction in the chain from primary to neutrino in which a meson is produced. The weight applied is based on the uncertainty associated with the hadron production at the parent and secondary particle energies of only this interaction.

Steps (1) and (2) involve making rather subjective decisions about the level and correlations of uncertainties from experiment and are described in detail in [8]. Various checks are made to validate the method by comparing alternatives.

Several effects have been neglected: the problems of extrapolating in atomic weight A; the uncertainties on neutron production and cross section uncertainties; the reduction in uncertainties which can be obtained by constraining the fluxes with cosmic ray muon flux measurements [23].

An example of various fluxes and errors is shown in figure 1. The fluxes of muon neutrinos are shown with full circles (red) using the right hand scale. All cosmic ray fluxes have the feature that they are steeply falling with energy. Above a few GeV, the fluxes take on a characteristic slope determined by the $E^{-\alpha}$ form of the primary spectrum where α is around 2.7. At lower energies, the spectra fall below the power law due to effects of the Earth's magnetic field and the solar wind. The uncertainties in the fluxes are shown on figure 1 both by the error bars on the (red) fluxes and by the solid lines (green) (left hand scale). They are around 15% and rise somewhat with

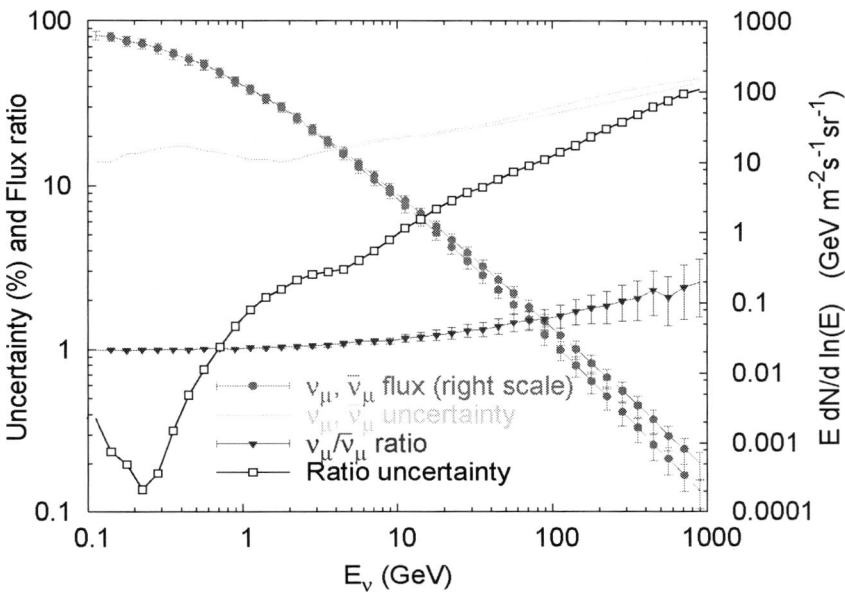

FIGURE 1. Summary of fluxes and flux uncertainties (see text)

energy, reaching around 40% at a neutrino energy of 1 TeV due to the increased effect of kaon production a high energies. [This is considerably more pronounced than simply the proportion of kaons produced in the interactions. Pions and muons are similar in mass whereas kaons are heavier, so there is more energy for the neutrino in the kaon centre of mass. Additionally, kaons are more likely to decay rather than interact due to their shorter lifetime.]

The effects of cancellation of errors is also shown on figure 1 — the example used is the ratio of muon neutrinos to muon antineutrinos where both the value of the ratio is shown (in solid triangles (blue)) with the uncertainties indicated as error bars and the uncertainties are shown separately (by the open squares (black)). In this case, at low energies, each μ^+ produces both a muon neutrino (at the point where it is produced, where the pion or kaon decays) and a muon antineutrino where it decays. This situation is reversed for a μ^-. Due to the decay kinematics, these two neutrinos coincidentally happen to (on average) have roughly the same energy. Therefore each time a muon is produced in an orientation where it will not hit the ground (any downward direction if it is below 2 GeV, more horizontal muons at higher energies) it will contribute to both numerator and denominator of the ratio. This is why the uncertainty on the $\nu_\mu/\overline{\nu}_\mu$ ratio is very low at low energies. The uncertainty from the hadron production has canceled to a high level. The $\nu_\mu/\overline{\nu}_\mu$ ratio uncertainty increases at higher energies partly because some of the muons hit the ground and the cancellation is less complete and partly because of the production of kaons.

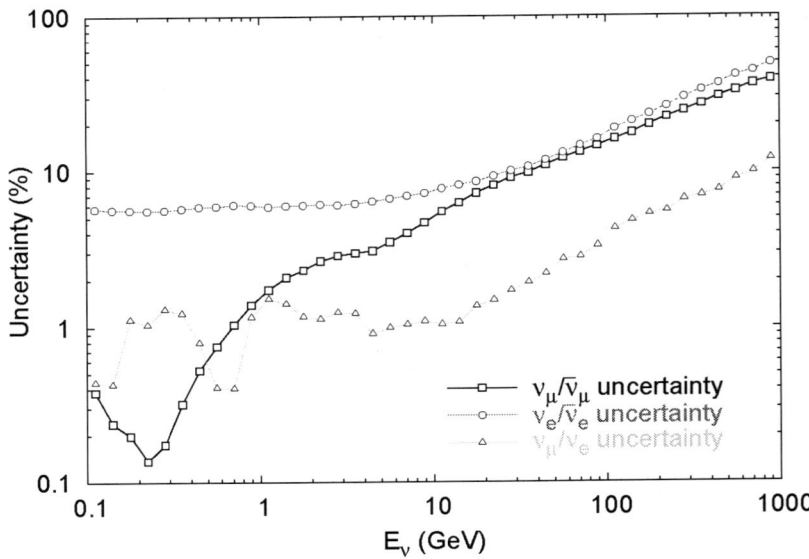

FIGURE 2. Flux uncertainties for flavour ratios (see text)

Figure 2 shows the uncertainties on the flavour ratios including the $\nu_\mu/\overline{\nu}_\mu$ ratio as repeated from figure 1 (open squares (black)). The open triangles (green) on figure 2 give the uncertainty on the $(\nu_\mu + \overline{\nu}_\mu)/(\nu_e + \overline{\nu}_e)$ flux. A similar level of cancellation of uncertainties is seen here. The open circles (red) give the uncertainties on the $\nu_e/\overline{\nu}_e$ ratio which is larger at low energy. This is because muons only associate themselves with one ν_e or $\overline{\nu}_\mu$ depending on their sign. The asymptotic low energy uncertainty simply reflects the 5% uncertainty on the π^+/π^- ratio used as input into the study.

Figure 3 shows the uncertainties on two types of directional ratios at Kamioka, the site of the SuperKamiokande neutrino detector. The ratio of upward going fluxes to downward going fluxes is shown (in squares (black) for ν_μ and in circles (red) for ν_e). The ratio of upward fluxes to horizontal fluxes is shown (in upward pointing triangles (green) for ν_μ and in downward triangles (blue) for ν_e). The features for the ν_μ and ν_e flavours are similar for both of these geometries. The up/down ratio has an uncertainty of around 7% at low energy which falls to zero around a few GeV. This effect is caused by the Earth's magnetic field — at high energies, the primary cosmic rays which produce the neutrinos are above the cutoff produced by the geomagnetic field and the hadron production uncertainty cancellation is exact. The up/horizontal ratio shows a different effect. This ratio behaves similarly at low energy, but then increases at higher energies. This is because the competition between decay and interaction of the mesons in the shower is occurring at different atmospheric densities for up and horizontal cosmic rays (at high energy, showers are highly relativistic and develop along a single direction). Horizontal showers develop higher up and therefore at lower density

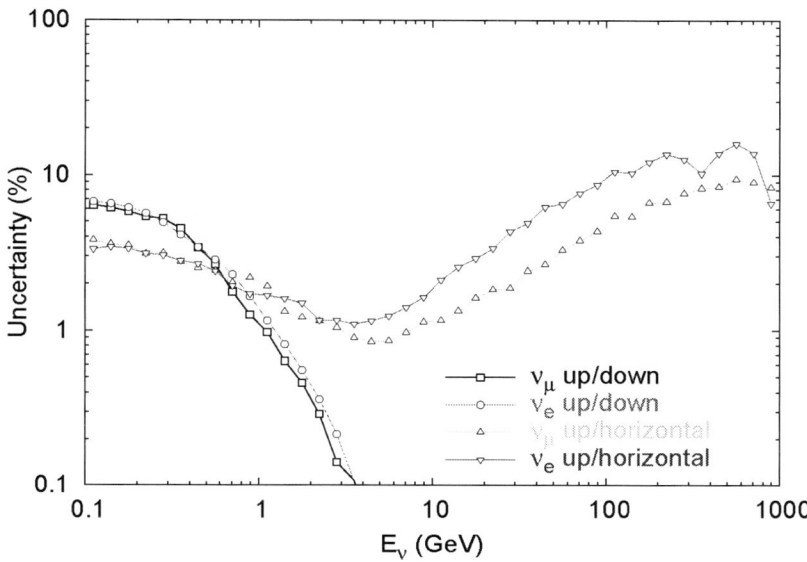

FIGURE 3. Flux uncertainties for directional ratios (see text)

than vertical showers.

The uncertainties on the fluxes can be decomposed to see which regions of phase space are more responsible for the uncertainties. It is interesting that the different ratios are affected differently by different regions of phase space. Full details are given in [8].

NA49 DATA

Since hadron production limits the accuracy to which atmospheric neutrino fluxes can be predicted, the best improvement can be obtained by doing new hadron production measurements. As discussed earlier, a wide range of primary and secondary energies are required on a target which is (as similar as possible to) nitrogen and/or oxygen. Since a single accelerator is not capable of providing beams over the entire primary energy range, several different measurements focusing on different primary energies are needed. The NA49 experiment [24] at CERN has carried out such a measurement with proton-proton interactions [25] and proton carbon interactions [9]. Details of these measurements are given in these publications. The results are also discussed [7] and compared with the main Monte-Carlos [26] in these proceedings.

NA49 is comprised of four large-volume time projection chambers as shown in figure 4. This configuration, which is similar to that used at HARP [27], E910 [28] and MIPP [29, 30] gives a large acceptance to secondary particles over a wide range of momenta and p_T. A magnetic field which bends the particles in the horizontal plane is used to measure the momenta of the particles. A track which traverses the whole

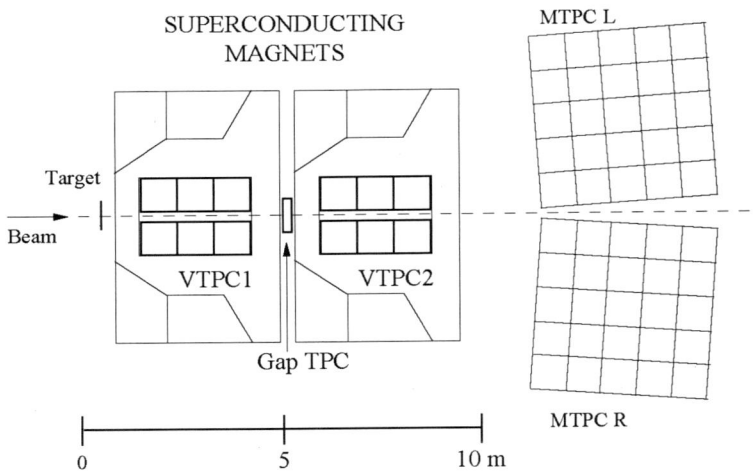

FIGURE 4. Layout of the detectors in the NA49 experiment

length of the TPCs has over 100 independent measurements of the energy deposition in the TPC gas along the track and these are combined using a truncated mean algorithm to distinguish e, π, K and p from the energy loss. Tracks with as few as 30 dE/dx measurements are easily identified and this allows tracks which exit the sides of the detector to be included in the analysis.

The data were collected with a 158 GeV/c tagged proton beam incident on a 1.5% interaction length carbon target. The beam was tagged in the trigger using a series of Cerenkov detectors and resulted in a beam which was pure protons to better than 1 part in 10^3. Carbon was selected as being particularly similar to air as it contains the same (even) ratio of the numbers of protons to neutrons as both nitrogen and oxygen and is more convenient to measure as it is a solid. A total of 377,000 inelastic events were collected. Measurements of π^+ and π^- yields in 270 bins each between x_F of -0.1 and $+0.5$ and p_T up to 2 GeV/c are described in Ref. [9] and discussed in detail in Ref. [31].

SUMMARY

An impressive amount has been learned about neutrinos from cosmic rays and this is despite the non-precision status of hadron shower simulation. Hadron production is certainly the most serious uncertainty in atmospheric neutrino flux calculations. There is even more information which could be learned using atmospheric neutrinos at higher precision [32].

This work has demonstrated that effects of the hadron production uncertainties are reduced substantially by measuring ratios as the neutrino experiments do. Values are given for the uncertainties in the fluxes and the ratios as a function of neutrino energy. The estimate presented here is not complete yet, for example, the effects of neutron production and interactions has not been considered carefully. However the technique used here to estimate the uncertainties on the fluxes could also be applied to other situations where hadron production uncertainties are important. New data on particle production has been obtained, a recent measurement of pion yields using the NA49 detector has been presented which gives precision yields in the x_F range between -0.1 and $+0.5$ and for a wide range of values of p_T.

REFERENCES

1. Y. Fukuda, et al., *Phys. Lett.* **B335**, 237–245 (1994).
2. Y. Ashie, et al., *Phys. Rev.* **D71**, 112005 (2005), hep-ex/0501064.
3. Y. Ashie, et al., *Phys. Rev. Lett.* **93**, 101801 (2004), hep-ex/0404034.
4. W. W. M. Allison, et al., *Phys. Rev.* **D72**, 052005 (2005), hep-ex/0507068.
5. M. Ambrosio, et al., *Eur. Phys. J.* **C36**, 323–339 (2004).
6. P. Adamson, et al., *Phys. Rev.* **D73**, 072002 (2006), hep-ex/0512036.
7. C. Meurer, *This symposium* (2006).
8. G. D. Barr, T. K. Gaisser, S. Robbins, and T. Stanev, *Phys. Rev.* **D74**, 094009 (2006), astro-ph/0611266.
9. C. Alt, et al., *Submitted to Eur. Phys. J.* (2006), hep-ex/0606028.
10. G. Battistoni, A. Ferrari, T. Montaruli, and P. R. Sala, *Astropart. Phys.* **19**, 269–290 (2003), hep-ph/0207035.
11. Y. Tserkovnyak, R. Komar, C. Nally, and C. Waltham, *Astropart. Phys.* **18**, 449–461 (2003), hep-ph/9907450.
12. J. Wentz, et al., *Phys. Rev.* **D67**, 073020 (2003), hep-ph/0301199.
13. J. Favier, R. Kossakowski, and J. P. Vialle, *Phys. Rev.* **D68**, 093006 (2003), astro-ph/0305460.
14. Y. Liu, L. Derome, and M. Buenerd, *Phys. Rev.* **D67**, 073022 (2003), astro-ph/0211632.
15. M. Honda, T. Kajita, K. Kasahara, and S. Midorikawa, *Phys. Rev.* **D70**, 043008 (2004), astro-ph/0404457.
16. G. D. Barr, T. K. Gaisser, P. Lipari, S. Robbins, and T. Stanev, *Phys. Rev.* **D70**, 023006 (2004), astro-ph/0403630.
17. V. Agrawal, T. K. Gaisser, P. Lipari, and T. Stanev, *Phys. Rev.* **D53**, 1314–1323 (1996), hep-ph/9509423.
18. M. Honda, T. Kajita, K. Kasahara, and S. Midorikawa, *Phys. Rev.* **D52**, 4985–5005 (1995), hep-ph/9503439.
19. G. Battistoni, et al., *Astropart. Phys.* **12**, 315–333 (2000), hep-ph/9907408.
20. P. Lipari, *Astropart. Phys.* **14**, 153–170 (2000), hep-ph/0002282.
21. S. Robbins, *D. Phil Thesis, University of Oxford, UK* (2004), CERN-THESIS-2005-005.
22. R. Engel, T. K. Gaisser, and T. Stanev, *Phys. Lett.* **B472**, 113–118 (2000), hep-ph/9911394.
23. M. Honda, T. Kajita, K. Kasahara, S. Midorikawa, and T. Sanuki (2006), astro-ph/0611418.
24. S. Afanasev, et al., *Nucl. Instrum. Meth.* **A430**, 210–244 (1999).
25. C. Alt, et al., *Eur. Phys. J.* **C45**, 343–381 (2006), hep-ex/0510009.
26. S. Striganov, *This symposium* (2006).
27. M. G. Catanesi, et al., *Nucl. Phys.* **B732**, 1–45 (2006), hep-ex/0510039.
28. J. Link, et al., *Presented at NuFact04, proceedings in Nucl. Phys. Proc. Suppl.* **149** (2005).
29. Y. Fisyak, et al., *The MIPP experiment, proposal 907, Fermilab, www.fnal.gov* (2000).
30. R. Raja, *This symposium* (2006).
31. G. Barr, et al. (2006), hep-ex/0606029.
32. T. Kajita, and K. Okumura, *Proceedings of the 5th RCCN Workshop on Sub-Dominant Oscillation Effects in Atmospheric Neutrino Experiments*, Universal Academy Press, Tokyo, 2005.

The muon component in extensive air showers and new p+C data in fixed target experiments

C. Meurer, J. Blümer, R. Engel, A. Haungs, M. Roth
and the HARP collaboration [1]

Institut für Kernphysik, Forschungszentrum Karlsruhe GmbH, Postfach 3640, 76021 Karlsruhe, Germany

Abstract. One of the most promising approaches to determine the energy spectrum and composition of the cosmic rays with energies above 10^{15} eV is the measurement of the number of electrons and muons produced in extensive air showers (EAS). Therefore simulation of air showers using electromagnetic and hadronic interaction models are necessary. These simulations show uncertainties which come mainly from hadronic interaction models. One aim of this work is to specify the low energy hadronic interactions which are important for the muon production in EAS. Therefore we simulate extensive air showers with a modified version of the simulation package CORSIKA. In particular we investigate in detail the energy and the phase space regions of secondary particle production, which are most important for muon production. This phase space region is covered by fixed target experiments at CERN. In the second part of this work we present preliminary momentum spectra of secondary π^+ and π^- in p+C collisions at 12 GeV/c measured with the HARP spectrometer at the PS accelerator at CERN. In addition we use the new p+C NA49 data at 158 GeV/c to check the reliability of hadronic interaction models for muon production in EAS. Finally, possibilities to measure relevant quantities of hadron production in existing and planned accelerator experiments are discussed.

Keywords: cosmic rays, extensive air showers, muon and hadron production, fixed target experiments
PACS: 13.85.Tp, 96.40.Pq

INTRODUCTION

The energy spectrum and composition of the cosmic rays with energies above 10^{15} eV are typically derived from measurements of the number of electrons and muons produced in extensive air showers at ground. However, the results of such a shower analysis are strongly dependent on the hadronic interaction models used for simulating reference showers [1]. Therefore it is important to study in detail the role of hadronic interactions and in particular the energy and secondary particle phase space regions that are most important for the observed characteristics of EAS.

The electromagnetic component of a shower is well determined by the depth of maximum and the energy of the shower. Due to the electromagnetic cascade, having a short radiation length of ~ 36 g/cm^2, any information on the initial distribution of photons produced in π^0 decays is lost. Therefore the electromagnetic shower component depends on the primary particle type only through the depth of shower maximum. In contrast, the

---

[1] See author list of [27]

CP896, *Hadronic Shower Simulation Workshop*
edited by M. Albrow and R. Raja
© 2007 American Institute of Physics 978-0-7354-0401-4/07/$23.00

muon component is very sensitive to the characteristics of hadronic interactions. Once the hadronic shower particles have reached an energy at which charged pions and kaons decay, they produce muons which decouple from the shower cascade. The muons propagate to the detector with small energy loss and deflection and hence carry information on hadronic interactions in EAS. Due to the competition between interaction and decay, most of the muons are decay products of mesons that are produced in low-energy interactions. Therefore it is not surprising that muons in EAS are particularly sensitive to hadronic multiparticle production at low energy [2, 3]. Recent model studies show that even at ultra-high shower energies the predictions on the lateral distribution of shower particles depend strongly on the applied low-energy interaction model [4, 5].

MUON PRODUCTION IN EXTENSIVE AIR SHOWERS

Motivated by the measurement conditions of the KASCADE array [6], we consider showers with a primary energy of 10^{15} eV and apply a muon detection threshold of 250 MeV. Using a modified version of the simulation package CORSIKA [7] we have simulated a sample of 1500 vertical proton induced showers. Below 80 GeV the low-energy hadronic interaction model GHEISHA 2002 [8] and above 80 GeV the high-energy model QGSJET 01 [9] are applied. Simulations with other models are in preparation.

In Fig. 1 (left) the energy distribution of muons at detector level (1030 g/cm^2) is shown for several lateral distance ranges. The maximum of this distribution shifts to lower energies for larger lateral distances. Most likely five consecutive hadronic interactions (number of generations) take place before a hadron decays into a muon, see Fig. 1 (right). Here and in the following we consider only those muons that reach the ground level with an energy above the detection threshold. The number of generations show only a weak dependence on the lateral distance.

To study the hadronic *ancestors* of muons in EAS, we introduce the terms *grandmother* and *mother particle* for each observed muon. The grandmother particle is the hadron inducing the *last* hadronic interaction that finally leads to a meson (mother particle) which decays into the corresponding muon. Most of the grandmother and mother particles are pions, but also about 20% of the grandmother particles are nucleons and a few are kaons. Details of the composition of mother and grandmother particles are given in Tab. 1.

TABLE 1. Particle types of mother and grandmother particles in a vertical proton induced shower at 10^{15} eV.

	mother	grandmother
pions	89.2%	72.3%
kaons	10.5%	6.5%
nucleons	–	20.9%

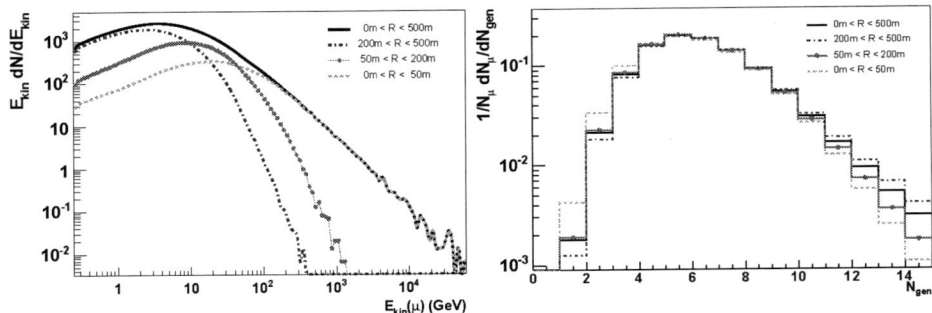

FIGURE 1. Left panel: Simulated energy distribution of muons for different lateral distances in vertical proton induced showers with a primary energy of 10^{15} eV. Right panel: Number of generations before producing a muon visible at ground level (shown for various lateral distances).

RELEVANT ENERGY RANGE

The energy spectra of different grandmother particles produced in vertical proton showers are shown in Fig. 2 (left). They cover a large energy range up to the primary energy with a maximum at about 100 GeV. The peak at 10^6 GeV in the nucleon energy spectrum shows that also a fraction of muons stems from decays of mesons produced in the first interaction in a shower. Furthermore, the step at 80 GeV clearly indicates a mismatch between the predictions of the low-energy model GHEISHA and the high-energy model QGSJET. In Fig. 2 (right) the grandmother particle energy spectrum is shown for different ranges of lateral muon distance. The maximum shifts with larger lateral distance to lower energies.

Comparing the *last* interaction in EAS with collisions studied at accelerators, one has to keep in mind that the grandmother particle corresponds to the beam particle and the mother particle is equivalent to a secondary particle produced in e.g. a minimum bias p-N interaction. The most probable energy of the grandmother particle is within the range of beam energies of fixed target experiments e.g. at the PS and SPS accelerators at CERN.

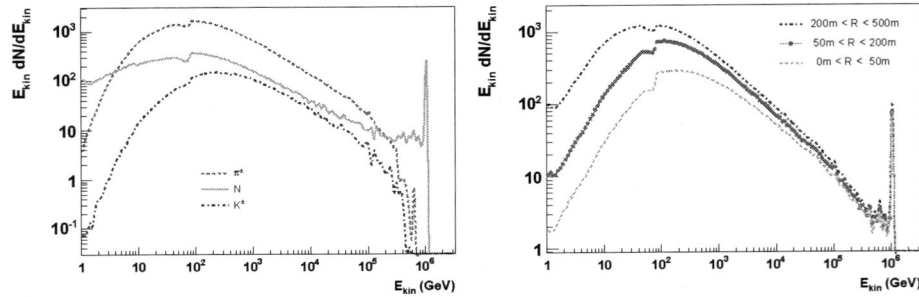

FIGURE 2. Energy distribution of grandmother particles in vertical proton showers. Left panel: different grandmother particle types for a muon lateral distance range of 0-500 m at ground level. Right panel: different lateral distances, all particle types are summed up.

TABLE 2. Energy and lateral distance ranges used for this analysis.

energy range	average energy	lateral distance range
80-400 GeV	160 GeV	50-200 m
30-60 GeV	40 GeV	200-600 m

RELEVANT PHASE SPACE REGIONS

The further study of the relevant phase space of the mother particles is done for two different grandmother energy ranges and muon lateral distance ranges at ground level, see Tab. 2. The lateral distance ranges are chosen to resemble typical lateral distances measured at KASCADE and KASCADE-Grande, respectively [10]. Motivated by the availability of protons as beam particles at accelerators we consider only those *last* interactions in EAS that are initiated by nucleons.

In Fig. 3 the rapidity spectra of mother particles (left: pions, right: kaons) are compared to the spectra of secondary particles of minimum bias proton-carbon and proton-air collisions with a fixed energy simulated with QGSJET labeled as *fixed target*. The spectra of mother particles in air showers are scaled to fit the falling tail of the fixed-energy collision spectra. No significant differences are found comparing the rapidity distributions of secondary particles in proton-carbon and proton-air collisions. As a consequence of the different selection criteria, the forward hemisphere in the mother rapidity spectra is clearly favoured compared to the spectra of secondaries in minimum bias collisions. The reason for this behaviour is the fact that the secondary particles (pions, kaons) are measured directly in fixed target experiments whereas in EAS an additional condition for the mother mesons is applied. In order to get the information of the mother meson, it has first to decay into a muon which is detectable. At low rapidity the mother mesons are missing in the EAS distribution because the energy of the daughter muon is lower than the applied detection threshold or the muon decays and it is not visible. The reason for the missing pions in the EAS distribution at the highest rapidities is the interplay between decay and further interactions, that means the pions do not decay but undergo further interactions. This effect is not much pronounced for the kaon distribution, because the energy, for which the probability of decay and interaction is of the same order, is eight times higher for kaons than for pions.

The phase space regions of mother particles produced in the *last* interaction in EAS are shown in Fig. 4 and 5. We choose a nucleon as the grandmother particle with a mean energy of 160 GeV (Fig. 4) and 40 GeV (Fig. 5), respectively. The transverse momentum of the mother particle is plotted vs. its rapidity divided by the rapidity of the grandmother particle which is equivalent to the beam rapidity in fixed target reactions. On the left hand side this distribution is shown for pions as mother particles, on the right hand side for kaons. The maximum of these distributions, which shows the most important phase space region for the muon production in EAS, is at $p_\perp \approx 0.1$ GeV and 0.7 in relative rapidity units (for a mean grandmother energy of 160 GeV) and shifts to slightly higher $p_\perp$ but stays at the same rapidity for a mean grandmother energy of 40 GeV. In both cases the distributions of pions and kaons are similar. For kaons higher particle transverse momenta are more important than for pions. The phase space regions of relevance to EAS are summarized in Tab. 3 and indicated with the dashed (red) boxes in Fig. 4 and 5.

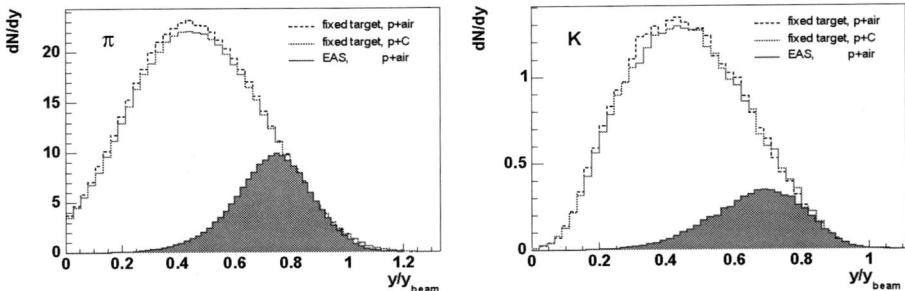

FIGURE 3. Rapidity distributions of mother particles (filled curves) compared with rapidity distributions of secondary particles in simulated single p+C (solid line) and simulated p+air (dashed line) collisions. Left panel: pions, right panel: kaons. The energy range of the grandmother particle is limited to 80-400 GeV and the lateral distance of the muons to 50-200 m to match experimentally accessible regions. The fixed target collision simulation is done at 160 GeV, corresponding approximately to the mean grandmother energy. The rapidity is normalized to the rapidity of the beam and grandmother particles, respectively.

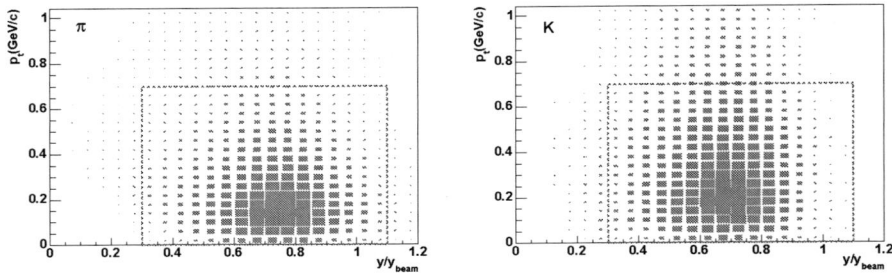

FIGURE 4. Phase space of mother particles. Grandmother energy range: 80-400 GeV. Left panel: pions, right panel: kaons. The dashed box (red) indicates the most interesting phase space region which includes more than 90% of this particles.

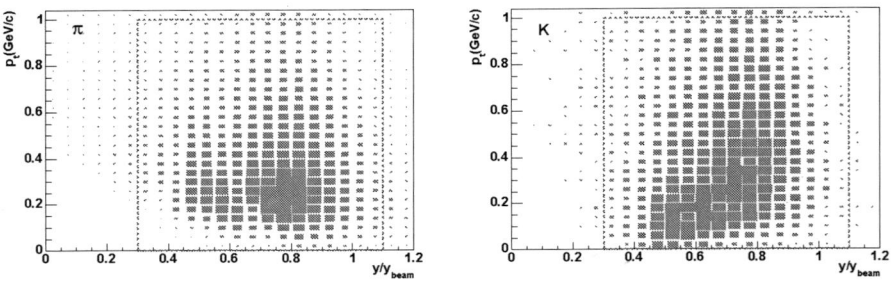

FIGURE 5. Same plots as in Fig. 4, but the grandmother energy range is 30-60 GeV.

TABLE 3. Phase space regions of hadronic interactions relevant for muon production in EAS.

average energy (GeV)	y/y_{beam}	$p_\perp$ (GeV/c)
160	0.3 - 1.1	0.0 - 0.7
40	0.3 - 1.1	0.0 - 1.0

PHASE SPACE COVERAGE BY FIXED TARGET DATA

p+Be data

In contrast to collider experiments, in fixed target setups it is more straight forward to measure particles in forward direction. The energy range of most of the existing fixed target experiments has a broad overlap with the energy range of relevance to EAS. There exist a number of fixed target measurements employing a proton beam with low energy (≤ 20 GeV) with a good phase space coverage, however, data at higher energy and for pion projectiles are very sparse. In Fig. 6 a compilation of p+Be target data is shown.

The existing data are indicated by shaded (colored) regions, whereas the beam momentum is given as abscissa and the secondary particle observable as ordinate. The relevant phase space is shown as box histogram. In this case the grandmother momentum is used instead of the beam momentum for the abscissa. The size of the boxes indicates the relative importance of the beam and secondary particle momenta.

The work of Eichten et al. [11] has become a widely used reference data set. This experiment used a proton beam with a beam momentum of 24 GeV and a beryllium target. The secondary particles (pions, kaons, protons) are measured in a broad angle range (17 mrad $< \theta <$ 127 mrad) and in a momentum region from 4 GeV/c up to 18 GeV/c. Other measurements cover only a smaller part of the phase space of interest to EAS [12, 13, 14, 15, 16]. A measurement for preparing the CERN neutrino beam experiments was performed by SPY/NA56 (NA52) [17, 18]. Therefore a 450 GeV proton beam and a thick beryllium target was used. Because of the very limited angular range ($\theta \approx 0°$) it is not included in Fig. 6.

p+C data

As simulations with hadronic interaction models show, the particle production in p+air collisions is much more similar to p+C collisions than to p+Be reactions because of the smaller difference in the atomic mass. Fig. 7 shows the p+C data which were available up to the time of these proceedings. In the past the only measurement of p+C collisions, which was not limited to a fixed angle, was the experiment done by Barton et al. [19]. These data were collected using the Fermilab Single Arm Spectrometer facility in the M6E beam line. A proton beam with a beam momentum of 100 GeV/c and a thin carbon target (1.37 g/cm^2) was used. The phase space of the secondary particles (pions, kaons, protons) covers only a small part of the phase space of interest to EAS. New p+C data at 12 GeV/c and 158 GeV/c, taken by the CERN experiments HARP [20] at the PS accelerator and NA49 [21] at the SPS accelerator, are available now. At both beam momenta the secondary particles (π^+, π^-) are measured in a broad momentum region and up to large angles. At 12 GeV/c the HARP data cover secondary momenta from 0.5 GeV/c to 8 GeV/c and an angular range from 30 mrad to 210 mrad. The NA49 data are taken in the momenta range $0.85\,\text{GeV/c} \lesssim p \lesssim 82.6\,\text{GeV/c}$ and at angles up to 440 mrad [22].

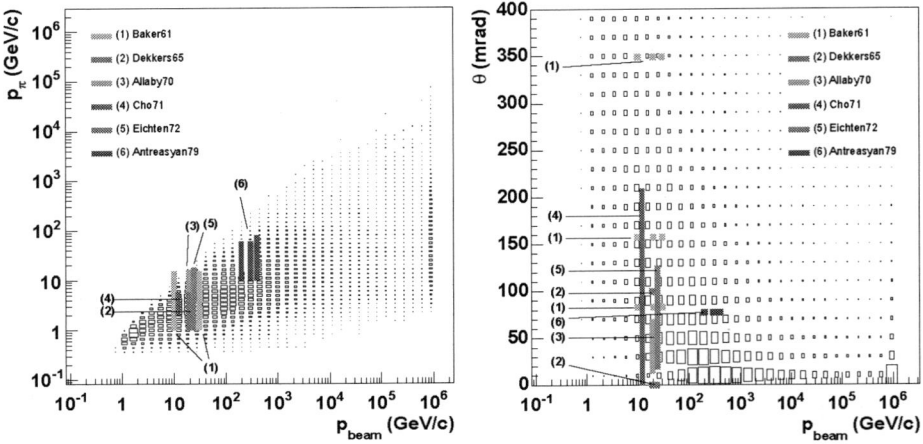

FIGURE 6. Coverage of the phase space regions of relevance to EAS (box histograms) by existing fixed target data using a proton beam and a beryllium target (shaded/colored regions). Left panel: total momentum of secondary pions vs. total momentum of proton projectiles. Right panel: angle between beam and secondary particle momentum vs. beam momentum.

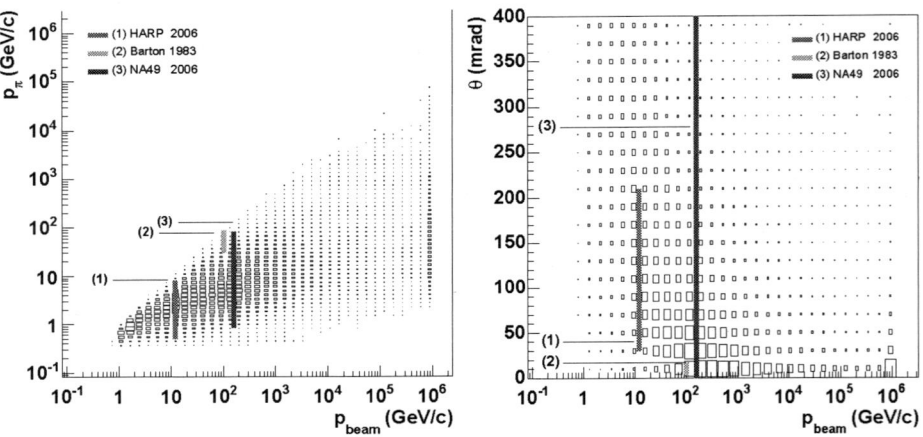

FIGURE 7. Coverage of the phase space regions of relevance to EAS (box histograms) by existing fixed target data using a proton beam and a carbon target (shaded/colored regions). Left panel: total momentum of secondary pions vs. total momentum of proton projectiles. Right panel: angle between beam and secondary particle momentum vs. beam momentum.

DATA FROM FIXED TARGET MEASUREMENTS

NA49 p+C data at 158 GeV/c

Pion spectra produced in p+C collisions with a beam momentum of 158 GeV/c measured by NA49 became available this summer [22]. In Fig.8 rapidity spectra of secondary pions are compared with simulations done with QGSJET-01, QGSJET-II [23, 24] and SIBYLL2.1 [25, 26]. SIBYLL2.1 and QGSJET-II predictions show a reasonable agreement with the data, but QGSJET-01 overestimates the measured spectra by a factor ∼1.5 in the central rapidity region. Comparing QGSJET-01 and QGSJET-II concerning the shape of the spectra, QGSJET-II predicts harder pion spectra than the older version.

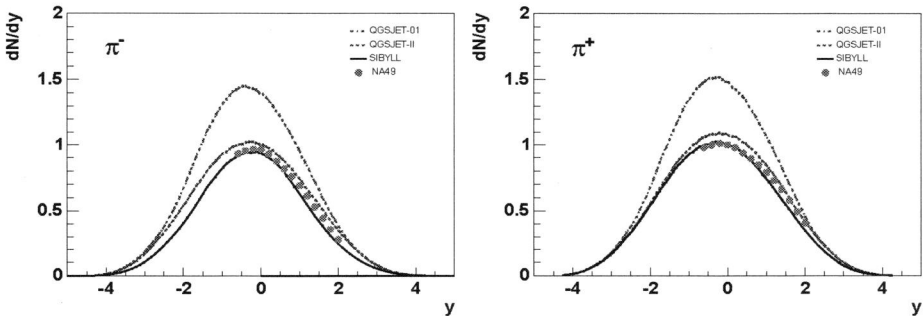

FIGURE 8. Rapidity spectra of pions produced in p+C collisions at 158 GeV/c. The filled circles indicate the NA49 measurements [22]. The three lines show the results of QGSJET-01 [9], QGSJET-II [23, 24] and SIBYLL2.1 [25, 26] simulations.

HARP p+C data at 12 GeV/c

In these proceedings, we present momentum spectra of secondary pions in p+C collisions with a beam momentum of 12 GeV/c measured by the HARP spectrometer [20]. The thickness of the carbon target is equivalent to 5% nuclear interaction length (3.56 g/cm^2). For the selection of the secondary particles (π^+, π^-) in the forward hemisphere four walls of drift chambers are used. The separation of particle types is done with different detector components: The time-of-flight measurement allows pion–kaon and pion–proton separation to be performed up to 3 GeV/c and beyond 5 GeV/c, respectively. Additional a Cherenkov detector is used to separate pions from protons and kaons above 2.5 GeV/c [27].

In Fig.9 the preliminary π^+ and π^- momentum spectra are presented in six different angular bins starting from 0.03 rad up to 0.21 rad. The cross section of the π^+ production is consistently higher for all angular bins than the cross section of π^-, especially in the higher momentum regions. This behaviour can be understood in terms of the leading particle effect which influences mainly positive pions. The error bars indicate the statistical and systematic errors. The size of both errors are of the same order of around 4 to 14% depending on the secondary momentum. Only for the highest momenta the statistical error of the π^- amounts more than 20%. Generally the statistical errors of the π^- are some what larger than that for the π^+. The dominant contributions to the systematic error are tertiary subtraction and momentum scale.

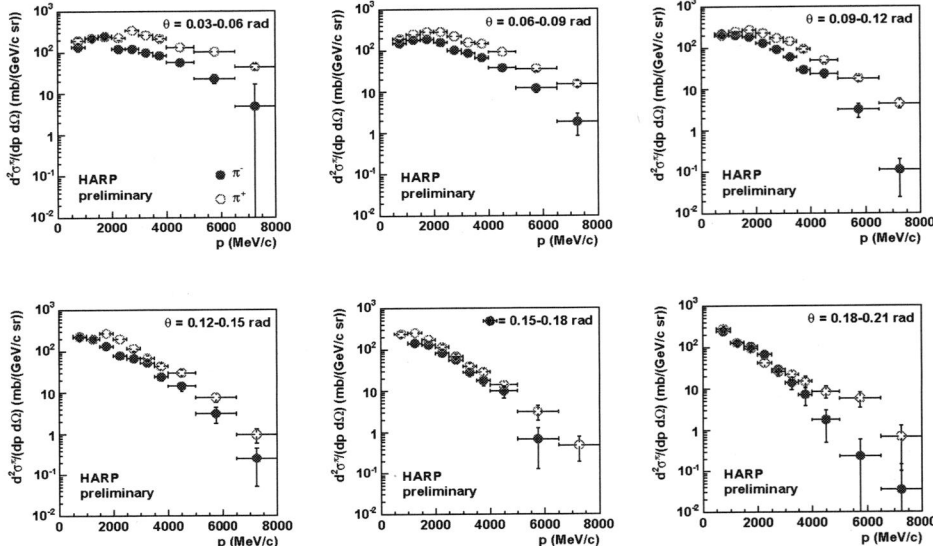

FIGURE 9. Preliminary momentum spectra of secondary π^- (filled circles) and π^+ (open circles) mesons in p+C reactions at 12 GeV/c measured with the HARP spectrometer at the PS accelerator at CERN. The six different panels show the spectra in different angular bins from 0.03 rad to 0.21 rad.

CONCLUSIONS AND OUTLOOK

New p+C data from NA49 and HARP are available now. They cover a broad range of phase space which is important for the EAS. However, only 20% of the *grandmother particles* in EAS are protons and most are pions. Currently an analysis of HARP π+C data is in progress to help tuning hadronic interaction models for primary pions. Future data of the recently proposed upgraded MIPP experiment at Fermilab [28, 29] of p+C, π+C and K+C collisions at 20, 60 and 120 GeV/c would give important input for hadronic interaction models at higher energies. As a future project, an upgrade of the NA49 detector is planned [30]. This will allow the measurement of reactions on p+C and π+C at the energies of 30, 40, 50 and 158 GeV.

REFERENCES

1. T. Antoni et al. (KASCADE Collab.), Astropart. Phys. 24 (2005) 1.
2. R. Engel, T.K. Gaisser, and T. Stanev, Proc. of ISMD, Providence, Rhode Island, August 9-13, 1999, World Scientific (2000), 457.

3. H.J. Drescher and G. Farrar, Astropart. Phys. 19 (2003) 235.
4. H.J. Drescher, M. Bleicher, S. Soff, H. Stöcker, Astropart. Phys. 21 (2004) 87.
5. D. Heck et al., Proc. of 28th ICRC, Tsukuba, Japan, (2003) 279.
6. T. Antoni et al. (KASCADE Collab.), Nucl. Instr. Meth. A 513 (2003) 490.
7. D. Heck et al., Report FZKA 6019, Forschungszentrum Karlsruhe, 1998.
8. H. Fesefeldt, Report PITHA-85/02, RWTH Aachen, 1985.
9. N.N. Kalmykov, S. Ostapchenko, and A.I. Pavlov, Nucl. Phys. B (Proc. Suppl.) 52B (1997) 17.
10. A. Haungs et al. (KASCADE Collab.), Czech. J. Phys. 56 (2006) A241.
11. T. Eichten et al., Nucl. Phys. B 44 (1972) 333.
12. W.F. Baker et al., Phys. Rev. Lett. 7 (1961) 101.
13. D. Dekkers et al., Phys. Rev. B 137 (1965) 962.
14. J.V. Allaby et al., CERN Yellow Report 70-12 (1970).
15. Y. Cho et al., Phys. Rev. D 4 (1971) 1967.
16. D. Antreasyan et al., Phys. Rev. D 19 (1979) 764.
17. G. Ambrosini et al. (NA56 Collab.), Phys. Lett. B 425 (1998) 208.
18. G. Collazuol, A. Ferrari, A Guglielmi and P. R. Sala, Nucl. Instr. Meth. A 449 (2000) 609.
19. D.S. Barton et al., Phys. Rev. D 27 (1983) 2580.
20. M. G. Catanesi et al. (HARP Collab.), CERN-SPSC-99-35, 1999.
21. S. V. Afanasiev et al. (NA49 Collab.), Nucl. Instr. Meth. A 430 (1999) 210.
22. C. Alt et al. (NA49 Collab.), hep-ex/0606028.
23. S. Ostapchenko, Nucl. Phys. Proc. Suppl. 151 (2006) 143, hep-ph/0412332
24. S. Ostapchenko, Phys. Rev. D 74 (2006) 014026, hep-ph/0505259.
25. R.S. Fletcher et al., Phys. Rev. D 50 (1994) 5710.
26. R. Engel et al., in Proc. 26-th Int. Cos. Ray Conf., Salt Lake City, 1999.
27. M.G. Catanesi et al. (HARP Collab.), Nucl. Phys. B 732 (2006) 1, hep-ex/0510039.
28. R. Raja et al. (MIPP Collab.), J. Phys. Conf. Ser. 9 (2005) 303.
29. D. Isenhower et al. (MIPP Collab.), Fermilab Proposal P-960, hep-ex/0609057.
30. N. Antoniou et al., CERN-SPSC-2006-001 (SPSC-P-329).

Results From MuScat

Malcolm Ellis on behalf of the MuScat Collaboration

Fermi National Accelerator Laboratory, Bataiva, IL, USA

Abstract. This paper presents the measurement of the scattering of 172 MeV/c muons in assorted materials and compares these measurements with the predictions of a number of simulations.

Keywords: Multiple scattering; Ionisation cooling.
PACS: 11.80.La

INTRODUCTION

The storage of intense muon beams is required for the construction of a neutrino factory or a muon collider. The ionisation cooling technique [1] is a leading contender to maximise the intensities for these facilities. This technique relies upon the cooling effect of ionisation energy losses in low-Z materials coupled with RF acceleration to reduce the transverse emittance of a beam. Multiple scattering heats the beam and the overall cooling is thereby reduced. The main purpose of this paper is to present a measurement of the multiple scattering of muons of around 200 MeV/c and compare these distributions with the predictions of some recent versions of the GEANT4 [2] simulation toolkit.

Experimental apparatus

The technique of the MuScat experiment was to prepare a narrow muon beam by collimation and to use it to illuminate a variety of targets. Te position of the outgoing particles was measured after traveling a sufficient distance that the position and angle were highly correlated. Details of the collimation system, tracking detectors, particle identification and analysis technique are described in more detail in [3].

M20 Beam-line

The experiment used the M20 beam-line at TRIUMF. The extracted proton beam from the cyclotron interacted in a target and produced pions which were then allowed to decay in a muon decay channel. The beam-line was tuned to optimise the capture of high momentum muons from forward decays of the pions. The mean beam momentum was determined to be 172 ± 2 MeV/c and the accepted momentum bite was approximately 1 MeV/c. The pion background was determined to be less than 1% and the trigger rate was 80 particles / s.

Collimation System

The collimator system consisted of two lead blocks with slits in them, augmented by other active and passive collimators. The system was behind a shield wall consisting of

CP896, *Hadronic Shower Simulation Workshop*
edited by M. Albrow and R. Raja
© 2007 American Institute of Physics 978-0-7354-0401-4/07/$23.00

8cm of steel with a 4 x 3 cm opening aligned with the lead collimators. During analysis of the data it was determined that one of the lead collimation blocks was not perfectly aligned. The description in the simulation of the experiment has been tuned to best reproduce the observed data when no scattering target was used. Figure 1 shows a comparison of the measured distribution in the first detector plane with that simulated with the assumed misalignment of the collimator

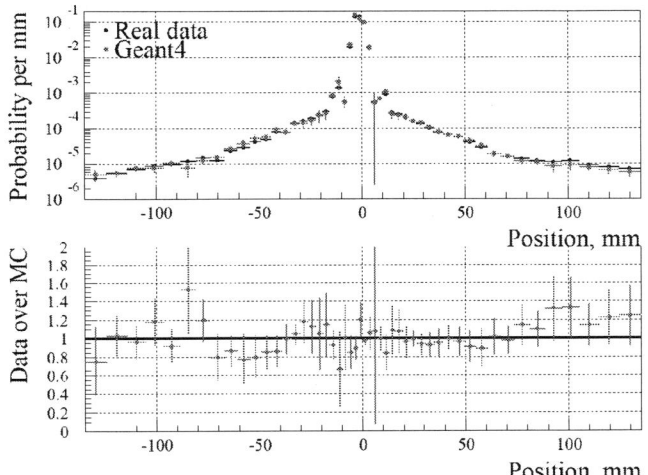

Figure 1. The distribution of particles at the detector plane 1 in data and simulation for no target.

Solid and Liquid Targets

There were two separate target systems: a wheel with a selection of solid targets and a vessel for liquid hydrogen. The target wheel was controlled by a stepper motor and presented a choice of 12 target positions to the beam. The LH_2 vessel was designed and built by the cryogenic targets group a the TRIUMF laboratory. The vessel was 150 mm by 100 mm by 100mm with 8cm diametre holes bored through the length and width allowing two lengths of LH_2 to be measured in addition to the empty target.

Scintillating Fibre Tracker

The scintillating fibre (Sci-Fi) tracker comprised three double fibre planes, readout by 24 HPK multi-anode PMTs. The planes were approximately 1.1 m from the target. The first and second planes were mounted together and displaced by 50 mm. The third plane was mounted to the rear of the vacuum vessel and was displaced 66.9 mm from the second plane. Each detector had an active area which was a 300 x 300 mm square.

The double planes were held in a Delrin frame that had 2048 holes drilled by a CNC machine to define the locations of the ends of the scintillating fibres. Each of the four faces of a detector were matched to a connector which took the 512 double-clad 1 mm fibres and took the light into two bundles of 256 fibres. These bundles were arranged in a

16x16 grid and matched to the sixteen 16 x 1 mm anodes on the multi anode PMTs. The multiplexing in the readout scheme was arranged such that the PMT reading out one side of a detector would determine which group of 16 fibres a hit was in and that on the opposite side would identify the fibre within the group of 16.

Due to difficulties in preparing exactly square bundles coupled with optical and electronic crosstalk, it was necessary to create a complete simulation of the readout of the tracker in order to remove the effects of crosstalk from the measured scattering distributions.

Particle Identification

The main method of selecting muons was through a measurement of time of flight. The arrival time of the beam was measured by a small scintillator upstream of the collimation system. This time was compared with a signal from the TRIUMF cyclotron. A fraction of the events were also able to be analysed by a NaI calorimeter (TINA) which was used to measured both the kinetic energy of the particle and in the case of the muon, the decay energy and decay time. Due to the size of the calorimeter not all particles were able to be analysed, so for the analysis particle selection was performed with time of flight.

Analysis Method

Event selection

The event selection criteria required that the high voltage must have been in a nominal state and the data-acquisition in a normal state when the event was recorded. There were some periods in which the time of flight system suffered unknown interference from an external source and these were removed from the sample.

The number of reconstructed space points on the first plane of the tracker was required to be one or two in order to reject events in which multiple hits, coupled with the effect of the multiplexing and cross-talk would make an unambiguous measurement of the muon difficult. The measured positions in the other two detector planes was required to be consistent with that in the first and the measured time in the upstream scintillator S1B was required to be within 0.2 ns of the nominal value.

Deconvolution technique

The deconvolution was performed using a GEANT4 simulation and measurements of the detector performance and beam to extract the scattering distribution Θ:

$$D = B + D_\pi + R \cdot \varepsilon \cdot \Theta \qquad (1)$$

In this equation D is the observed scattering distribution, B is the background of particles that pass the selection criteria but which did not pass through the target, D_π is the small contamination from pions, R is the response of the detector to a particle deflected through angle Θ and ε is the efficiency of the detector for particles deflected through that angle.

The background was found through simulation and the pion contamination is taken from the data. The response and efficiency matrices were determined from simulations.

The equation (1) is solved for Θ, imposing a requirement of symmetry about $\Theta = 0$ using MINUIT which also finds the errors and correlations. It has been checked that the technique is mathematically correct, in that if given simulated data as input the scattering distribution used in the simulation is recovered as the deconvolved distribution.

Results

In [2] the full set of deconvolved scattering distributions are presented as well as a comparison with GEANT4 version 7.0.p01. Several changes to the multiple scattering simulation in GEANT4 version 8 and later have resulted in the agreement between simulation and the MuScat results getting better, however there are still differences, particularly for the low Z materials that are of most interest to ionisation cooling. The following figures will compare the unfolded distributions from MuScat with several versions of GEANT4.

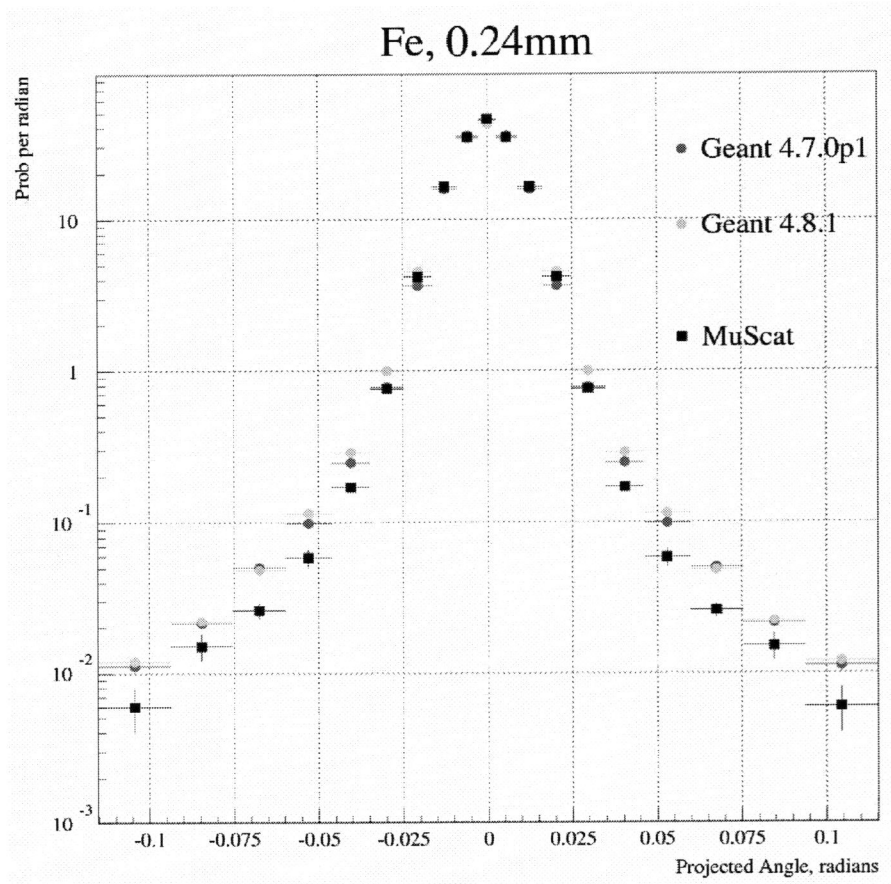

Figure 2. The scattering distribution through the thin Iron target measured by MuScat and simulated by two versions of GEANT4

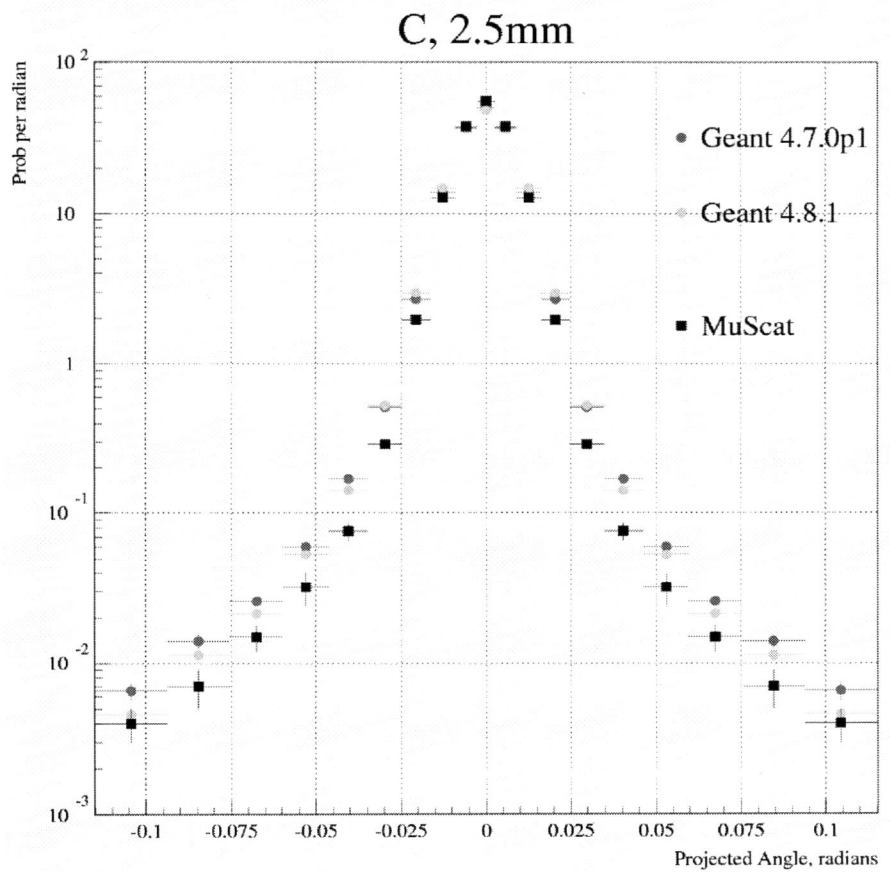

Figure 3. The scattering distribution through the Carbon target measured by MuScat and simulated by two versions of GEANT4

173

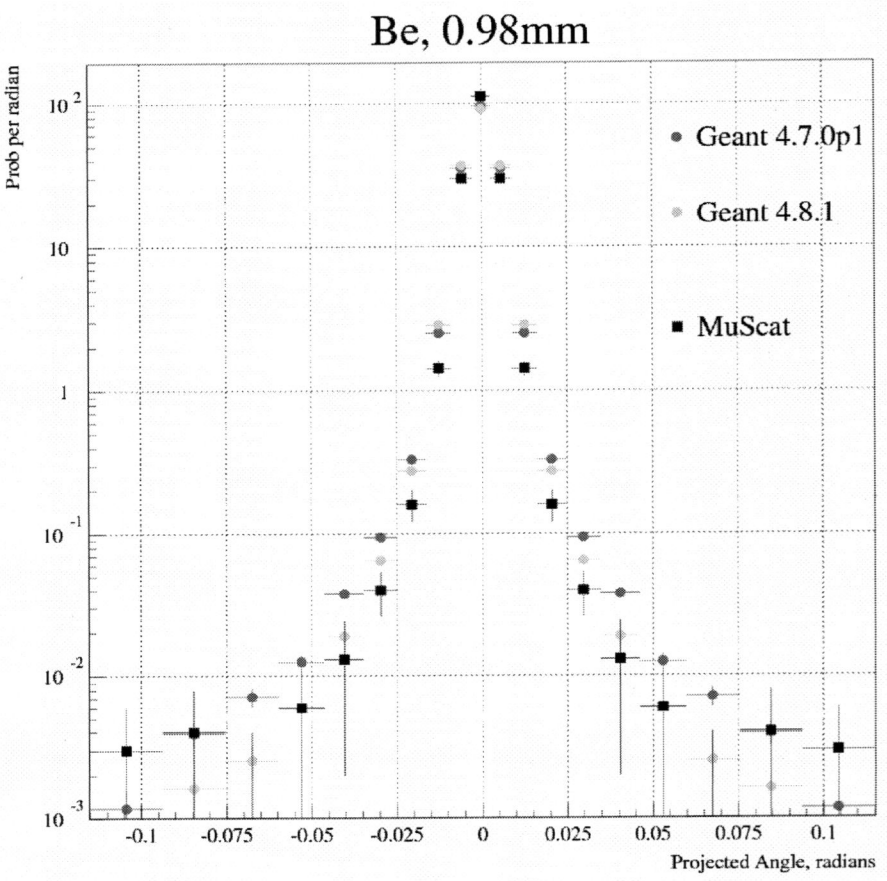

Figure 4. The scattering distribution through the thin Beryllium target measured by MuScat and simulated by two versions of GEANT4

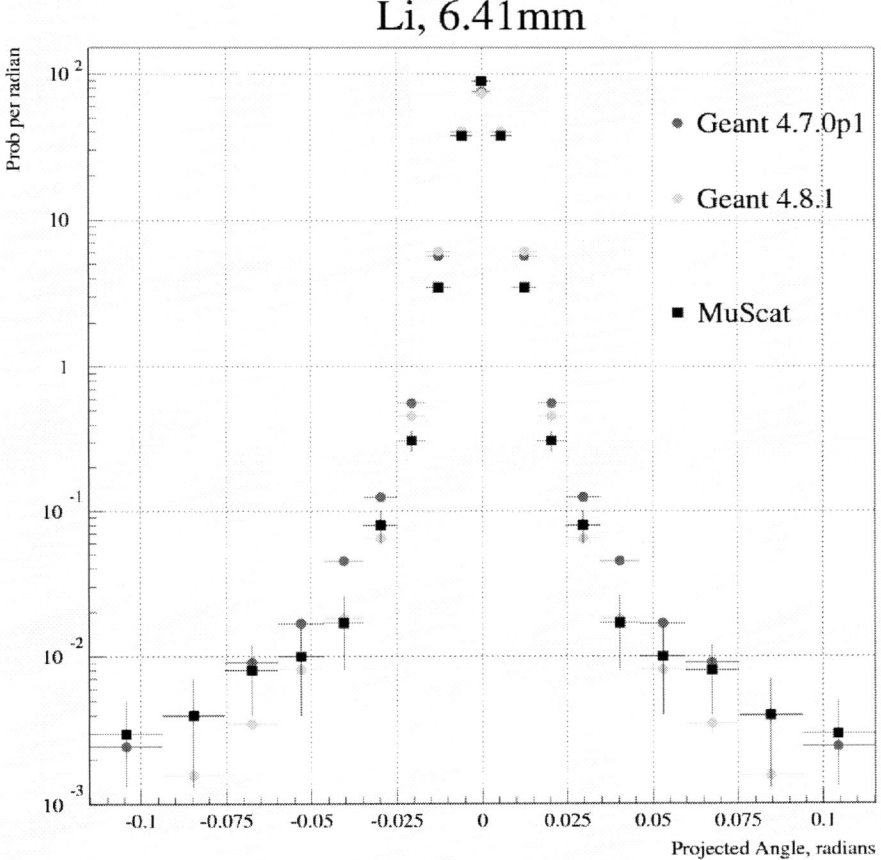

Figure 5. The scattering distribution through the thin Lithium target measured by MuScat and simulated by two versions of GEANT4

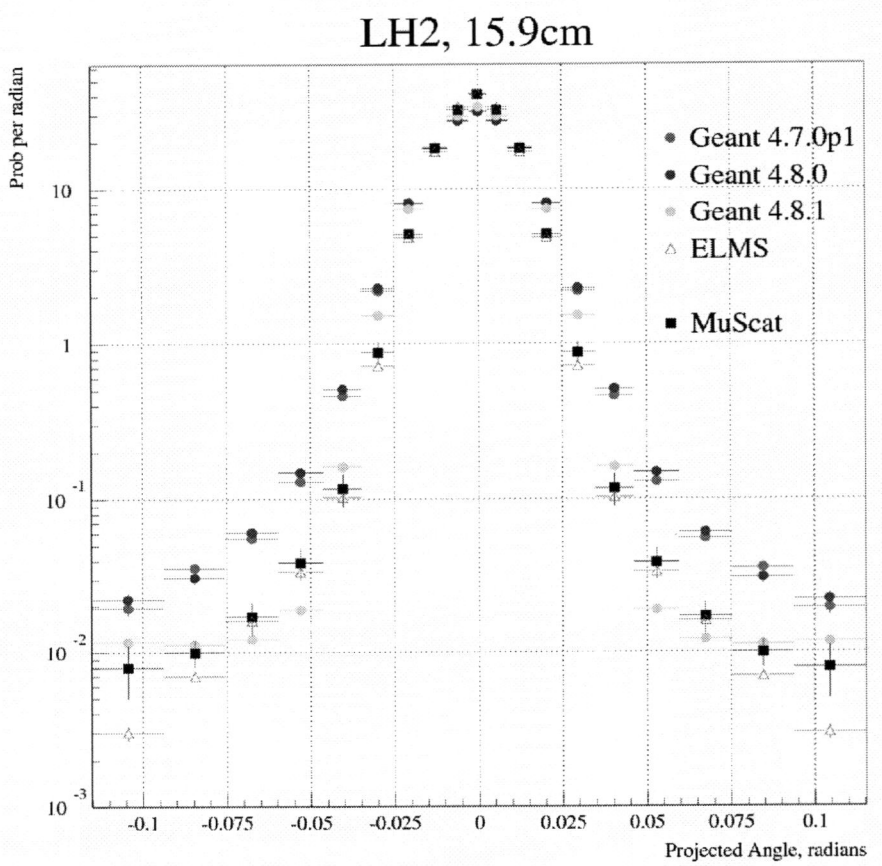

Figure 6. The scattering distribution through the long Liquid Hydrogen target measured by MuScat and simulated by three versions of GEANT4 and the ELMS [4] program.

ACKNOWLEDGMENTS

We would like to thank all the staff of TRIUMF laboratory for the helpful and friendly way in which we were welcomed to the laboratory and for all the facilities that were put at our disposal. The following all lent support, for which we are most grateful: M.-P. Boudet, R. Barlow, S. Burge, J. Flynn, R. Hampton, D. Laihem, S. Malton, C. Marshall, E. O'Neill, P. Rock, W. Sievers, D. Wade, P. Zubko. We also acknowledge the financial support from PPARC and CCLRC.

REFERENCES

1. D.V. Neuffer, Muon cooling and applications, in: Proc. Workshop on Beam Cooling, Montreux 1993, CERN Rep. 94-03, 1994. p. 49.
2. J. Allison et. al., GEANT4 developments and applications, in IEEE Transactions on Nuclear Science 53 No 1 (2006) 270-278.
3. D. Attwood et. al., The scattering of muons in low-Z materials, Nucl. Instrum. Meth. B251 (2006) 41-55.
4. W.W.M. Allison, Calculations of energy loss and multiple scattering (ELMS) in molecular hydrogen, J. Phys G; Nucl. Part. Phys. 29 (2003) 1701. Available from: <http://www.adams-institute.ac.uk/ELMS/>

Neutrino Event Generators

Steven Dytman

Dept. of Physics and Astronomy
University of Pittsburgh
Pittsburgh, PA 15260

Abstract. A brief description of recent progress in NEUGEN improvements is given. Comparison of various neutrino event generators based on a previous study is also made. Existing event generator models differ most strongly in the treatment of nuclear fsi, the focus of the paper.

Keywords: <neutrino event generator fsi>
PACS: 13.15.+g,14.60.Pq

INTRODUCTION

This conference was dominated by presentations on the major codes particle physicists use for simulations of experiments. Of course, neutrino experiments use those same codes. Every experiment needs an event generator which contains the correct physics for the interaction to be studied. For neutrino experiments, the event generator is particularly important. The beam energy is a key part of the neutrino oscillation formula and it must be determined by the detector because each beam has a broad energy spectrum. Since the cross sections and the beam fluxes are low, detectors are massive. In many cases, they are coarse-grained calorimeters. MINOS will be used as the example in this talk because I am a member of that collaboration. It has a near (far) detector with 15m (30m) of alternating steel (1 in. thick) and scintillator (1 cm) sheets. Understanding the energy deposition in these calorimeters (both target and detector) is one of the important issues in determining the beam energy of each event,

$$E_\nu = E_\mu + \sum_i E_{hadron,i}.$$ (1)

All the energy in the final state is summed to get the beam energy. Charged current events are guaranteed by the detection of a muon which is typically well-characterized. The hadrons are more difficult to accurately identify and measure. The typical experiment today is most interested in neutrinos of about 1 GeV energy because that is where the oscillation effects are largest. At energies of interest, the primary reactions are quasielastic and pion production.

There are 4 codes in common use- NEUGEN [1], NEUT [2], NUANCE [3], and NUX [4]. Each is the work of a small number of people and has been developed over many years. The general structure of an event generator must have 3 elements- the initial interaction, the formation zone (where the hadrons have no fsi), and the final state interactions (fsi). They are designed for a particular experiment. The Nuint (neutrino-nucleus interactions) series of conferences has been a valuable way to pull

CP896, *Hadronic Shower Simulation Workshop*
edited by M. Albrow and R. Raja
© 2007 American Institute of Physics 978-0-7354-0401-4/07/$23.00

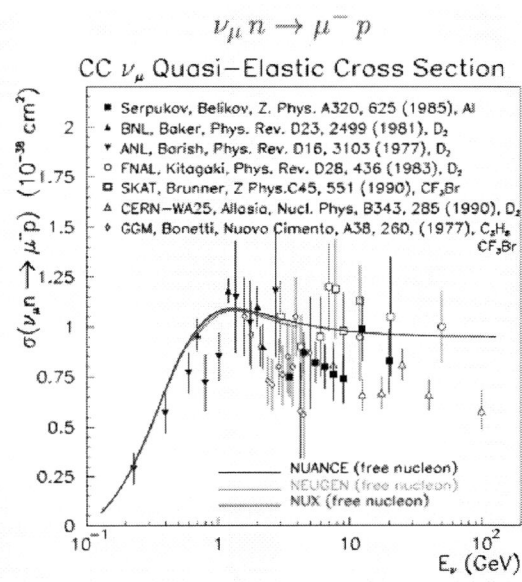

$$\nu_\mu n \longrightarrow \mu^- p$$

CC ν_μ Quasi—Elastic Cross Section

- Serpukov, Belikov, Z. Phys. A320, 625 (1985), Al
- BNL, Baker, Phys. Rev. D23, 2499 (1981), D$_2$
- ANL, Barish, Phys. Rev. D16, 3103 (1977), D$_2$
- FNAL, Kitagaki, Phys. Rev. D28, 436 (1983), D$_2$
- SKAT, Brunner, Z Phys.C45, 551 (1990), CF$_3$Br
- CERN—WA25, Allasia, Nucl. Phys. B343, 285 (1990), D$_2$
- GGM, Bonetti, Nuovo Cimento, A38, 260, (1977), C$_3$H$_8$ CF$_3$Br

NUANCE (free nucleon)
NEUGEN (free nucleon)
NUX (free nucleon)

FIGURE 1. Total cross section for $\nu_\mu n \to \mu^- p$. The values used for 3 models are also shown [5].

results together in ways not otherwise possible. In 2002, S. Zeller [5] made a comparison with known cross sections for the elementary processes. The total cross section for $\nu_\mu n \to \mu^- p$ is shown in Fig. 1 along with values used by NEUGEN, NUANCE, and NUX. They agree well because the same elementary model [6] is used in each case. Similar agreement is found for other elementary cross sections. In 2004, H. Gallagher [7] coordinated a more general comparison. The standard case was $\overline{\nu}_\mu$ ^{16}O. The choice of anti-neutrino was required to include all models. Since there is no data for comparison, this choice is not critical. Various inclusive distributions were then compared. Since all models us the same KNO prescription [8] to choose particles in the initial interaction, distributions that characterize the entire event (such as the total hadronic mass, W, and x_B) are also very similar. We show the proton multiplicity and momentum distribution for 5 GeV neutrinos [7] in Fig. 2 and the same quantities for π^+ in Fig. 3. At this energy, the full range of possible final states is sampled. With the wide variation between models seen in these figures, we make the interesting conclusion that the fsi provides the widest differences among them.

BUILDING A MODEL

The propagation of a hadron through a nucleus is a complicated problem which can only be described with approximations. All models use the intranuclear cascade model (INC).

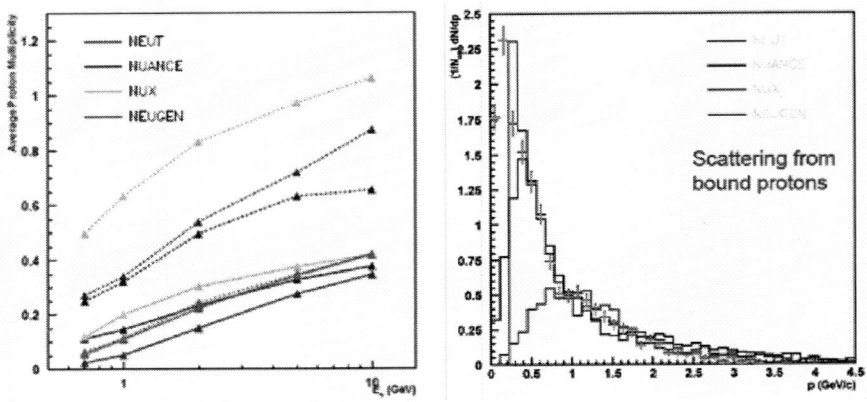

FIGURE 2. (left) Proton multiplicity distribution for $\bar{\nu}_\mu$ ^{16}O (dashed) and free proton (solid) for different models at 5 GeV [7]; (right) Proton momentum distribution for $\bar{\nu}_\mu$ ^{16}O at 5 GeV.

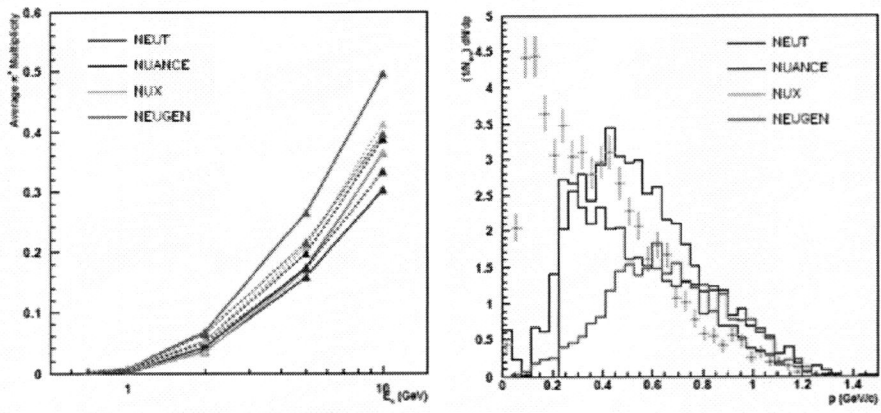

FIGURE 3. (left) Positive pion multiplicity distribution for $\bar{\nu}_\mu$ ^{16}O (dashed) and free proton (solid) for different models at 5 GeV [7]; (right) Positive pion momentum distribution for $\bar{\nu}_\mu$ ^{16}O at 5 GeV.

Each nucleon has a Fermi momentum and each hadron-nucleon collision is described by the free cross section. There are fortunately many data sets and accompanying partial wave amplitude fits [9] for many elementary interactions such as $\pi^- p \rightarrow \pi^0 n$ and $pp \rightarrow pp$. The literature has a good history of using the INC to describe inelastic hadron-nucleus interactions [10]. Since the primary variation between models seems to be in the fsi, we focus on that aspect here. Since the author is in the middle of a rewrite of the fsi section of the NEUGEN model, that will be described.

In developing an fsi model, the lack of neutrino-nucleus data is a problem. The hadron-nucleus data becomes the primary resource for validation. Most of the existing pion-nucleus data came from LAMPF [11] which had a limited energy region of study. The most significant data is the pion-nucleus interaction for kinetic energy less than ~350 MeV. The proton-nucleus inclusive interaction has been well-studied at ITEP and GSI. An alternate source is the INC models which have been constructed, e.g. the CEM03 model [12]. Stepan Mashnik (see his separate talk in this workshop) is a primary author of this code. For proton-nucleus interactions, CEM03 appears to be an excellent guide; we use it to fill in the gaps where there is no data. The primary new feature this model provides is the large multi-nucleon knockout cross sections at low energy. This causes a significant change in the nucleon distributions.

The best information for pions is the total cross section data of Rutherford Lab from the 1970's [13]. We use the Carroll data at low energies and the Clough data (extrapolated to higher A) at high energies. However, this must be subdivided into the component processes (elastic and inelastic scattering, charge exchange, absorption, and pion production). Various experiments at low energy [11] provide good separation of the processes. At higher energies, the CEM03 calculations are more important. Since the calculations are not in good agreement with the data at low energies, scaling factors are required for overall consistency. This is not a simple process; to some extent, all hadronic models must face this issue. It is surprising how seldom it is discussed.

NEW MODEL RESULTS

The process briefly discussed above is used to develop a new fsi model for NEUGEN. The longer term goal is to write a good INC code which matches the hadron-nucleus data at low energy and is then extrapolated to higher energy where there is good hadron-nucleon data. PWA amplitudes based on πp, πn, pp, and np extend to beyond 1.5 GeV kinetic energy. The first steps toward that model are complete. The existing code reduces the fsi to a single interaction which is based on the appropriate total cross section. Many processes can then be taken into account. As the goal is to understand the response of the MINOS detector, processes with more than 5 nucleons in the final state (a significant portion of the cross section) are represented by 5 nucleons. The final state particles are then distributed in angle and energy by phase space. Existing data shows this is a reasonable but not highly accurate choice.

We now show the same quantities as were studied in Gallagher's Nuint04 talk [7]- π^+ multiplicity and momentum distribution in Fig. 4 and proton multiplicity and momentum in Fig. 5. In each case, we show the evolution of NEUGEN. In 2005, pion absorption effects were added. These effects were improved and numerous other processes added in 2006, the developments described above. The pion distributions are not strongly affected. However, pion absorption and multinucleon knockout are important sources of low energy protons. Each code improvement brings more of them and shifts the average to lower energy. For MINOS, this should have a strong effect on simulations. The increase in proton multiplicity for the recent code is not in agreement with the results shown in Fig. 2. Although there is unfortunately no data to test this prediction, the results are firmly attached to existing hadron-nucleus data.

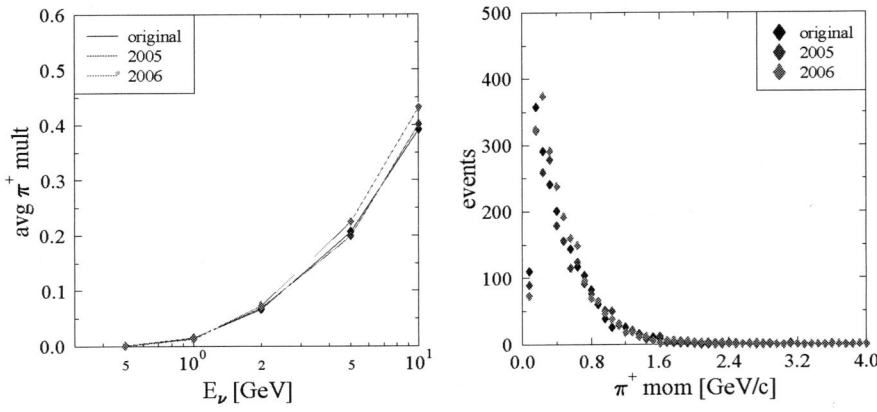

FIGURE 4.   (left) Positive pion multiplicity distribution for $\overline{\nu}_\mu$ ^{16}O for different NEUGEN models at 5 GeV; (right) Positive pion momentum distribution for $\overline{\nu}_\mu$ ^{16}O at 5 GeV. See text for details.

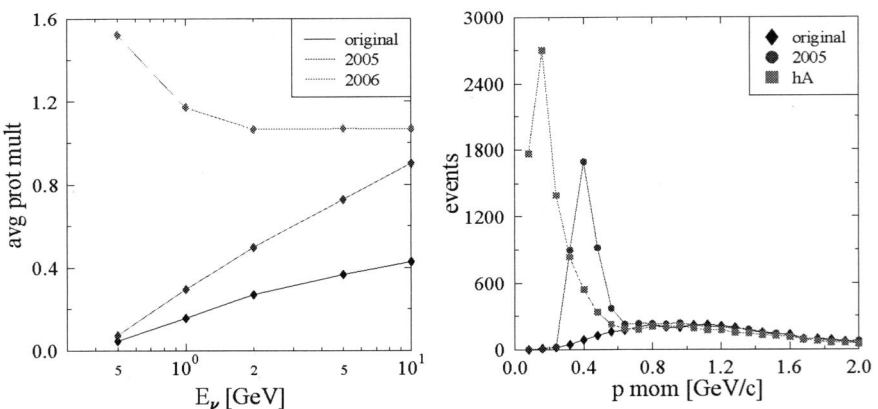

FIGURE 5.   (left) Proton multiplicity distribution for $\overline{\nu}_\mu$ ^{16}O for different NEUGEN models at 5 GeV; (right) Proton momentum distribution for $\overline{\nu}_\mu$ ^{16}O at 5 GeV. See text for details.

FUTURE

As neutrino experiments become more sophisticated, the need for better Monte Carlo simulations increases. In the present round of experiments using coarse-grained calorimeters, simulations are required to get detector response correct. Future experiments will measure the distributions of particles in the final state models can then be tuned much better. However, expectations for the accuracy of fundamental measurements in new experiments will also create more pressure to improve simulations.

Neutrino event generators contain many kinds of physics- neutrino-nucleon cross sections, hadronization, and hadronic fsi. Better understanding of neutrino-nucleon cross sections will require new data. Significant progress has come in the latter 2 topics because of recent multiplicity data and improved nuclear modeling. It is interesting to note that the most significant difference among the leading neutrino event generators is in the fsi. However, this conclusion is a little disingenuous because codes tend to use the same models.

The main topic of this note is a discussion of one attempt to describe nuclear modeling. A good description of pion-nucleus and proton-nucleus reactions are required at energies up to about 2 GeV. That was difficult for the nuclear physics community and won't be simple for this application either. With the lack of data above $\sim$800 MeV, bootstrapping will be required. The INC models will be good for this; the models can be calibrated at lower energy where the nuclear effects are stronger and then extrapolated to higher energies where there is still good hadron-nucleon data. The elastic channel is special as an INC model is unlikely to get this correct because it is largely a wave effect. Here, nuclear models will be very useful.

This progress report shows how much a particular model can change with new inputs. This work will continue as the full INC model is developed. However, the main effect has probably already been realized.

Future experimental results from the MINERνA experiment will be critical to really pin down event generator models. Existing neutrino-nucleus cross section data largely comes from bubble chamber experiments and is therefore limited in accuracy. Thus, continued improvement of these models is guaranteed.

ACKNOWLEDGMENTS

A large number of interesting conversations with Hugh Gallagher were an important part of this work.

REFERENCES

1. H. Gallagher, *Nucl. Phys. Proc. Suppl.* **112**, 188 (2002).
2. Y. Hayato, *Nucl. Phys. Proc. Suppl.* **112**, 271 (2002).
3. D. Caspar, *Nucl. Phys. Proc. Suppl.* **112**, 161 (2002).
4. F. Battistoni, P.R. Sala, and A. Ferrari, *Acta Phys. Polon. B***37**, 2361 (2006).
5. G.P. Zeller, arXiv:hep-ex/0312061.
6. C.H. Llewellyn-Smith, *Phys. Repts.* **3**, 261 (1972).

7. H. Gallagher, in *Proceedings of the Third International Workshop on Neutrino-Nucleus Interactions in the Few-GeV Region*, Nucl. Phys. Proc. Suppl. **139**, 278 (2005)..

8. Z. Koba, H.B. Nielsen, and P.L Olesen, *Nucl. Phys. B* **40**, 317 (1972); N. Schmitz, in *Proc. Intl. Symp. on Lepton and Photon Int. at High Energy*, Bonn, W. Germany, pg. 527 (1981).

9. R.A. Arndt, W.J. Briscoe, I.I. Strakovsky, R.L. Workman, and M.M. Pavan, *Phys. Rev. C* **65**, 025213 (1994).

10. Z. Fraenkel, E. Piastsky, and G. Kalbermann, *Phys. Rev. C***26**, 195 (1982); G.D. Harp, *Phys. Rev. C***10**, 2387 (1974).

11. D. Ashery, *et al.*, *Phys. Rev. C* **23**, 2173 (1981); T.J. Bowles, *et al.*, *Phys. Rev. C* **23**, 439 (1981), I. Navon, *et al.*, *Phys. Rev. C* **28**, 2548 (1983), R.D. Ransome, *et al.*, *Phys. Rev C* **45**, R509 (1992).

12. Stepan G. Mashnik, Konstantin K. Gudima, Arnold J. Sierk, Mircea I. Baznat, and Nikolai V. Mokhov, LANL Report LA-UR-05-7321 (2005), RSICC Code Package PSR-532; S. G. Mashnik, A. J. Sierk, K. K. Gudima, M. I. Baznat, Journ. Phys.: Conference Series, 41 (2006) 340-351 (E-print: nucl-th/0510070).

13. A.S. Carroll, *et al.*, *Phys. Rev. C* **14**, 635 (1976), A.S. Clough, *et al.*, *Nucl. Phys. B* **76**, 15 (1974).

Hadronic Interaction Modelling in MINOS

Michael Kordosky

For the MINOS collaboration

Department of Physics and Astronomy
University College London
Gower Street
London WC1E6BT
United Kingdom

Abstract.
The Main Injector Neutrino Oscillation search (MINOS) uses two detectors separated by 735 km to measure a beam of neutrinos created by the Neutrinos at the Main Injector (NuMI) facility at Fermi National Accelerator Laboratory. The experiment has recently reported an observation [1] of ν_μ disappearance consistent with neutrino oscillations. We describe the manner in which the experiment's results depend on the correct understanding and modeling of hadronic systems.

Keywords: hadron calorimeter, final state interactions, neutrino beam
PACS: 13.75.-n,13.85.-t,24.10.Lx,29.40.Vj,29.25.-t

INTRODUCTION

The Main Injector Neutrino Oscillation Search (MINOS) is a long baseline, two-detector neutrino oscillation experiment that will use a muon neutrino beam produced by the Neutrinos at the Main Injector (NuMI) facility at Fermi National Accelerator Laboratory (FNAL) [2, 3]. The measurement is conducted by two functionally identical detectors, located at two sites, the Near Detector (ND) at FNAL and the Far Detector (FD) in the Soudan Underground Laboratory in Minnesota. The experiment began collecting beam data in March 2005 and has recently reported ν_μ disappearance consistent with quasi-two-neutrino oscillations according to $|\Delta m_{32}^2| = 2.74^{+0.44}_{-0.26} \times 10^{-3} \, \text{eV}^2/\text{c}^4$ and $\sin^2 2\theta > 0.87$ at 68% CL [1].

The MINOS detectors are tracking-sampling calorimeters, optimised to measure neutrino interactions in the energy range $1 \lesssim E_\nu \lesssim 50 \, \text{GeV}$. The active medium comprises 4.1 cm-wide, 1.0 cm-thick plastic scintillator strips arranged side by side into planes. Each scintillator plane is encased within aluminum sheets to form a light-tight module and then mounted on a steel absorber plate. The detectors are composed of a series of these steel-scintillator planes hung vertically at a 5.94 cm pitch with successive planes rotated by $90°$ to measure the three dimensional event topology. Wavelength-shifting and clear optical fibers transport scintillation light from each strip to Hamamatsu multi-anode photomultiplier tubes that reside in light-tight boxes alongside the detector. Both detectors are magnetized so as to measure muon charge-sign and momentum via curvature.

The ν_μ disappearance measurement is done by using the event topology to identify interactions as $\nu_\mu-$CC, rather than NC/ν_e-CC, then reconstructing the neutrino energy

CP896, *Hadronic Shower Simulation Workshop*
edited by M. Albrow and R. Raja

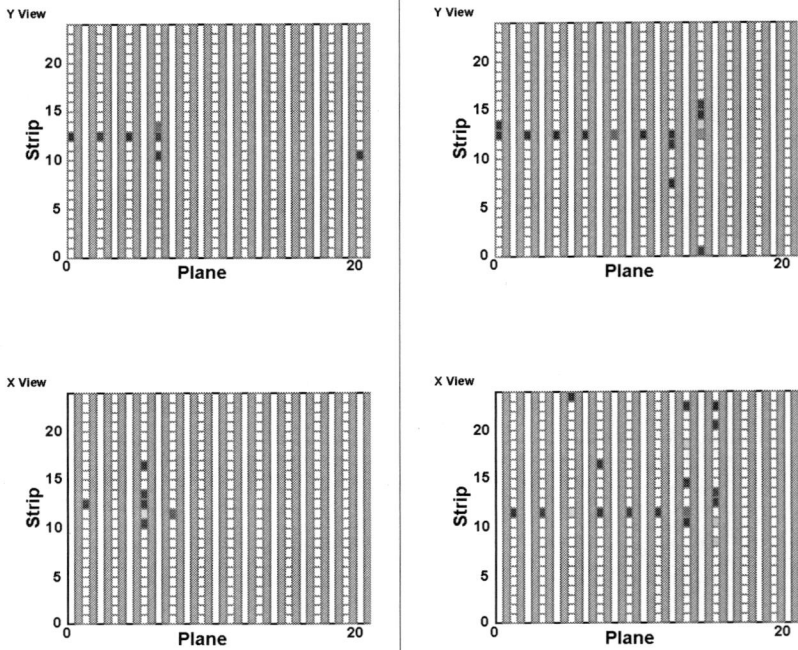

FIGURE 1. π^+ events measured in the MINOS calibration detector. The pions are incident from the left and two (orthogonal) views of the event are shown. Cells demarcate the detector's scintillator strips, with shading/color indicating the pulse-height of hits. The event on the left was collected at 1 GeV/c beam momentum while the one on the right is from a 2 GeV/c run. To remove hits due to optical cross-talk in the multi-anode PMTs, only strips registering a pulse-height larger than 1.5 PE (photoelectrons) are shown.

as the sum of the muon energy and the energy transferred to the nucleus : $E_\nu = E_\mu + \nu$ where $nu = y \times E_\nu$. The oscillation hypothesis is tested by comparing the measured neutrino energy spectrum to the spectrum expected in the absence of oscillations. The latter spectrum is anchored to observations made with the Near Detector. MINOS endeavors to measure Δm^2 and $\sin^2 2\theta$ with an accuracy of better than 10%. The results depend upon a reliable knowledge of the event selection efficiency and the energy scale, both of which are affected by the modeling of the hadronization process and the detector's signal and topological response to those hadrons. We will discuss the way that MINOS has dealt with these issues. In addition, uncertainties in the production of hadrons in the NuMI target result in uncertainties in the neutrino flux and poor agreement with data. We close by describing the manner in which data from the Near Detector was used to constrain the simulation of the NuMI beam.

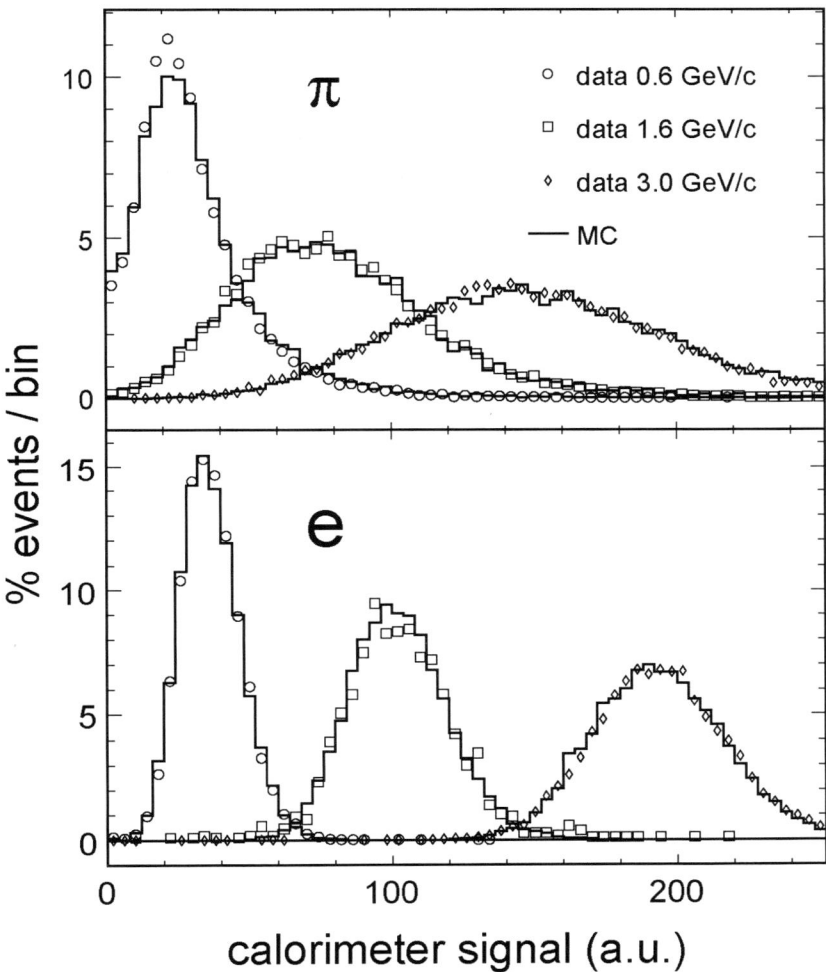

FIGURE 2. Pion and electron line shapes as measured in the MINOS calibration detector [4, 5]. The Monte Carlo calculation is that of the experiment's GEANT3/GCALOR simulation. The same program is used to model neutrino induced interactions in the MINOS Near and Far Detectors.

HADRONIC ENERGY SCALE

MINOS data suggests that (assuming quasi-two-neutrino oscillations) the largest suppression of ν_μ occurs for $1 < E_\nu < 2\,\mathrm{GeV}$ and that the suppression slowly shrinks as the energy increases to about 10 GeV. This is a relatively low energy range, but not low enough to make the experiment dominated by quasi-elastic $\nu_\mu - \mathrm{CC}$ for which E_ν may

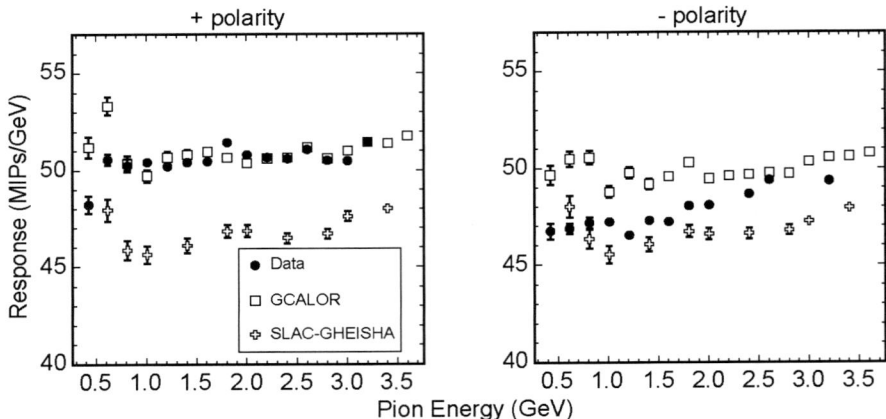

FIGURE 3. MINOS detector response to $\pi^\pm$ induced showers [4]. Data measured in the MINOS calibration detector are compared with the result of GEANT3/GCALOR and GEANT3GHEISHA simulations. The π^+/π^- response asymmetry is absent above 4 GeV.

be reconstructed from only the muon energy and scattering angle. Instead, events with $1 < E_\nu < 10$ GeV have, after experimental cuts, an average inelasticity of ≈ 0.3. We are therefore interested in hadronic showers in the few GeV range and in which the energy is carried by a relatively few particles. The response of calorimeters to these low energy showers is notoriously difficult to model. Experiments generally remedy this by measuring the response to single-particles, either to perform a direct calibration or as a way of constraining/tuning the shower model. Since the MINOS Near and Far Detectors are large and have been constructed underground, direct exposure to a calibration beam was not possible. Instead a dedicated Calibration Detector (CalDet) was built to establish the energy scale and develop the calibration technique.

The CalDet, a scaled-down but functionally equivalent model of the MINOS Far and Near detectors, was exposed to test beams in the CERN PS East Area during 2001–3 to establish the response of the MINOS calorimeters to hadrons, electrons and muons in the momentum range 0.2–10 GeV/c. The detector consisted of sixty 2.5 cm × 1 m × 1 m steel plates each of which supported a plane of twenty-four 100 cm-long scintillator strips. Alternating planes were rotated ±90° as in the larger MINOS detectors. The sampling fraction is 6.4% and the detector has a longitudinal (transverse) granularity of $0.15\,\lambda_I$ $(0.25\,\lambda_I)$. Identification of $\pi + \mu, e$ and p was accomplished using a time of flight system and several threshold Čerenkov counters. Pions and muons were discriminated using the event topology. Fig. 1 shows two pion induced showers measured in the detector. Data were collected in 200 MeV/c steps (1 GeV/c for $p_{\text{beam}} > 4$ GeV/c) in both positive and negative beam polarity and in two beamlines (T11 and T7).

The detector was calibrated using a procedure similar to that employed in the Near and Far detectors [6]. Gain and light-output non-uniformities were corrected using an LED based light injection system [7] and through-going cosmic-ray muons [8]. The absolute scale, in "muon equivalent units", was defined as the average signal per-plane induced by muons with momenta between 0.5–1.1 GeV/c at the time of plane crossing [9]. The momentum was determined by requiring that the muons stop in the detector ($p_\mu \lesssim 2.2\,\text{GeV/c}$), then working backward along the track, inverting the range-energy relation[10].

Figure 2 shows π^+ and electron line-shapes measured in the calibration detector and compared to the GEANT3/GCALOR [11, 12, 13] Monte Carlo. The simulated response to electrons was in good agreement with the data after upstream energy loss (e.g., via bremsstrahlung and ionization) was included in the simulation. The e/π response ratio is energy dependent but always > 1 and ≈ 1.27 for momenta above a few GeV. The line-shapes display an asymmetric, high-side tail characteristic of non-compensating calorimeters. The agreement between the data and the MC is surprisingly good, even at the lowest momentum settings. The energy resolution may be parameterised as $56\%/\sqrt{E} \oplus 2\%$ for $\pi^\pm$ showers and $21.4\%/\sqrt{E} \oplus 4\%/E$ for electrons.

The detector's response to $\pi^\pm$ is shown as a function of energy in Fig. 3. The data are in somewhat better agreement with the GCALOR based simulation than they are with the GHEISHA [14, 15] based one. The agreement is excellent for π^+ induced showers but poorer for π^- showers. This reflects a measured asymmetry, as much as 6% for few-GeV energies, in the detector's response to π^+ and π^-. This asymmetry was not present in the response to $e^\pm$ nor in the range of $\mu^\pm$ and vanishes for $E_\pi \gtrsim 3\,\text{GeV}$. Its origin may lie in differences, between $\pi^+ + N$ and $\pi^- + N$, in the multiplicity of secondaries produced in inelastic interactions at these low energies [16].

Hadronic showers produced in neutrino interactions are simulated with the version of GEANT/GCALOR shown here. Because these showers consist of multiple particles, and because the response to e, π and p all differ (the latter two below $\sim 1\,\text{GeV}$ where protons tend to range out before interacting hadronically) we cannot derive the relation between detector signal and energy carried by the shower particles directly from the CalDet data. Instead we derive the relationship from the MC but assign a somewhat pessimistic 6% systematic error on the response. This error is small enough to have a negligible effect on the precision of results from the low-statistics first year MINOS dataset. The uncertainty may be reduced by introducing data based correction factors into the relation or, more ideally, by improving the hadronic shower model. Over the long term MINOS plans to transition to a GEANT4 based simulation in the hopes of improving the hadronic shower treatment.

FINAL STATE INTERACTIONS

Test beam measurements, such as the one discussed above, are used to determine a detector's hadronic and electromagnetic response to single particle induced showers. The single particle based calibration may be used for energy measurements of multi-particle showers (e.g., "jets") if the energy response is sufficiently linear and if the particle content of the shower is well known or, especially, if the calorimeter's response to different particles of the same energy is identical (e.g.,a compensating calorimeter) [17]. These concerns apply to neutrino induced showers (which bear some similarity to jets) but there is an additional complication owing to the fact that the neutrino interacts in the dense nuclear medium of the target nucleus. Hadrons produced by the neutrino must

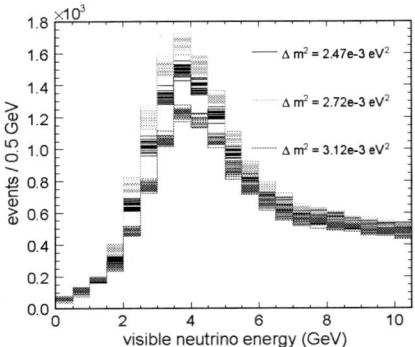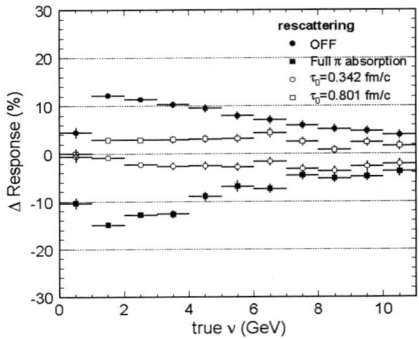

FIGURE 4. **Left:** Energy distribution of simulated $\nu_\mu - CC$ events at the MINOS Far Detector. The effect of neutrino oscillations is included for $\Delta m^2 = 2.47$, 2.72 and 3.12×10^{-3} eV^2 (68% CL limits of Ref. [1]) and $\sin^2 2\theta = 1$. Individual curves correspond to uncorrelated $\pm 1\sigma$ variations in the formation time, pion absorption cross-section and the neutrino induced 2π production cross-section. The MC sample corresponds to the "high-statistics limit", approximately $\times 10$ of the planned MINOS exposure. **Right:** The variation in the calorimeter's response as a function of the energy transfer ν and under different final state interaction scenarios. Zero on the vertical scale corresponds to the $\pi \to npnp$ model of pion absorption and $\tau_0 = 0.52$ fm/c.

escape the nucleus to create a signal in the detector but, at MINOS energies, often suffer a hadronic interaction before doing so. It is important to account for these "final state interactions" because neutrino oscillation experiments seek to measure the energy, ν, transferred by the neutrino to the target, so as to reconstruct $E_\nu = E_\mu + \nu$. The presence of final state interactions causes $\nu \neq E_{\text{vis}}$, where the latter is the visible energy of shower particles exiting the nucleus. We will describe the way in which final state interaction are modeled and how uncertainties are accounted for.

Neutrino interactions in MINOS are modeled by NEUGEN-v3 [18]. NEUGEN includes a final state interactions (also referred to as "intranuclear re-scattering") package known as INTRANUKE. The code is anchored to a comparison of final states in $\nu + d$ and $\nu + Ne$ interactions as measured in the BEBC and ANL−12 ft bubble chambers [19]. The library includes a treatment of pion elastic and inelastic scattering, single charge exchange and absorption in a cascade simulation of the final state. The relative probabilities for these processes are approximately 35:50:10:5 (40:30:7:23) at a pion kinetic energy of 1 GeV (250 MeV). The formation zone concept is included by suppressing interactions that would have occurred before the pion has traveled a distance $l = \tau_0 p/m \; \tau_0 = 0.52$ fm/c [20]. The simulation includes a treatment of pion absorption inspired by [21]. Absorbed pions transfer their energy to a $npnp$ cluster and the individual nucleons are then tracked in GEANT. A more comprehensive description of the INTRANUKE code and it's effect in MINOS is given in Ref. [22].

We derive the effect that parametric uncertainties in the final state interaction model have on the oscillation result as follows. First, the model contains a number of parameters for which the uncertainty may be estimated, however roughly, based on external data. The dependence of the reconstructed shower energy on ν was studied as a function of these model parameters and the most influential were identified. The parameters were then varied in an uncorrelated way over multiple trials and the visible $\nu_\mu - CC$ energy spectra were constructed, accounting for the detector energy smearing. This was done without oscillations so as to represent measurements made in the Near Detector and with oscillations at varying Δm^2 and $\sin^2 2\theta$ to represent measurements at the Far Detector. An example of simulated Far Detector spectra is shown on the left in Fig. 4. The Δm^2 values were chosen according to the best-fit and $\pm 1\sigma$ uncertainties on the initial MINOS result [1]. For each trial a fit was performed to extract Δm^2 and $\sin^2 2\theta$ and the differences with respect to the input values were histogrammed. The mean and width of the distributions is an estimate of the uncertainty attributed to the parameter uncertainties in the final state interaction model. We find $\delta \Delta m^2 \approx 3 \times 10^{-5} \, \mathrm{eV}^2/\mathrm{c}^4$ and $\delta \sin^2 2\theta \approx 0.01$. These errors are small compared to the overall uncertainty as may be noticed from the fact that the curves for different values of Δm^2 in Fig. 4 are easily distinguished.

A much larger uncertainty occurs as a consequence of the rather simple way in which the final state interaction code simulates pion absorption and inelastic collisions. Absorption is modeled, by default, as solely due to $\pi \rightarrow npnp$ but we know that processes such as $\pi \rightarrow npp$ and $\pi \rightarrow nnp$ may also occur and that the nucleons are themselves attenuated inside the nuclear medium (this effect is not accounted for). In addition, $\pi + N$ inelastic collisions are modeled but the energy lost by the π is discarded and secondary π are not created. These model based deficiencies result in a large uncertainty on the amount of visible energy emitted from the nucleus.

The model based uncertainties cannot be accounted for by the parametric variation technique described above. Instead, we study simulations done (a) without re-scattering, (b) with pion absorption according to $\pi \rightarrow npnp$ (the default) and (c) in which the energy of absorbed pions is discarded. The differences $a - b$ and $c - b$ in the reconstructed shower energy are shown as a function of ν on the right-hand side of Fig. 4. The curve labeled "OFF" corresponds to $a - c$ and represents the change in visible energy due to inelastic pion interactions. The curve labeled "Full π absorption" corresponds to $c - b$ and represents the change in visible energy due to pion absorption. For comparison the (much smaller) effect of $\pm 1\sigma$ variations in the formation time is also shown. To estimate the uncertainty on the oscillation parameters, the $\nu \leftrightarrow E_{\mathrm{shower}}$ relation is modified according to Fig. 4 and the analysis is repeated. We find that the parameters shift by $\delta \Delta m^2 \approx 6 \times 10^{-5} \, \mathrm{eV}^2/\mathrm{c}^4$ and $\delta \sin^2 2\theta \approx 0.05$. When drawing the oscillation contours, uncertainties in both the single particle energy scale and final state interactions are included as an 11% penalty term ("nuisance parameter") on the hadronic energy scale.

The experiment is currently working on improvements in the final state interaction model [23] to remedy the deficiencies mentioned above. In addition, the model is being re-implemented to facilitate re-weighting of MC events. A re-weightable model is extraordinarily important because it vastly reduces the number of MC events needed to estimate systematic uncertainties. Re-weighting machinery would also allow the parametric variation technique described above to be used directly in the oscillation fit by including the final state interaction model parameters as nuisances. The importance of event by event re-weighting is one reason that experiments may prefer home-grown codes, for which the model parameters can be altered and uncertainties understood, over more sophisticated but also more opaque general purpose codes.

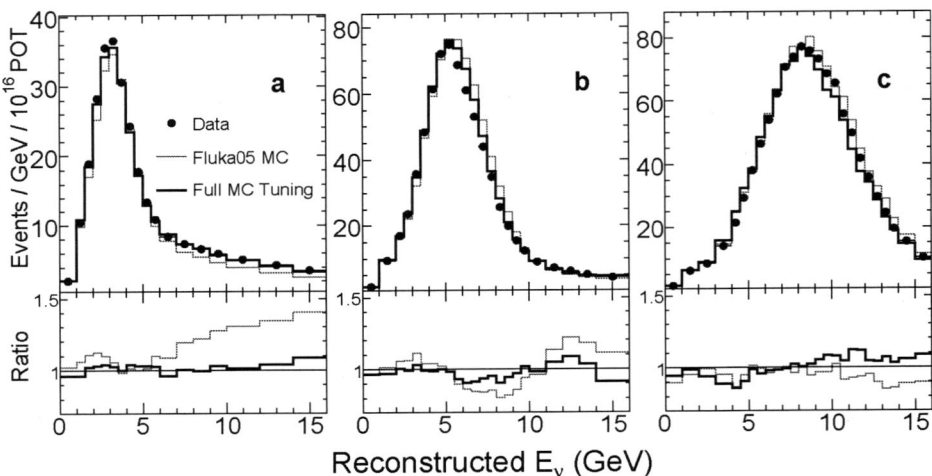

FIGURE 5. The reconstructed energy distribution of ν_μ–CC events observed in MINOS. The data were collected in the (a) low (b) medium and (c) high energy beam configurations and are compared to the predictions of the neutrino beam Monte Carlo, using FLUKA05 to simulate hadron production in the NuMI target. We show the MC prediction before and after the tuning procedure described in the text.

ESTIMATION OF THE NEUTRINO FLUX

The neutrino beamline model is separated into three parts: (a) a simulation of the hadrons produced by 120 GeV/c protons incident on the NuMI target and (b) the propagation of those hadrons and their progeny through the magnetic focusing elements, along the 677 m decay pipe, and into the primary beam absorber and (c) decay of the mesons and calculation of the probability that any neutrino progeny traverses the Near and Far detectors. Hadronic production is simulated using a detailed model of the target geometry and material composition. By default, interactions are generated with FLUKA05 [24, 25, 26] but output from the MARS-v15 [27, 28, 29] and GEANT-FLUKA [30] codes, as well as calculations based on the parameterisations of BMPT [31] and Malensek [32], are used as cross-checks. The produced hadrons are propagated in a GEANT3 simulation of the NuMI beamline. Decays in which a neutrino is produced are saved and later used as input for neutrino event simulation in the Near and Far Detectors.

The neutrino fluxes predicted by the different models vary by 10-20%. This is unsurprising given the relatively paucity of hadron production data in the x_F, p_T region interesting to NuMI, not to mention at 120 GeV/c and on a thick Carbon target [33]. Therefore, the rather good agreement between the ν_μ–CC energy spectrum measured in the Near Detector and the MC prediction based on FLUKA05 (Fig. 5), comes as a pleasant surprise. The data/MC agreement is, however, not perfect and can be improved by taking advantage of the flexibility of the NuMI beamline. The NuMI target may be remotely moved along the beam axis, changing the x_F, p_T region focused by the magnetic horns, and thereby the beam energy. Figure 5 shows data taken in three target positions:

$z = 10\,\mathrm{cm}, 100\,\mathrm{cm}$ and $250\,\mathrm{cm}$ where z=0 cm is the location of the target when fully inserted into the first horn. These data, along with data taken at $z = 10\,\mathrm{cm}$ with the horns off and with the horn current modified by $\pm 8\%$, are used in a parametric fit to constrain the beam model.

The fit is based on a parametric model [34], similar to that of BMPT, which is able to describe the $d^2N/dx_F dp_T$ distribution predicted by FLUKA. The model parameters were allowed to float in a fit to the ν_μ−CC energy spectra. The fit also included nuisance parameters to account for uncertainties in beam focusing, intensity, and detector response. The mean transverse momentum of pions, $\langle p_T \rangle$ was constrained to the FLUKA05 value of 364 MeV/c, with a penalty term of 15 MeV/c, obtained from the variation of hadron production models. The results of the fit are shown as the "Full MC Tuning" curve in Fig. 5. The fitting procedure dramatically improves the data/MC agreement, particularly in the lowest energy configuration in which most of the oscillation dataset was recorded. The oscillation analyses use these results as the first step in the prediction of the neutrino spectrum at the Far Detector.

SUMMARY

We have attempted to highlight some of the ways in which MINOS depends upon the correct understanding and modeling of hadronic interactions from 120 GeV (hadron production) to a few tens of MeV and less (calorimetry). Hadronic interactions are difficult to model correctly and whenever possible we have attempted to benchmark and/or constrain simulations using internal or external datasets. One expects that models will never completely agree with the data so much effort is concerned with quantifying the effect that disagreements have upon the oscillation results. Uncertainties are generally included as nuisance parameters and fitting is greatly facilitated by the ability to re-weight events for model changes. Finally, in the interest of brevity we have been forced to omit some important topics, such as the effect of shower modeling uncertainties on the NC background estimated in the reconstructed ν_μ−CC sample as well as the sensitivity of the ν_e−CC event selection to the fragmentation/hadronisation treatment.

The author would like to thank J. Morfin, H. Gallagher, R. Ransome and S. Dytman for advice and comments regarding final state interactions and N. Mokhov for comments about low energy pion nucleus interactions. The hadron production fitting is due to S. Kopp, Ž. Pavlović and P. Vahle.

REFERENCES

1. D. G. Michael, et al., *Phys. Rev. Lett.* **97**, 191801 (2006), hep-ex/0607088.
2. E. Ables, et al. (1995), FERMILAB-PROPOSAL-0875.
3. The MINOS Collaboration (1998), The MINOS Detectors Technical Design Report.
4. M. A. Kordosky, *Hadronic Interactions in the MINOS Detectors*, Ph.D. thesis, University of Texas (2004), FERMILAB-THESIS-2004-34.
5. P. L. Vahle, *Electromagnetic Interactions in the MINOS Detectors*, Ph.D. thesis, University of Texas (2004), FERMILAB-THESIS-2004-35.
6. P. Adamson, et al., *Nucl. Instrum. Meth.* **A556**, 119–133 (2006).

7. R. Nichol, *Calibration of the MINOS Detectors*, Ph.D. thesis, University College London (2003), FERMILAB-THESIS-2003-41.
8. C. Smith, *Calibration of the MINOS Detectors and Extraction of Neutrino Oscillation Parameters*, Ph.D. thesis, University College London (2002), FERMILAB-THESIS-2002-58.
9. J. J. Hartnell, *Measurement of the MINOS Detectors' Relative Calorimetric Energy Response*, Ph.D. thesis, St. John's College, Oxford (2005), FERMILAB-THESIS-2005-51.
10. D. E. Groom, N. V. Mokhov, and S. I. Striganov, *Atom. Data Nucl. Data Tabl.* **78**, 183–356 (2001).
11. R. Brun, et al., GEANT Detector Description and Simulation Tool, CERN Program Library Long Writeup W5013 (1994).
12. T. A. Gabriel, J. E. Brau, and B. L. Bishop, *IEEE Trans. Nucl. Sci.* **36**, 14–22 (1989).
13. C. Zeitnitz, and T. A. Gabriel, *Nucl. Instrum. Meth.* **A349**, 106–111 (1994).
14. H. Fesefeldt, *GHEISHA The Simulation of Hadronic Showers: Physics and Applications*, As reproduced by CERN, 1985.
15. G. Bower, and R. Cassel (c. 2002), SLAC-GHEISHA refers to a version of the GHEISHA code improved to fix several coding errors present in the original.
16. N. Mokhov, private communication (2006).
17. R. Wigmans, *Calorimetry : Energy Measurement in Particle Physics*, Oxford Science Publications, 2000.
18. H. Gallagher, *Nucl. Phys. Proc. Suppl.* **112**, 188–194 (2002).
19. R. Merenyi, et al., *Phys. Rev.* **D45**, 743–751 (1992).
20. V. Ammosov, Low multiplicity final states in neutrino interaction at low energy – skat results, Talk given at The First International Workshop on Neutrino-Nucleus Interactions in the Few GeV Region (NuINT01) (2001), In our work the formation time has been modified to account for an effective nuclear radius used our simulation.
21. R. D. Ransome, *Nucl. Phys. Proc. Suppl.* **139**, 208–212 (2005).
22. M. Kordosky, *Nucl. Phys. Proc. Suppl.* **159**, 223–228 (2006), hep-ex/0602029.
23. S. Dytman (2006), Please refer to the talk given in this workshop.
24. G. Collazuol, A. Ferrari, A. Guglielmi, and P. R. Sala, *Nucl. Instrum. Meth.* **A449**, 609–623 (2000).
25. A. Ferrari, and P. Sala, "," in *Proceedings of the Workshop on Nuclear Reaction Data nd Nuclear Reactors Physics, Design and Safety*, edited by A. Gandini, and G. Reffo, World Scientific, 1996, aTLAS Internal Note ATL-PHYS-97-113.
26. P. Sala (2000), talk given at the Second International Workshop on Neutrino Beams and Instrumentation.
27. N. V. Mokhov (1995), fERMILAB-FN-0628.
28. O. E. Krivosheev, and N. V. Mokhov (1997), talk given at the 3rd Workshop on Simulating Accelerator Radiation Environments (SARE3), Tsukuba, Japan, 7-9 May 1997.
29. N. V. Mokhov, et al. (1998), talk given at the 4th Workshop on Simulating Accelerator Radiation Environments (SARE4), Knoxville, TN, 13-15 Sept 1998, nucl-th/9812038.
30. A. Fasso, A. Ferrari, J. Ranft, and P. R. Sala (????), given at 4th International Conference on Calorimetry in High-energy Physics, La Biodola, Italy, 19-25 Sep 1993.
31. M. Bonesini, A. Marchionni, F. Pietropaolo, and T. Tabarelli de Fatis, *Eur. Phys. J.* **C20**, 13–27 (2001), hep-ph/0101163.
32. A. J. Malensek (1981), fERMILAB-FN-0341.
33. R. Raja (2006), FNAL-E-907 (MIPP) is attempting to measure hadron production off of a prototype NuMI target. These measurements are expected to significantly improve the NuMI flux model. Please refer to the talk given in this workshop.
34. S. Kopp, Žarko Pavlović, and P. Vahle (2006), MINOS internal document 1650.

CMS validation Experience:
Test-beam 2004 data vs Geant4

Stefan Piperov

Fermilab, Batavia, Illinois, USA
INRNE-BAS, Sofia, Bulgaria

Abstract. A comparison between the Geant4 Monte-Carlo simulation of CMS Detector's Calorimetric System and data from the 2004 Test-Beam at CERN's SPS H2 beam-line is presented. The overall simulated response agrees quite well with the measured response. Slight differences in the longitudinal shower profiles between the MC predictions made with different Physics Lists are observed.

Keywords: Monte-Carlo Simulation, Geant4 Validation, Physics Lists, Calorimeters, Test-Beam, Hadronic Showers, Longitudinal Shower Profiles
PACS: 29.40.Vj,07.05.Tp

INTRODUCTION

A series of Test-Beam measurements has been performed on the calorimetric system of CMS over the last few years in order to optimize the design and study its performance. Detailed Monte-Carlo simulations of the test-beam configuration in 2004 have been made and the results compared with the test-beam measurements. Presented here is a comparison between the results from the 2004 Test-Beam and the Geant4-based Monte-Carlo simulations performed at the same time.

The CMS detector

The Compact Muon Solenoid detector (CMS)[1] is one of the general purpose detectors for the Large Hadron Collider (LHC) that is being assembled at CERN. A cross-section of the detector (Fig.1) identifies the major sub-systems of the apparatus. The Pixel Detector, Silicon Tracker, Preshower, Electromagnetic Calorimeter (ECAL) and Hadronic Calorimeter (HCAL) are positioned inside the superconducting solenoidal magnet generating the strong 4T magnetic field. Outside of the solenoid are the Muon Detectors embedded into the magnet's return yoke, and the Very-Forward Calorimeters.

Following the cylindrical symmetry of the apparatus, most sub-detectors consist of a Barrel and End-cap parts, labeled with a "B" or "E" respectively in further references.

CP896, *Hadronic Shower Simulation Workshop*
edited by M. Albrow and R. Raja
© 2007 American Institute of Physics 978-0-7354-0401-4/07/$23.00

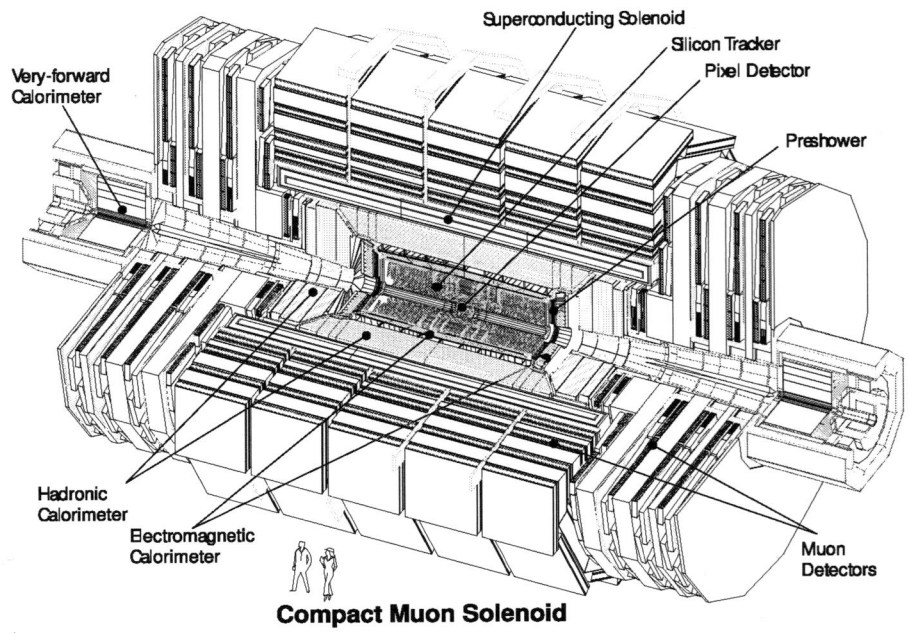

FIGURE 1. CMS - detector subsystems

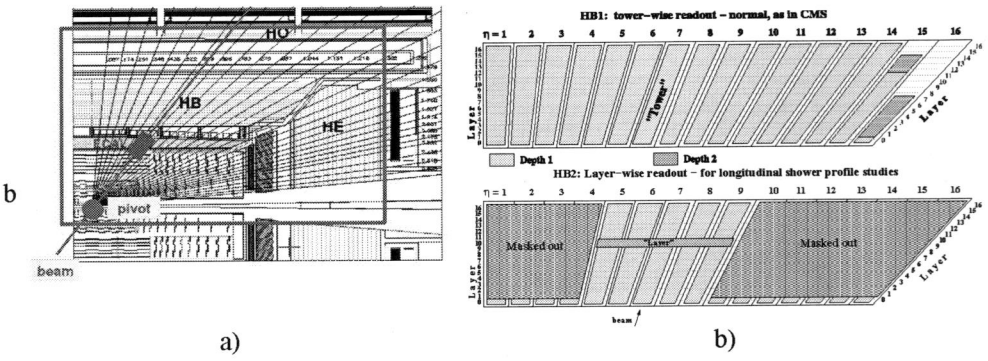

a)   b)

FIGURE 2. a) Calorimetric systems present on main moving table of Test-beam 2004 (framed). Pivot point corresponds to beam-crossing point in CMS; b) Two different read-out schemes for the HB wedges: Tower-wise (top) and Layer-wise (bottom)

CMS Calorimetric System

The Calorimetric System of CMS consists of the Electromagnetic Calorimeter (ECAL) Barrel (EB) and Endcap (EE) parts, Hadronic Calorimeter (HCAL) Barrel

(HB), Endcap (HE) and Outer (HO)[1] parts, and the Very-Forward Calorimeter (HF).

ECAL

The Electromagnetic Calorimeter [2] is a homogeneous calorimeter made of over 80,000 lead-tungstate ($PbWO_4$) crystals equipped with avalanche photo-diodes (APDs) for readout. The crystals in the barrel have a front face of $\sim 22x22mm^2$ and are 23 cm (~ 26 radiation lengths) long. Following the overall 18-fold ϕ-symmetry of the barrel part of CMS, the EB crystals are organized into 36 Super-Modules (18 in each positive and negative Z-direction) covering 20° in ϕ-direction and pseudo-rapidity range of $|\eta| \lesssim 1.5$. Each ECAL crystal is read-out independently by two Avalanche Photo Diodes (APDs).

HCAL

The Hadronic Calorimeter [3] is a sampling calorimeter with $\sim 50mm$ thick copper absorber plates interleaved with $4mm$ thick scintillator sheets (barrel part). Again, following the overall 18-fold ϕ-symmetry of the barrel part of CMS, the HB is organized into 36 "wedges" (18 in each positive and negative Z-direction) covering 20° in ϕ-direction and pseudo-rapidity range of $|\eta| \lesssim 1.5$. Scintillator tiles are optically grouped together in towers covering equal surface (0.087x0.087) in $\eta - \phi$ space, and are read-out together (single electronics channel per tower, as shown on Fig.2b) - top) by Hybrid Photo Diodes (HPDs).

Thickness of HB in the central region ($|\eta| = 0$) is $\sim 90cm$ (or ~ 6 nuclear interaction lengths λ), which is somewhat thin. For that reason HB is complemented with scintillator tiles embedded in the first muon absorber layer just outside the magnet coil, thus forming the Outer Hadron Calorimeter (HO).

HF

The Very Forward Calorimeter (HF) is a Cherenkov quartz-fiber, steel absorber calorimeter, covering the very high pseudo-rapidity region ($3 < |\eta| < 5$) in CMS. Its design is driven by the requirements for extreme radiation hardness necessary in this part of the detector.

---

[1] Not labeled in Fig.1, HO is situated in the immediate outside of the magnet coil, embedded in the first layer of muon absorber, and consists of Barrel part only

The 2004 Test-beam setup

The following elements of the calorimetric system of CMS were present in the 2004 test-beam:

- Two wedges of HB.
- One wedge of HE.
- A 7x7 matrix of prototype ECAL crystals, read-out by individual photo-multipliers.
- One wedge of HO.
- One wedge of HF.

All detector elements, except HF wedge, were mounted on a moving table (see Fig.2a)), allowing for beam particles to be sent to different (η, ϕ) sections of the calorimeter. The HF wedge was mounted on a separate table and was positioned in the beam independently downstream of the main calorimetric system.

The H2 beam-line of CERN's SPS accelerator was arranged as shown in Fig.3 to allow production of the following particle beam types:

- hadrons (mainly $\pi^{\pm}$) with momenta: $2 - 300 GeV/c$
- muons with momenta: $80, 150 GeV/c$
- electrons with momenta: $9 - 100 GeV/c$

The Very Low Energy ("VLE Setup") part of the beam-line was used to produce the beam particles below $10 GeV/c$. The following detectors were used for particle identification (PID) and beam cleanup:

- Three Wire Chambers: WC A,B,C - used for tagging of interactions in the beam-line;
- Two Cherenkov counters: CK2 - used for electron tagging, and CK3 used for pion/proton tagging;
- Three scintillators: V3, V6 and VM - used for muon tagging.

HB readout

By re-arranging the Optical Decoupling Units (ODUs) of the two HB wedges two different readout schemes were implemented: HB1 was read out tower-wise as usual (top of Fig.2b), while HB2 was read out layer-wise (bottom of Fig.2b) thus allowing for measurement of the longitudinal shower profiles in the calorimeter.

MONTE-CARLO SIMULATION

The complete TB2004 setup was simulated using the CMS software, which internally employs the Geant4 toolkit[4, 5]. A very detailed simulation geometry was used, as presented in Fig.4. The version of Geant4 used was 6.2_p02. The simulations were

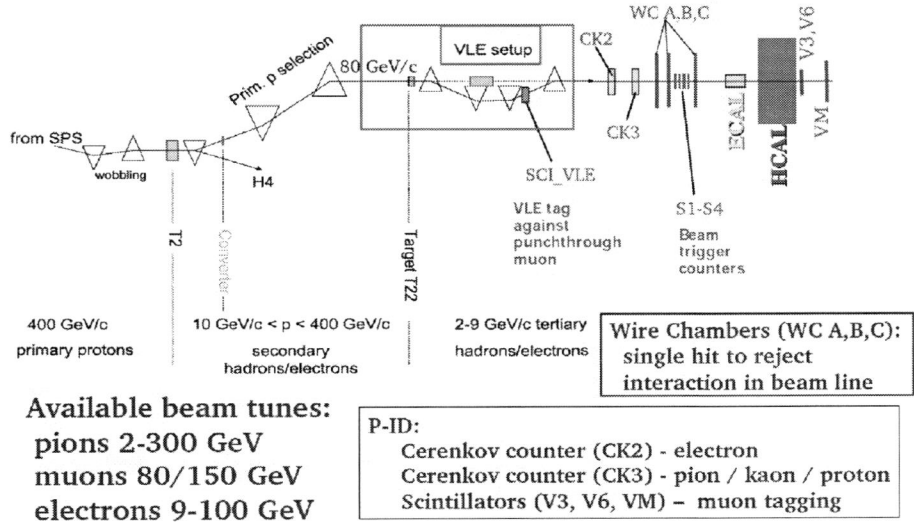

FIGURE 3. SPS's H2 beam-line. The VLE section is used for production/selection of very low beam energies (2 - 10 *GeV*)

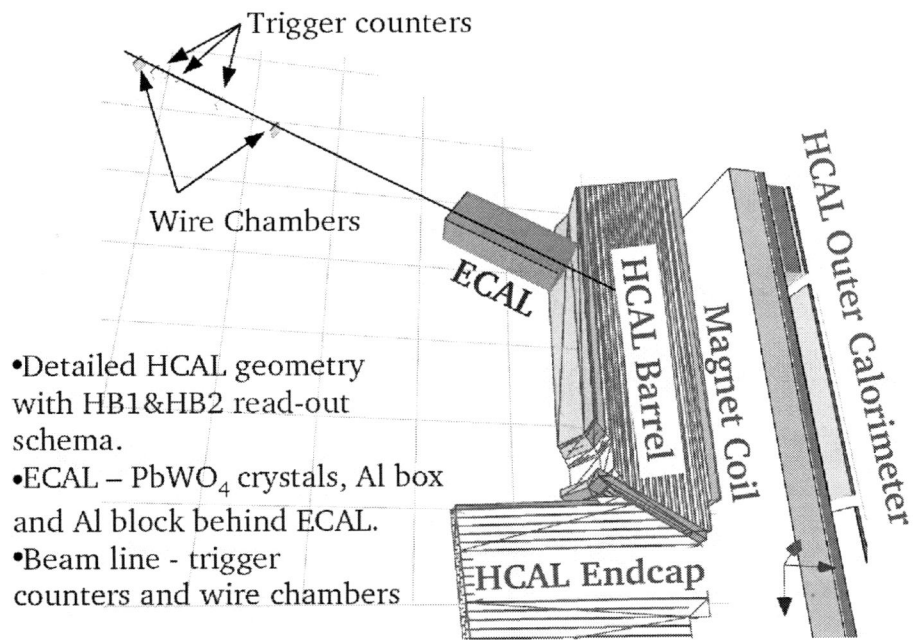

FIGURE 4. MC simulation geometry

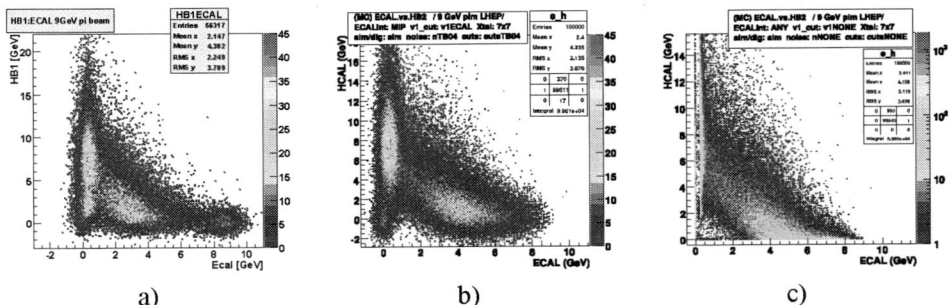

FIGURE 5. The "banana" plot - HCAL signal vs. ECAL signal - of a 9 GeV/c pion. a)Test-beam data; b) MC Simulation with included Gaussian noise; c) MC Simulation without noise.

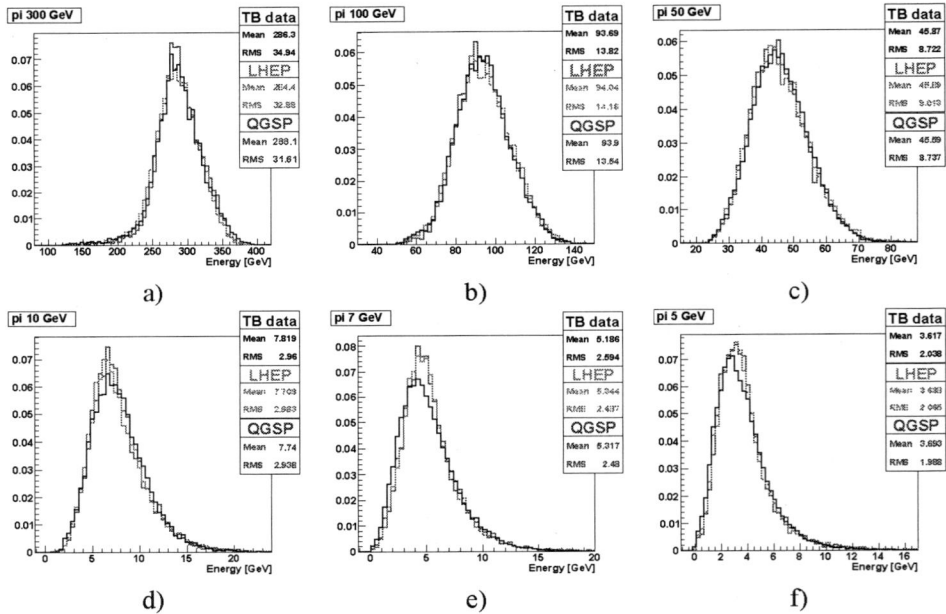

FIGURE 6. Comparison of detector response to pion beam of various momenta with the Monte-Carlo predictions obtained with two physics lists

repeated with all four physics lists (LHEP, QGSP, QGSC, FTFP) for High Energy Physics Calorimetry available in PACK 2.5[2].

Simulated detector response was studied at the simulated hit level - i.e. energy deposited in active materials. The digitization step was not simulated. Also, no detailed

---

[2] http://geant4.web.cern.ch/geant4/physics_lists/

TABLE 1. TB2004 data sets

TB Data	Simulation
Very Low Energies (VLE)	
2,3,5,7,9 GeV mainly $\pi^{\pm}$ beam with/without ECAL HB1/HB2 Full particle identification	2,3,5,7,9 GeV $e^{\pm}$, $\pi^{\pm}$, p, $K^{\pm}$ beam with/without ECAL HB2
Medium Energies	
10,15,20 GeV $e^{\pm}$, $\pi^{\pm}$ beam with/without ECAL HB1/HB2 Partial particle identification	10,15,20 GeV $e^{\pm}$, $\pi^{\pm}$, p, $K^{\pm}$ beam with/without ECAL HB2
High Energies	
30,50,100,150,300 GeV $e^{\pm}$, $\pi^{\pm}$ beam with/without ECAL HB1/HB2	30,50,100,150,300 GeV $e^{\pm}$, $\pi^{\pm}$, p, $K^{\pm}$ beam with/without ECAL HB2

simulation of the detector noise was performed. Simple, gaussian-distributed noise with amplitudes matched to the ones observed in the real detectors was used where necessary.

RESULTS

We consider here only the response of the ECAL and HCAL detectors to various beam particles. HO and HF detectors were not used in this study. On the Monte-Carlo side we have found that all three "calculated" (or "model-based") physics lists (QGSP, QGSC and FTFP) show similar results, so we only use QGSP as an example of the "calculated" physics lists, and we compare it to the "parametrized" physics list LHEP.

Detector response

Fig.5 shows the so-called "banana plot" - HCAL vs. ECAL response - of the combined system to a $9GeV/c$ pion beam. The left plot shows the response of the detectors measured in TB2004. Middle plot is the simulated response with included gaussian noise. Right plot shows the simulated response without noise. Several features of the TB2004 setup become evident from this comparison:

- There is electron contamination in the pion beam (the spot at $E_{ECAL} = 9$, $E_{HCAL} = 0$ in Fig.5a);
- Small fraction of the pions - perhaps due to scattering in the beam-line components - miss completely the ECAL crystals, leaving signal only in the HCAL;
- There are muons in the beam, leaving only a minimum-ionizing-particle signal in both calorimeters. These could be the result of pion decay in flight, or beam contamination (the spot at $E_{ECAL} = 0.3GeV$, $E_{HCAL} = 2GeV$ in Fig.5c).

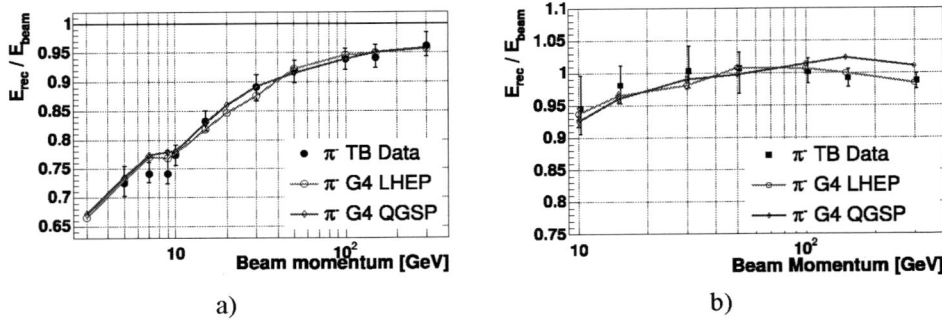

FIGURE 7. Linearity of pion response: a) Combined system ECAL + HCAL; b) HCAL alone.

These findings clearly indicate inefficiencies in the beam particle identification detectors and the muon vetoes. To eliminate the effect of these contaminations the following (geometrical) cuts were applied:

- Events in the vicinity of the ($E_{ECAL} = 0$, $E_{HCAL} = 0$) region were excluded from the analysis;
- Pion events with $E_{ECAL} > 0.8 \times E_{beam}$ were excluded from the analysis;
- the exact cut values were optimized for each beam type to maximize the cleaning efficiency;
- the same cuts were used both for TB data and MC simulation.

With these cuts applied, a reasonably good agreement (see Fig.6) was achieved between the reconstructed energy spectra of pions in wide momentum range. The cleaning process was not efficient in the lowest momentum range ($2, 3 GeV/c$), so those points were excluded from the comparison.

Another important quantity - the linearity of response - also shows good agreement with the predictions obtained with both physics lists (Fig.7). The left plot shows the linearity of response for the whole system (ECAL+HCAL), while the right plot compares the linearity of response for the HCAL alone. This was accomplished by requiring only a MIP signal in ECAL and in the first readout layer (L0) of HCAL. In this second configuration (HCAL alone), the calorimeter is clearly too thin to completely contain the showers of high-energy pions, and a significant leakage is observed for energies above $50 GeV$. As we can see, the QGSP physics list does not reproduce this leakage very precisely, which could be an indication of differences in the predicted/simulated longitudinal shower profiles (see next section).

Longitudinal shower profiles

The only significant difference observed among the MC simulations performed with different physics lists (LHEP, QGSP, QGSC, FTFP) was the prediction of the longitudinal shower profiles. Fig.8 compares the longitudinal shower profiles for pions of

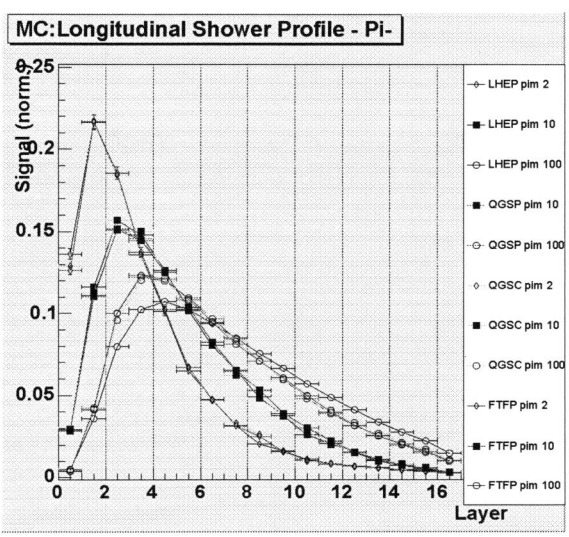

FIGURE 8. Longitudinal profiles of simulated pion showers. Three different energies and four Physics lists are shown.

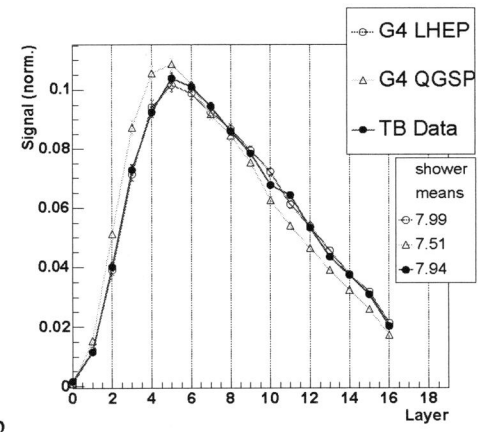

b

FIGURE 9. Longitudinal profiles of $300 GeV$ pion showers - MC simulation compared to Test Beam 2004 data

3 different energies (2,10,100 GeV) obtained with the different physics lists. Clearly, at high energies (100 GeV) the predictions of parametrized LHEP list start to differ from the predictions of the calculated (QGSP, QGSC, FTFP) lists. Comparison of these shower profiles with the one measured in TB2004 in Fig.9 indicates that the prediction of the LHEP physics list agrees better with the data. This result is in agreement with the

differences in the amounts of energy leakage seen in Fig.7.

CONCLUSION

The measured response of the combined ECAL+HCAL calorimetric system in Test-Beam 2004 agrees quite well with the Monte-Carlo simulations based on Geant4. The parametrized Physics List (LHEP) shows better agreement with data, while the model-based QGSP list seems to predict shorter showers than we measured.

REFERENCES

1. The CMS Collaboration, Technical Proposal, Tech. rep., CERN/LHCC 94-38 (1994).
2. The CMS Collaboration, CMS ECAL Technical Design Report, Tech. rep., CERN/LHCC 97-33 (1997).
3. The CMS Collaboration, The HCAL Technical Design Report, Tech. rep., CERN/LHCC 97-31 (1997).
4. J. Allison, et al., *IEEE Trans. Nucl. Sci.* **53**, 270 (2006).
5. S. Agostinelli, et al., *Nucl. Instrum. Meth.* **A506**, 250–303 (2003).

Hadronic Shower Validation Experience for the ATLAS End-Cap Calorimeter

A.E. Kiryunin and D. Salihagić

Max-Planck-Institut für Physik, Werner-Heisenberg-Institut,
Föhringer Ring 6, 80805 München, Germany

(presented by D. Salihagić on behalf of the ATLAS Liquid Argon Collaboration)

Abstract. Validation of GEANT4 hadronic physics models is carried out by comparing experimental data from beam tests of modules of the ATLAS end-cap calorimeters with GEANT4 based simulations. Two physics lists (LHEP and QGSP) for the simulation of hadronic showers are evaluated. Calorimeter performance parameters like the energy resolution and response for charged pions and shapes of showers are studied. Comparison with GEANT3 predictions is done as well.

Keywords: simulation, liquid argon calorimeter, testbeam
PACS: 07.05.Tp, 07.20.Fw

INTRODUCTION

An accurate description of hadronic interactions and hadronic showers is required for any package to be used for Monte Carlo (MC) simulations of experiments at the Large Hadron Collider (LHC). The validation of physics models of different simulation packages is going on within the "Simulation Physics Validation Project" (running in the framework of the "LHC Computing Grid Project").

A part of the validation project is the evaluation of hadronic physics models of the GEANT4 simulation package [1], carried out for liquid argon (LAr) calorimeters of the ATLAS detector. The validation is based on the comparison of GEANT4 predictions with experimental results obtained during various beam tests of calorimeter modules. The validation includes comparisons of relevant calorimeter performance parameters like the energy dependence of the response, resolution and shower shapes for charged pions.

Different hadronic physics lists, provided by the GEANT4 toolkit, are studied. It is a long-term project. Therefore the follow-up of GEANT4 versions is carried out as well.

The comparison with predictions of GEANT3 [2] (a simulation package, which was successfully used over many years) is also part of the project.

In this talk results obtained for the following testbeams are presented:

- Beam tests of serial production modules of the ATLAS hadronic end-cap calorimeter (HEC)
- Combined beam tests of modules of the HEC and of the ATLAS electromagnetic end-cap calorimeter (EMEC)

CP896, *Hadronic Shower Simulation Workshop*
edited by M. Albrow and R. Raja
© 2007 American Institute of Physics 978-0-7354-0401-4/07/$23.00

GEANT PACKAGES

GEANT4 [1] is a toolkit for simulating the passage of particles through matter. It is based on software engineering techniques and object-oriented technology, and is implemented in the C++ programming language.

Different physics lists are provided by the GEANT4 package for simulations of hadronic showers [3]. For the comparison with the ATLAS end-cap calorimeter testbeam data, two lists have been selected, namely: LHEP and QGSP. The LHEP physics list uses the low- and high-energy pion parametrization models for inelastic scattering. The QGSP physics list is based on theory-driven models: it uses the quark-gluon-string model for interactions of the projectile with nucleons and a pre-equilibrium decay model to describe the behavior of the damaged nucleus.

In GEANT4, charged particles are tracked to zero kinetic energy. The production of secondaries is controlled by a minimum range requirement: secondary particles with a range below a specified value ("range cut") are not tracked. The range cut is translated into a minimum production energy threshold for a given particle and material.

Different versions of GEANT4 and different values of the range cut are used for testbeam simulations (as presented later).

Concerning the GEANT3 package [2], version 3.21 is used for the calorimeter test-beam simulations, with the G-CALOR code [4] for the hadronic shower development. The energy threshold of particle tracking is set to 100 keV and of secondary production of photons and electrons to 1 MeV.

COMPARISON WITH DATA OF BEAM TESTS OF HEC SERIAL MODULES

HEC beam tests

The ATLAS hadronic end-cap calorimeter [5] is a LAr sampling calorimeter with parallel copper absorber plates. The HEC is structured in two wheels (HEC1 and HEC2), each wheel consisting of 32 modules. Annular spacers define 8.5 mm gaps between the absorber plates. The thickness of the copper absorber plates is 2.5 cm for HEC1 and 5.0 cm for HEC2. The calorimeter has four longitudinal layers (combined 8 and 16 gaps of LAr in the HEC1 wheel, and 8 and 8 gaps in the HEC2 wheel). The total thickness of the calorimeter is $\sim$103 radiation lengths X_0 or $\sim$10 absorption lengths.

During the years 2000-2001, the serial production calorimeter modules were exposed to test beams as part of the standard quality control procedure during HEC construction. The beam tests have been carried out in the H6 beam line of the CERN SPS, using secondary and tertiary beams. Pion beams of different energies (from 10 to 200 GeV) were used. More details, as well as data analysis results can be found in [6].

Simulation and analysis

For the validation of hadronic physics models of GEANT4, energy scans with negatively charged pions have been simulated for the HEC testbeam in the interval from 10 to 200 GeV. Versions of GEANT4 and hadronic physics lists, used for those simulations, are summarised in Table 1. A range cut of 20 μm was used for all simulations. In addition, simulations with version 8.0 were done with 100 and 1000 μm range cut. The statistics was typically 5000 events per energy, version and range cut.

TABLE 1. Versions of GEANT4 and corresponding physics lists

Version	5.2 patch-02	6.2 patch-02	7.0 patch-01	8.0 patch-01
Release date	October 2003	October 2004	February 2005	February 2006
Physics lists	LHEP 3.6	LHEP 3.7	LHEP 3.7	LHEP 4.0
	QGSP 2.7	QGSP 2.8	QGSP 2.8	QGSP 3.0

The analysis of the simulated data followed rather closely the experimental procedure [6]. An energy independent electromagnetic (EM) scale factor (obtained by the analysis of electron samples, simulated with the corresponding GEANT4 versions and range cuts) was applied for the energy reconstruction. Clusters of fixed size were used, i.e. the same set of cells was selected for energy measurements in all events and for all beam energies. The use of clusters of fixed size for the energy reconstruction allowed the evaluation of the electronics noise (including possible correlations between channels). To compare the experimental and simulated data, the noise was subtracted from the experimental data.

Detailed results on the validation of physics models of GEANT4 (obtained for version 6.2) with the HEC testbeam data can be found in [7].

Results

Energy response and resolution. To get the energy resolution and response for pions, Gaussian curves were fitted to the reconstructed energy distributions in the interval $\pm 3\sigma$ around the peak value E_0. The parameters E_0 and σ from this fit were used to determine the response E_0/E_{BEAM} and the resolution σ/E_0.

The relative response (response normalised to one at E_{BEAM}=200 GeV) as a function of the beam energy is shown in Figure 1 for the experimental data and for the GEANT4 and GEANT3 simulations. The ratio between simulated and experimental data is shown there as well. The non-linear response reflects the non-compensating nature of the HEC calorimeter. The variation with energy differs slightly for different simulation codes. The GEANT3 predictions are closest to the experimental values, but within errors LHEP and QGSP are consistent with the experimental data as well.

Values of the energy response, obtained for different versions of GEANT4 and with different range cuts, are very close.

The energy dependence of the resolution can be parametrized by the following two-term formula:

$$\frac{\sigma}{E_0} = \frac{A}{\sqrt{E_{BEAM}}} \oplus B, \tag{1}$$

207

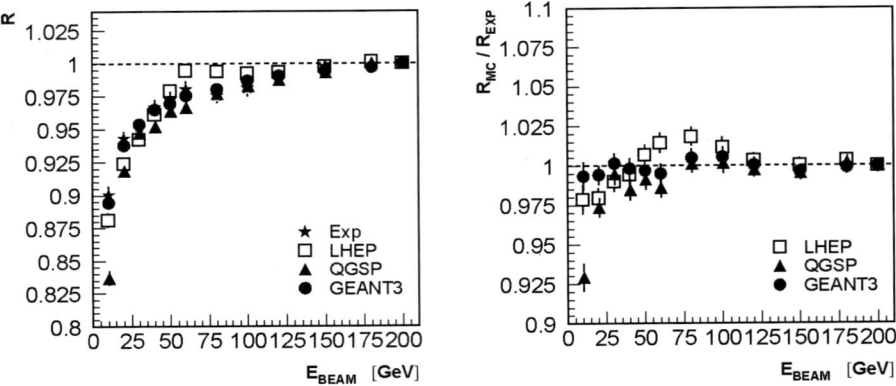

FIGURE 1. Relative response $R = (E_0/E_{BEAM})/(E_0/E_{BEAM})_{E_{BEAM}=200\ GeV}$ for pions (left-hand plot) and ratio between simulated and experimental data for the pion response (right-hand plot) as functions of the beam energy. The GEANT4 points correspond to version 8.0 with 20 μm range cut.

where A and B are the sampling and the constant terms, respectively.

In Figure 2, the parameters A and B are shown for different GEANT4 versions and hadronic physics lists. Both GEANT4 hadronic lists describe the pion energy resolution quite well (where QGSP is somewhat better). GEANT3 systematically predicts a too good energy resolution. No significant differences between GEANT4 versions are observed. In Figure 3, the terms of the energy resolution are presented for different range cuts. Within errors the values are constant.

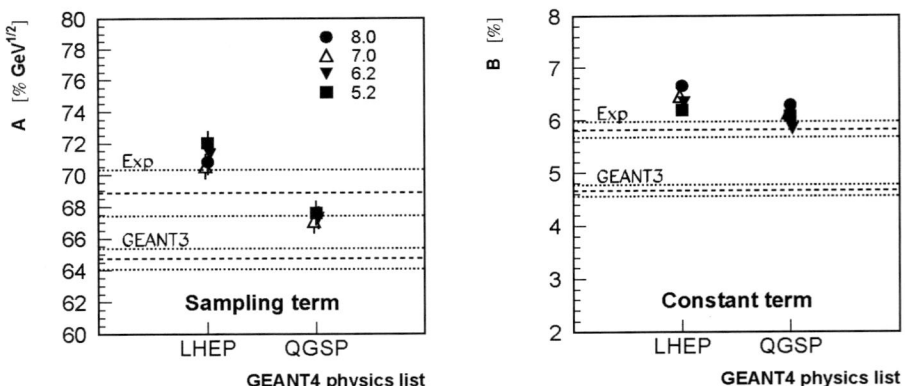

FIGURE 2. Sampling and constant terms of the pion energy resolution for different GEANT4 versions with 20 μm range cut. Lines mark the $\pm 1\sigma$ band of experimental values and of GEANT3 predictions.

Longitudinal shape of hadronic showers. Longitudinally the HEC calorimeter is segmented into four layers, which allows to study the longitudinal development of hadronic showers. An appropriate variable is the fraction of energy in a HEC layer with

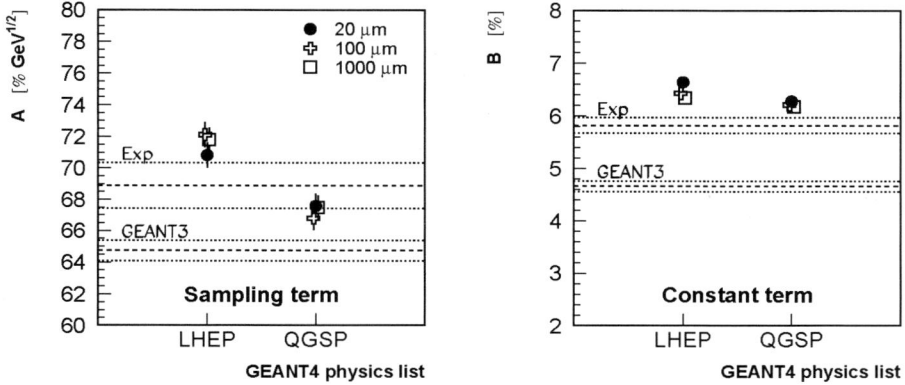

FIGURE 3. Sampling and constant terms of the pion energy resolution for GEANT4 version 8.0 with different range cuts. Lines mark the $\pm 1\sigma$ band of experimental values and of GEANT3 predictions.

respect to the total energy reconstructed in the pion cluster. It can be defined as an average energy in a layer divided by the sum of average energies in four layers. These energy fractions are presented in Figure 4 as functions of the beam energy. The ratio between simulation and data is shown as well (Figure 5).

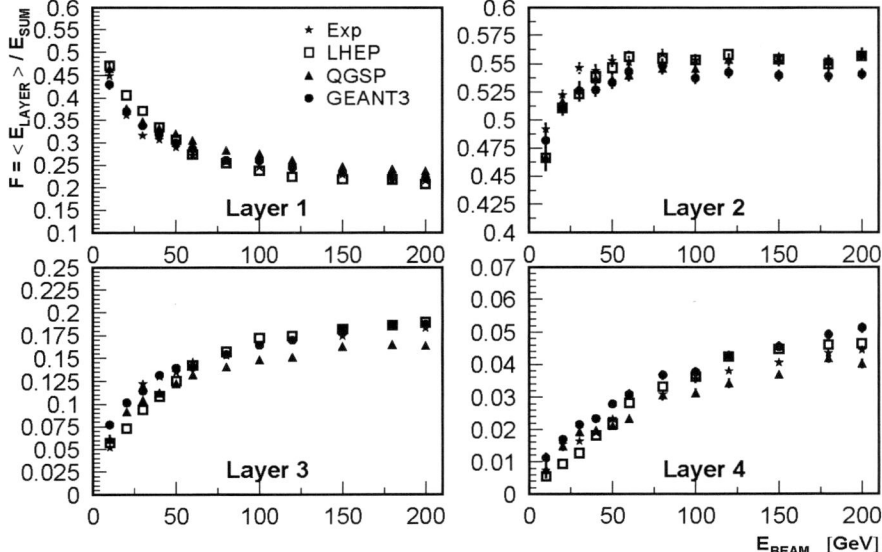

FIGURE 4. Fractions of the pion shower energy in the various longitudinal layers as a function of the beam energy. The GEANT4 points correspond to version 8.0 with 20 μm range cut.

Approximately half of the pion signal is deposited in the second layer, which is rather well described by all three simulation codes. On the other hand, for the energy fractions

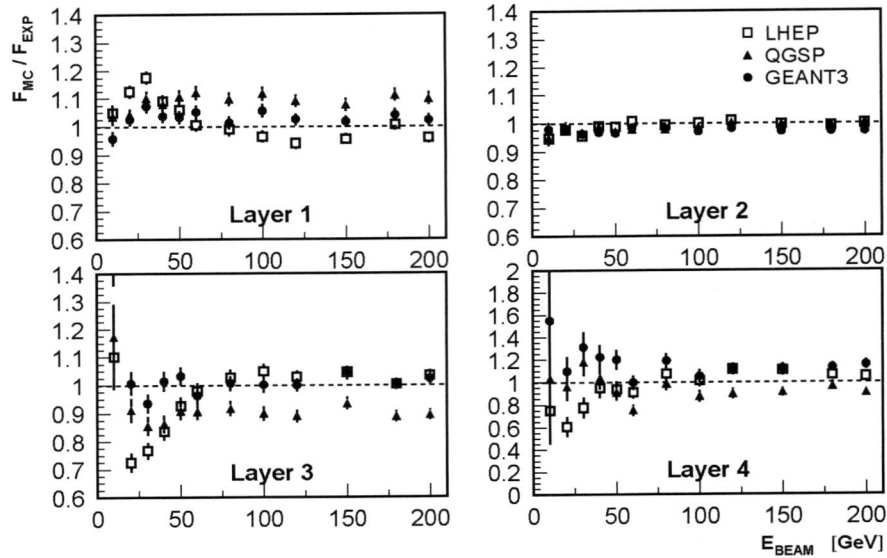

FIGURE 5. Ratio between simulated and experimental data for the pion energy fractions in the various longitudinal layers as a function of the beam energy. The GEANT4 points correspond to version 8.0 with 20 μm range cut.

in the first and last layers the agreement between experimental data and simulation is quite modest. QGSP predicts more energy in the front part of the calorimeter and less energy in the rear part than the data for almost all beam energies. This means that hadronic showers in this model start earlier and are more compact. LHEP shows different trends at different beam energies: at small energies it is rather close to QGSP, whereas at high energies it predicts a later start of the shower. In general, GEANT3 describes the longitudinal profiles of hadronic showers better than GEANT4.

Comparison of longitudinal profiles, predicted by different versions of GEANT4, shows that for the QGSP physics list a small improvement is observed for versions 7.0 and 8.0. Results for LHEP stay unchanged from version to version. No dependence of the longitudinal shape of hadronic showers on the range cut is seen.

Ratio e/π. The e/π ratio in the HEC is defined as the ratio of the average energies reconstructed in the electron and pion clusters. This ratio is presented in Figure 6 as a function of energy for the data and for the GEANT4 and GEANT3 simulations. The experimental values shown here result from averaging of e/π as obtained for different impact points and with different calibration methods; their errors (as RMS values) include the systematical uncertainties. The ratios of the simulations relative to the data are shown in Figure 6, too. The QGSP simulation describes the data very well. LHEP predicts larger values of e/π, while GEANT3 is systematically lower.

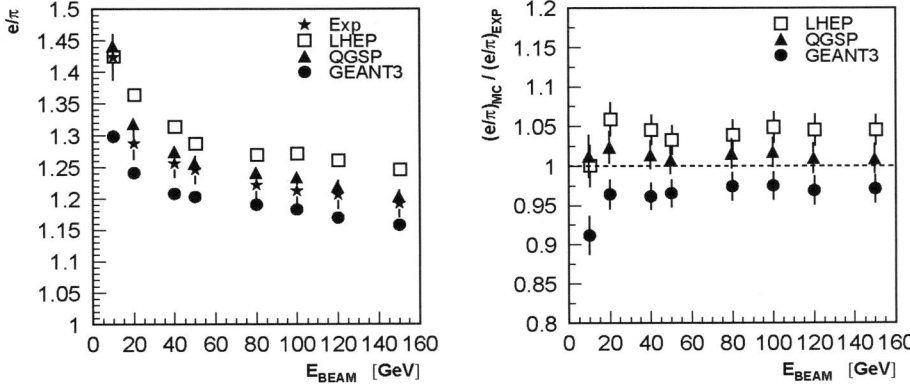

FIGURE 6. Ratio e/π (left-hand plot) and ratio between simulated and experimental data for e/π (right-hand plot) as functions of the beam energy. The GEANT4 points correspond to version 8.0 with 20 μm range cut.

Comparison of results, obtained with different GEANT4 versions (Figure 7) and with different range cuts (Figure 8), show no significant differences for the e/π ratio.

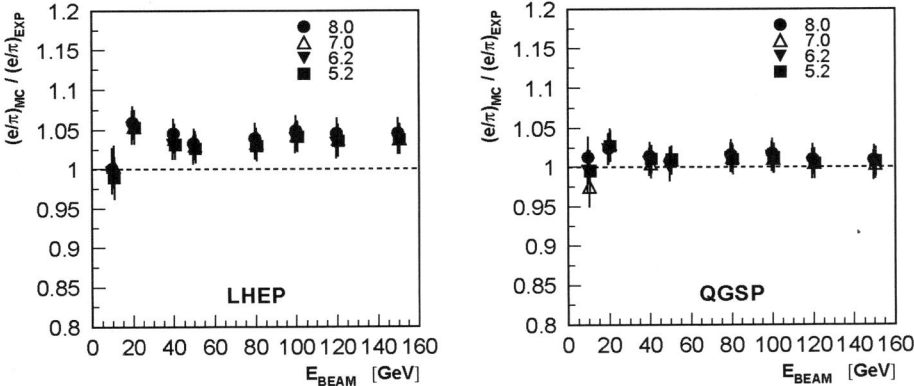

FIGURE 7. Ratio between simulated and experimental data for e/π as a function of the beam energy for different GEANT4 versions with 20 μm range cut.

COMPARISON WITH DATA OF COMBINED BEAM TESTS OF EMEC AND HEC MODULES

Combined beam tests of modules of the ATLAS end-cap calorimeters took place at CERN in 2002. Modules of electromagnetic (EMEC) and hadronic (HEC) calorimeters in the configuration of the final ATLAS set-up were exposed to beams.

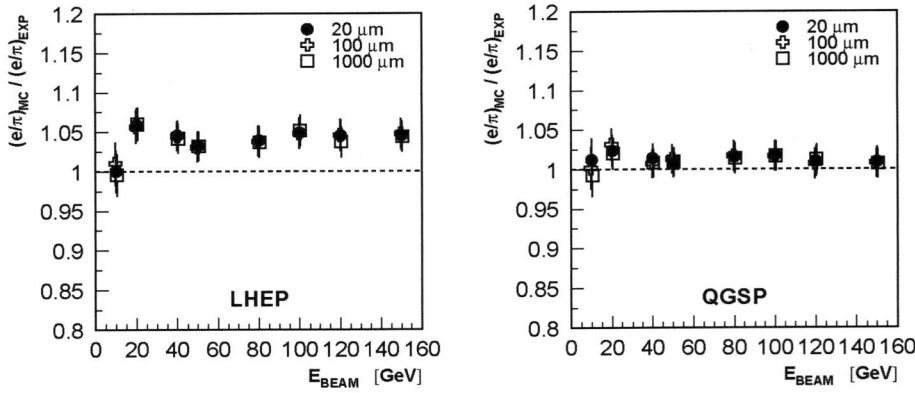

FIGURE 8. Ratio between simulated and experimental data for e/π as a function of the beam energy for GEANT4 version 8.0 with different range cuts.

Testbeam set-up

The EMEC [5] is a LAr detector with accordion-shaped kapton electrodes and lead absorbers. The set of calorimeter modules for the combined tests consisted of one full azimuthal wedge of EMEC (1/8 of a wheel), three full wedges of the front HEC (3/32 of a HEC1 wheel) and two wedges of the rear HEC (2/32 of a HEC2 wheel). The last modules were only half of the nominal depth. The limitation was due to the size of the cryostat available for beam tests. In the region exposed to beams, EMEC had three longitudinal layers: the first section of about 4 radiation lengths, the second one extends up to 22 X_0 (in total) and the last one adds a thickness 4-12 X_0. The HEC was also segmented into three longitudinal layers. The total thickness of the calorimeter set-up was about 8.8 absorption lengths.

Pion beams of different energies (from 10 to 200 GeV) were used. A more detailed description of the testbeam set-up, of available experimental data and results of the analysis can be found in [8].

Simulation and analysis

For simulations of the combined testbeam, version 5.0 (with patch-01) of GEANT4 was used. Again two hadronic physics lists: LHEP (version 3.3) and QGSP (version 2.3) were selected. The range threshold for production was set to 0.7 mm.

For the comparison of experimental and simulated data, signals at the same scale have to be used. In the simulation dimensionless ratios of the beam energy and the total visible energy of electrons have been obtained. For the HEC this EM-scale calibration parameter equals to 23.7 (23.3) for GEANT4 (GEANT3) based simulations (these values were obtained for the HEC stand-alone testbeam simulations). Simulations of electrons for the combined testbeam were used to get the EMEC scale factor: 12.05

(11.35) for GEANT4 (GEANT3). The noise has been simulated using real data. The procedure of the reconstruction and analysis of simulated data follows very closely the reconstruction and analysis of the experimental data [8].

Results

In Figure 9 the average fractions of energy in the EMEC and HEC (with respect to the total reconstructed energy) are presented as functions of the beam energy. The GEANT3 and QGSP models describe the energy sharing between the two calorimeters rather well.

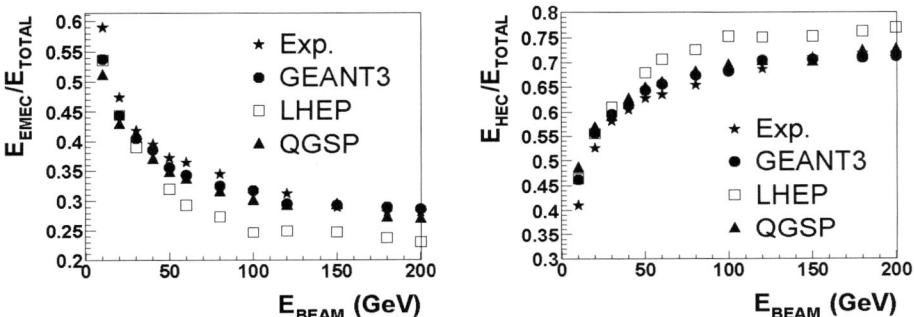

FIGURE 9. Average fraction of energy in the EMEC (left-hand plot) and in the HEC (right-hand plot) as a function of the beam energy for charged pions.

More detailed studies of the energy sharing between the six longitudinal layers however show that none of the MC models gives a sufficient good description of the longitudinal development of hadronic showers. For the total signal the LHEP model gives the best agreement with experimental data, as shown, for example, for 200 GeV pions in Figure 10.

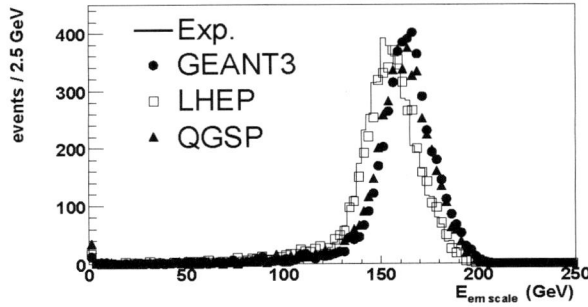

FIGURE 10. Total response to 200 GeV charged pions in the EMEC and HEC at EM-scale.

The work on the evaluation of GEANT4 physics models with the combined testbeam data of EMEC and HEC is going on.

CONCLUSIONS

A long-term program to evaluate hadronic physics models of GEANT4 has been established in ATLAS. It is based on the comparison of the simulation results with experimental data obtained during numerous beam tests of ATLAS calorimeters. New GEANT4 releases and physics lists are systematically validated as soon as they become available and reflect changes.

In general, it is observed that important hadronic signal characteristics like the e/π ratio and the energy dependence of the resolution and response for charged pions are described by GEANT4 reasonably well.

There are still some open questions, related to the description of shapes of hadronic showers, which require further improvement of the GEANT4 simulation code.

Experimental data, obtained during combined beam tests of ATLAS end-cap and forward calorimeters, are to be used soon for the evaluation of the GEANT4 hadronic physics models as well.

ACKNOWLEDGMENTS

We would like to thank all our colleagues from the ATLAS Liquid Argon Collaboration for their work on building and testing calorimeter modules and analysing testbeam data.

Especially we are grateful to S. Menke, H. Oberlack, P. Schacht and P. Strizenec for their help to prepare this talk, as well as to P. Loch for his valuable comments and suggestions.

We also want to thank the GEANT4 team, in particular H.-P. Wellisch, J. Apostolakis and G. Folger for the help and fruitful discussions.

REFERENCES

1. S. Agostinelli et al., *Nucl. Instr. and Meth.* A506, 250–303 (2003).
2. R. Brun et al., "GEANT3", CERN DD/EE/84-1 (1986).
3. J.P. Wellisch, http://geant4.web.cern.ch/geant4/physics_lists/
4. C. Zeitnitz and T.A. Gabriel, *Nucl. Instr. and Meth.* A349, 106–111 (1994).
5. The ATLAS Liquid Argon Collaboration, "Liquid Argon Calorimeter Technical Design Report", CERN/LHCC/96-41 (1996).
6. B. Dowler et al., *Nucl. Instr. and Meth.* A482, 94–124 (2002).
7. A.E. Kiryunin et al., *Nucl. Instr. and Meth.* A560, 278–290 (2006).
8. C. Cojocaru et al., *Nucl. Instr. and Meth.* A531, 481–514 (2004).

Particle Flow Calorimetry at the ILC

M. A. Thomson

Cavendish Laboratory, Univ. of Cambridge, JJ Thomson Av., Cambridge CB3 0HE, UK

Abstract. One of the most important requirements for a detector at the ILC is good jet energy resolution. It is widely believed that the particle flow approach to calorimetry is the key to achieving the goal of $0.3/\sqrt{E(\mathrm{GeV})}$. In contrast to the traditional approach to calorimetry, potentially the performance of particle flow calorimetry is sensitive to the detailed structure of hadronic showers. This paper describes the current performance of the PANDORAPFA particle flow algorithm. For 45 GeV jets in the Tesla TDR detector concept, the ILC jet energy resolution goal is reached. First detector optimisation studies are presented and the aspects of hadronic showers which are most likely to impact particle flow performance are discussed.

Keywords: calorimetry, particle flow
PACS: 07.05.Kf, 29.40.Vj, 29.85.+c

INTRODUCTION

Many of the interesting physics processes at the ILC will be characterised by multi-jet final states, often accompanied by charged leptons and/or missing transverse energy associated with neutrinos or the lightest super-symmetric particles (LSP). The reconstruction of the invariant masses of two or more jets, and identification of W and Z bosons, will provide a powerful tool for event reconstruction and background rejection. At LEP, kinematic fitting[1] enabled precise jet-jet invariant mass reconstruction almost independent of the intrinsic jet energy resolution of the detectors. At the ILC kinematic fitting will be less powerful due to missing particles (neutrinos/LSP) and beamstrahlung; invariant mass reconstruction will rely on the detector having excellent jet energy resolution. The ILC goal is to achieve a mass resolution for $W \rightarrow q'\overline{q}$ and $Z \rightarrow q\overline{q}$ decays which is comparable to their natural widths, *i.e.* ~ 2 GeV. A jet energy resolution of of the form $\sigma_E/E = \alpha/\sqrt{E(\mathrm{GeV})}$ leads to a di-jet mass resolution of roughly $\sigma_m/m = \alpha/\sqrt{E_{jj}(\mathrm{GeV})}$, where E_{jj} is the energy of the di-jet system. At the ILC operating in the centre-of-mass energy range $0.5 - 1.0$ TeV, typical di-jet energies will be in the range $150 - 350$ GeV, suggesting the goal of $\sigma_E/E = 0.3/\sqrt{E(\mathrm{GeV})}$. This is more than a factor two better than the best jet energy resolution achieved at LEP, $\sigma_E/E = 0.6(1+|\cos\theta|)/\sqrt{E(\mathrm{GeV})}$ [2]. Meeting the jet energy resolution goal is a major factor in the overall design of a detector for the ILC. It seems unlikely that this goal can be reached using traditional calorimetric techniques.

CP896, *Hadronic Shower Simulation Workshop*
edited by M. Albrow and R. Raja
© 2007 American Institute of Physics 978-0-7354-0401-4/07/$23.00

THE PARTICLE FLOW APPROACH TO CALORIMETRY

It is widely believed that the most promising strategy for achieving a jet energy resolution of $\sigma_E/E = 0.30/\sqrt{E(\text{GeV})}$ at the ILC is the particle flow analysis (PFA) approach to calorimetry. In contrast to a purely calorimetric measurement, particle flow requires the reconstruction of the four-vectors of all visible particles in an event. The reconstructed jet energy is the sum of the energies of the individual particles. The momenta of charged particles are measured in the tracking detectors, while the energy measurements for photons and neutral hadrons is performed with the calorimetric system.

Measurements of jet fragmentation at LEP have provided detailed information on the particle composition of jets (*e.g.* [3, 4]). On average, after the decay of short-lived particles, roughly 62% of the energy of jets is carried by charged particles (mainly hadrons), around 27% by photons, about 10% by long-lived neutral hadrons (*e.g.* n/K_L^0), and around 1.5% by neutrinos. Assuming calorimeter resolutions of $\sigma_E/E = 0.15/\sqrt{E(\text{GeV})}$ for photons and $\sigma_E/E = 0.55\sqrt{E(\text{GeV})}$ for hadrons, a jet energy resolution of $0.19/\sqrt{E(\text{GeV})}$ is obtained, with the contributions from tracks, photons and neutral hadrons shown in Tab. 1. In practice it is not possible to reach this level of performance for two main reasons. Firstly, particles travelling at small angles to the beam axis will not be detected. Secondly, and more importantly, it is not possible to perfectly associate all energy deposits with the correct particles. For example, if a photon is not resolved from a charged hadron shower, the photon energy is not counted (*missed energy*). Similarly, if some of the energy from a charged hadron is identified as a separate cluster (*a neutral fragment*) the energy is effectively double-counted. On an event-by-event basis fluctuations in the amount of missed energy and the energy of neutral fragments contributes to the overall jet energy resolution. This *confusion* is the limiting factor in determining particle flow performance. The crucial aspect of particle flow is therefore the ability to correctly assign calorimeter energy deposits to the correct reconstructed particles, requiring efficient separation of nearby showers. This places stringent requirements on the granularity of electromagnetic and hadron calorimeters. Consequently, particle flow performance is one of the main factors driving the overall ILC detector design.

TABLE 1. Contributions from the different particle components to the jet-energy resolution (all energies in GeV). The table lists the approximate fractions of charged particles, photons and neutral hadrons in a jet and the assumed single particle energy resolution.

Component	Detector	Energy Fraction	Energy Res.	Jet Energy Res.
Charged Particles ($X^{\pm}$)	Tracker	$\sim 0.6E_{\text{jet}}$	$10^{-4}E_{X^{\pm}}^2$	$< 3.6 \times 10^{-5}E_{\text{jet}}^2$
Photons (γ)	ECAL	$\sim 0.3E_{\text{jet}}$	$0.15\sqrt{E_{\gamma}}$	$0.08\sqrt{E_{\text{jet}}}$
Neutral Hadrons (h^0)	HCAL	$\sim 0.1E_{\text{jet}}$	$0.55\sqrt{E_{h^0}}$	$0.17\sqrt{E_{\text{jet}}}$

THE ILC DETECTOR CONCEPTS

The work on detectors for the ILC is currently concentrated in four detector design groups. Three[1] of these, **GLD**[5], **LDC**[6] and **SiD**[7], are optimised for particle flow calorimetry. The high granularity electromagnetic and hadronic calorimeters required for particle flow are significant cost drivers for the overall detector design. The detector concepts share a number of features. All consist of a vertex detector, a large central tracking volume (either a time projection chamber (TPC) or a silicon tracker) and sampling electromagnetic (ECAL) and hadronic (HCAL) calorimeters. The calorimeters are located inside the solenoid and have high transverse and longitudinal segmentation. The electromagnetic calorimeters (ECAL) have transverse segmentation of between $0.5 \times 0.5 \, cm^2$ and $2.0 \times 2.0 \, cm^2$, and the hadronic calorimeters (HCAL) have transverse segmentation of between $1.0 \times 1.0 \, cm^2$ and $5.0 \times 5.0 \, cm^2$. The main parameters of the different detector concepts are listed in Tab. 2.

TABLE 2. Main features of the GLD, LDC and SiD detector concepts. The table lists the tracker technology, the approximate outer radius of the tracker, the passive and active media of the sampling calorimeters and the magnetic field.

Concept	Tracker	Tracker Radius	ECAL	HCAL	B-Field
GLD	TPC	$\sim 2.0 \, m$	Pb-Scint	Pb-Scint	3 T
LDC	TPC	$\sim 1.6 \, m$	Si-W	Steel-Scint/RPC	4 T
SiD	Silicon	$\sim 1.3 \, m$	Si-W	Steel-RPC	5 T

It should be noted that the jet energy resolution obtained for a particular detector concept is the combination of the intrinsic detector performance and the performance of the PFA software used to reconstruct the energy deposits in the calorimeter. In addition the Monte Carlo (MC) studies used to optimise the detectors, the choice of hadron shower model will also impact the calorimetric performance obtained. Results from any detector optimisation studies should be considered in the light of the the potential impact of imperfect PFA software and hadron shower simulation.

THE PANDORAPFA PARTICLE FLOW ALGORITHM

PANDORAPFA[10] is a C++ implementation of a PFA algorithm running in the MARLIN[11, 12] framework. It was designed to be sufficiently generic for ILC detector optimisation studies and was developed and optimised using events generated with the MOKKA[13] program, which provides a GEANT4[14] simulation of the Tesla TDR[15] detector concept (the predecessor if the LDC concept). The PANDORAPFA algorithm performs both calorimeter clustering and particle flow in a single stage. The algorithm has six main stages:

i) Tracking: for the studies presented in this paper, the track pattern recognition is performed using Monte Carlo information[11]. The track parameters are then extracted using a fit to a helix. The projections of tracks onto the front face of the ECAL are calculated using helical fits (which do not take into account energy loss along the track).

---

[1] The fourth design study[8] adopts the DREAM[9] approach to calorimetry.

Neutral particle decays resulting in two charged particle tracks (V^0s) are identified by searching for pairs of tracks which do not originate from the interaction point and that are consistent with coming from a single point in space. Kinked tracks from charged particle decays to a single charged particle and a number of neutrals are also identified. When a kink is identified the parent track is usually removed for the purposes of forming the reconstructed particles.

ii) Calorimeter Hit Selection and Ordering: isolated hits, defined on the basis of proximity to other hits, are removed from the initial clustering stage. This has the effect of removing energy deposits from soft neutrons. The remaining hits are ordered into *pseudo-layers* which follow the detector geometry so that particles propagating outward from the interaction region will cross successive pseudo-layers. In most of the calorimeter the pseudo-layers follow the physical layers of the calorimeters except in the barrel-endcap overlap region and where the ECAL stave structure[15] results in low numbered layers which are far from the front face of the calorimeter. The assignment of hits to pseudo-layers removes the dependence of the algorithm on the explicit detector geometry whilst following the actual geometry as closely as possible. Within each pseudo-layer hits are ordered by decreasing energy.

iii) Clustering: the main clustering algorithm is a forward projective method working from innermost to outermost pseudo-layer. In this manner hits are added to clusters or are used to seed new clusters. Throughout the clustering algorithm clusters are assigned a direction (or directions) in which they are growing. The algorithm starts by *seeding* clusters using the projections of reconstructed tracks onto the front face of the calorimeter. The initial direction of a track-seeded cluster is obtained from the track direction. The hits in each subsequent pseudo-layer are then looped over. Each hit, i, is compared to each clustered hit, j, in the previous layer. The vector displacement, $\mathbf{r}_{ij}$, is calculated and is used to calculate the parallel and perpendicular displacement of the hit with respect to the unit vector(s) $\hat{\mathbf{u}}$ describing the cluster propagation direction(s), $d_{\parallel} = \mathbf{r}_{ij} \cdot \hat{\mathbf{u}}$ and $d_{\perp} = |\mathbf{r}_{ij} \times \hat{\mathbf{u}}|$. Associations are made using a cone-cut, $d_{\perp} < d_{\parallel} \tan\alpha + \beta D_{\text{pad}}$, where α is the cone half-angle, D_{pad} is the size of a sensor pixel in the layer being considered, and β is the number of pixels added to the cone radius. Different values of α and β are used for the ECAL and HCAL with the default values set to $\{\tan\alpha_{\text{E}} = 0.3, \beta_{\text{E}} = 1.5\}$, and $\{\tan\alpha_{\text{H}} = 0.5, \beta_{\text{H}} = 2.5\}$ respectively. Associations may be made with hits in the previous 3 layers. If no association is made, the hit is used to seed a new cluster. This procedure is repeated sequentially for the hits in each pseudo-layer (working outward from ECAL front-face).

iv) Topological Cluster Merging: by design the initial clustering errs on the side of splitting up true clusters rather than clustering energy deposits from more than one particle. The next stage of the algorithm is to merge clusters from tracks and hadronic showers which show clear topological signatures of being associated. A number of track-like and shower-like topologies are searched for including looping minimum ionising tracks, back-scattered tracks and showers associated with a hadronic interaction. Before clusters are merged, a simple cut-based photon identification procedure is applied. The cluster merging algorithms are only applied to clusters which have not been identified as photons.

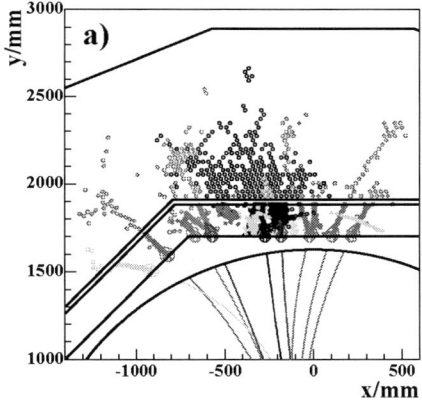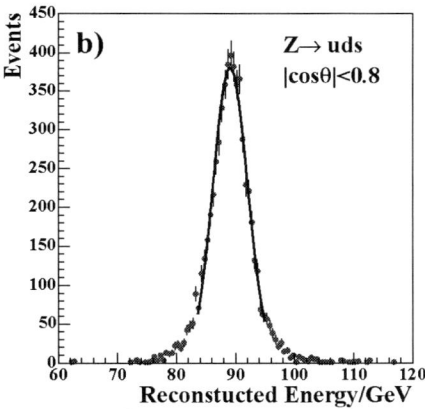

FIGURE 1. a) PANDORAPFA reconstruction of a 100 GeV jet in the MOKKA simulation of the Tesla TDR detector. b) The total reconstructed energy from reconstructed PFOs in Z → uds events for initial quark directions within the polar angle acceptance $|\cos\theta_{q\bar{q}}| < 0.8$. The solid line shows a Gaussian fit to the peak region with a standard deviation of 2.9 GeV.

v) Statistical Re-clustering: The previous four stages of the algorithm were found to perform well for 50 GeV jets. However, at higher energies the performance degrades rapidly due to the increasing overlap between hadronic showers from different particles. To address this, temporary associations of tracks with reconstructed calorimeter clusters are made. If the track momentum is incompatible with the energy of the associated cluster re-clustering is performed. If $E_{CAL} - E_{TRACK} > 3.5\sigma_E$, where σ_E is the energy resolution of the cluster, the clustering algorithm, described in *iii)* and *iv)* above, is reapplied to the hits in that cluster. This is repeated, using successively smaller values of the αs and βs in the clustering finding algorithm until the cluster splits to give an acceptable track-cluster energy match. Similarly, if $E_{TRACK} - E_{CAL} > 3.5\sigma_E$ the algorithm attempts to merge additional clusters with the cluster associated with the track. In doing so high energy clusters may be split as above.

vi) Formation of Particle Flow Objects: The final stage of the algorithm is to create Particle Flow Objects (PFOs) from the results of the clustering. Tracks are matched to clusters on the basis of the distance closest approach of the track projection into the first 10 layers of the calorimeter. If a hit is found within 50 mm of the track extrapolation an association is made. The reconstructed PFOs are written out in LCIO[11] format.

PERFORMANCE

Fig. 1a) shows an example of a PANDORAPFA reconstruction of a 100 GeV jet from a Z → uū decay at $\sqrt{s} = 200$ GeV using the Tesla TDR detector model. The ability to track particles in the high granularity Tesla TDR calorimeter can be seen clearly. Fig. 1b) shows the total PFA reconstructed energy for Z → uds events with $|\cos\theta_{q\bar{q}}| < 0.8$,

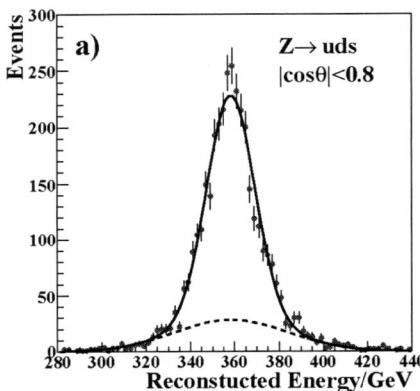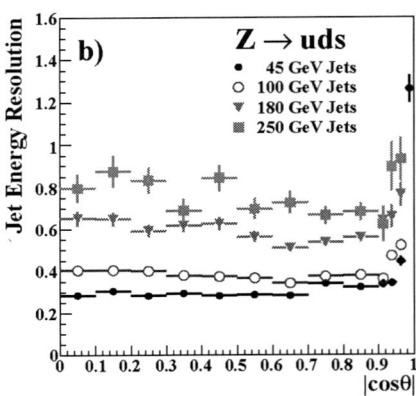

FIGURE 2. a) The total reconstructed energy from reconstructed PFOs in Z → uds at $\sqrt{s} = 360\,\mathrm{GeV}$ for initial quark directions within the polar angle acceptance $|\cos\theta| < 0.8$. The solid line shows a results of the fit to two Gaussians and the dashed line indicates the contribution from the broader Gaussian which is constrained to contain 25 % of the events. b) The jet energy resolution, defined as the α in $\sigma_E/E = \alpha\sqrt{E(\mathrm{GeV})}$, plotted versus $\cos\theta_{q\bar{q}}$ for four different values of $\sqrt{s}$.

where $\theta_{q\bar{q}}$ is the polar angle of the generated $q\bar{q}$ system. These events were generated at $\sqrt{s} = 91.2\,\mathrm{GeV}$ using the Tesla TDR detector model. The root-mean-square deviation from the mean (rms) of the distribution is 4.0 GeV. However, quoting the rms as a measure of the performance over-emphasises the importance of the tails. For example, in this figure, the central peak is well described by a Gaussian of width 2.9 GeV, equivalent to a resolution of $\sigma_E/E = 0.31/\sqrt{E(\mathrm{GeV})}$. In this paper two measures of the performance are quoted. The first measure, rms_{90}, is the rms in the smallest range of reconstructed energy which contains 90 % of the events. The second performance measure is obtained from a fit to the reconstructed energy distribution. The fit function is the sum of two Gaussian distributions with a common mean but different widths. The width of the narrower Gaussian, which is constrained to contain 75 % of the events, gives a measure of the resolution in the peak, σ_{75}. For the data shown in Fig. 1b) both methods give a resolution of $\sigma_E/E = 0.3/\sqrt{E(\mathrm{GeV})}$; the ILC goal. However, this is of little consequence to ILC physics where, in general, the jets will be higher in energy.

The majority of interesting ILC physics will consist of final states with at least six fermions, setting a "typical" energy scale for ILC jets as approximately 85 GeV and 170 GeV at $\sqrt{s} = 500\,\mathrm{GeV}$ and $\sqrt{s} = 1\,\mathrm{TeV}$ respectively. Fig. 2a shows the reconstructed total energy in Z →uds events (generated without ISR or beamstrahlung effects) at $\sqrt{s} = 360\,\mathrm{GeV}$. The fit to the sum of a double Gaussian gives $\sigma_{75} = 10.8\,\mathrm{GeV}$, equivalent to a resolution of $\sigma_E/E = 0.57/\sqrt{E(\mathrm{GeV})}$, significantly worse than that obtained for lower energy jets. Fig. 2b shows the jet energy resolution for Z →uds events plotted against $|\cos\theta_{q\bar{q}}|$ for four different values of $\sqrt{s}$.

TABLE 3. Jet energy resolution, expressed as both rms$_{90}$ and σ_{75}, for $Z \to$ uds events with $|\cos\theta_{q\bar{q}}| < 0.8$.

Jet Energy	rms$_{90}$	rms$_{90}/\sqrt{E(\text{GeV})}$	σ_{75}	$\sigma_{75}/\sqrt{E(\text{GeV})}$
45 GeV	2.8 GeV	0.30	2.8 GeV	0.30
100 GeV	5.3 GeV	0.38	5.2 GeV	0.37
180 GeV	11.0 GeV	0.58	10.8 GeV	0.57
250 GeV	16.8 GeV	0.76	16.8 GeV	0.75

The results described above are summarised in Tab. 3. The observed jet energy resolution for simulated events is not described by the expression $\sigma_E/E = \alpha/\sqrt{E(\text{GeV})}$. This is not surprising, as the particle density increases it becomes harder to correctly associate the calorimetric energy deposits to the particles and the confusion term increases. Empirically it is found that the total energy resolutions in Tab. 3 can be described by a jet energy resolution of $\sigma_E/E = 0.265/\sqrt{E(\text{GeV})} + 1.2 \times 10^{-4}E(\text{GeV})$, where E is the energy of the jet. This expression represents the current performance of the PANDORAPFA algorithm and should not be be considered as anything more fundamental. It should be noted that in the current MOKKA simulation of the Tesla TDR detector the muon chambers are not included. In principle these can be used as a "tail-catcher" to improve the energy measurement for high energy hadronic showers which may not be fully contained in the HCAL. In the current version of PANDORAPFA no attempt is made to correct for this energy leakage. It is noticeable in Fig. 2b that the energy resolution improves with increasing polar angle in the barrel region of the detector. This is due to increasing shower containment which becomes important for jets of energy above 100 GeV.

ILC DETECTOR OPTIMISATION STUDIES

The ultimate goal of the ILC detector design groups is to produce a detector design based on cost-performance analysis. The main performance requirements for an ILC detector are summarised below:

- *Momentum resolution:* $\sigma_{1/p} \sim 5 \times 10^{-5}\,\text{GeV}^{-1}$ (a factor of ten better than that achieved at LEP). Good momentum resolution is important for the reconstruction of the leptonic decays of Z bosons.
- *Impact parameter resolution:* Efficient b and c quark tagging which implies good impact parameter (d_0) resolution:

$$\sigma_{d_0}^2 < (5.0\,\mu\text{m})^2 + \left(\frac{5.0\,\mu\text{m}}{p(\text{GeV})\sin^{\frac{3}{2}}\theta}\right)^2.$$

This is a factor of three better resolution than obtained at SLD.
- *Jet energy resolution:* $\sigma_E/E \sim 0.3/\sqrt{E(\text{GeV})}$.

The vertex detector is essentially a standalone system whose design has minimal impact on the design of the rest of the detector. Of the above requirements, it is the jet energy

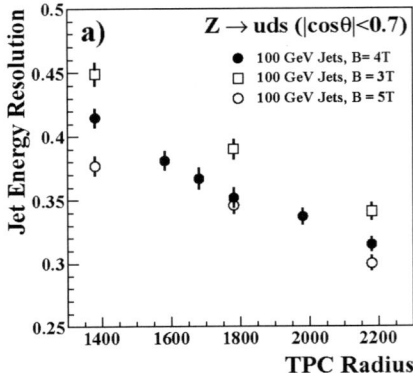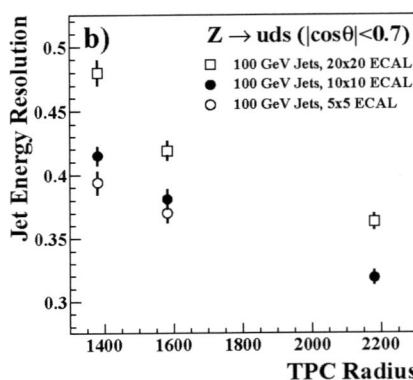

FIGURE 3. a) The jet energy resolution obtained with PANDORAPFA and the Tesla TDR detector model plotted as a function of TPC outer radius and magnetic field. b) The jet energy resolution obtained with PANDORAPFA and the Tesla TDR detector model plotted as a function of TPC outer radius and ECAL transverse segmentation (mm^2) for a magnetic field of 4 T. For both plots jet energy resolution is defined as the α assuming the expression $\sigma_E/E = \alpha\sqrt{E(\text{GeV})}$.

resolution goal that has the largest single impact on the overall detector design parameters such as the size of the TPC outer radius detector and the magnetic field. In turn, both the size and the magnetic field are significant driving factors for the cost of the detector. Optimising an ILC detector from the point of view of particle flow performance is therefore of vital importance.

A number of optimisation studies have recently been performed with the PANDORAPFA particle flow algorithm and the Tesla TDR detector model. For example, Fig. 3a shows how the jet energy resolution depends on the TPC radius and magnetic field. As expected, the resolution improves with increasing radius and increasing magnetic field (both of which increase the mean transverse separation of particles at the front face of the ECAL). For 100 GeV jets it is found that the dependence of jet energy resolution is approximately described by

$$\frac{\sigma_E}{E} \propto B^{-\frac{1}{4}} R^{-\frac{3}{5}}.$$

This relation represents the dependence on the jet energy resolution obtained with the current version of the PANDORAPFA algorithm. Fig. 3b shows how the jet energy resolution depends on the transverse segmentation of the ECAL for a number of different TPC outer radii. As expected, higher granularity gives better resolution and it is apparent that a transverse segmentation of 20×20 mm^2 is insufficient. The improvement in going from 10×10 mm^2 segmentation to 5×5 mm^2 is not particularly large because for 100 GeV jets the confusion of clusters in the ECAL does not contribute significantly to the overall jet energy resolution in either case.

DISCUSSION

The above results give a flavour of the simulation studies necessary for the design and optimisation for an ILC detector. However, two caveats apply: i) the results are combination of the intrinsic particle flow performance of the detector model being considered and the performance of the particle flow software; ii) the particle flow performance obtained from Monte Carlo simulation will depend on the modelling of hadronic showers. At this stage no strong conclusions should be drawn from the results.

An interesting question is which aspects of hadron shower simulation are most likely to impact particle flow performance. At this stage it is not possible to provide a definitive answer. However, it is possible to identify the limitations of the PANDORAPFA algorithm by isolating the three contributions to the resolution arising from confusion:

1. "lost photons": energy deposits from photons which are merged into charged particle clusters;
2. "lost neutral hadrons": energy deposits from neutral hadrons which are merged into charged particle clusters;
3. "shower fragments": isolated energy deposits from hadronic showers of charged particles which are reconstructed as separate particles.

The relative importance of each contribution has been investigated by associating calorimeter energy deposits to generated Monte Carlo particles and, on an event-by-event basis, determining the amount of calorimetric energy in the above three classes. For 100 GeV jets with the Tesla TDR detector model the contribution from "shower fragments" is by far the largest contribution to the jet energy resolution; these fragment clusters mostly have low reconstructed energy, typically less than 2 GeV. The performance of the current PANDORAPFA algorithm is limited by isolated clusters produced in hadronic showers and, consequently, the resolution obtained depends on the number of such clusters produced in the simulation. It is not only the gross features of hadronic showers (transverse and longitudinal profiles) that are relevant but also the detailed substructure of the showers.

CONCLUSIONS

Particle flow calorimetry is widely believed to be the key to reaching the ILC jet energy resolution goal of $\sigma_E/E = 0.3/\sqrt{E(\text{GeV})}$. Consequently, the design and optimisation of detectors for the ILC depends both on hardware and on sophisticated software reconstruction. For the Tesla TDR detector concept, the PANDORAPFA particle flow algorithm achieves good performance, $< 0.4/\sqrt{E(\text{GeV})}$, for jet energies upto about 100 GeV. For higher energies the performance degrades significantly reaching the equivalent of $0.6/\sqrt{E(\text{GeV})}$ for 200 GeV jets. With further optimisation of the algorithm the performance is expected to improve. First detector optimisation studies are presented. However, to use the results of such studies to optimise an ILC detector design requires an accurate simulation of the sub-structure of hadronic showers. At this stage it is far from clear whether existing hadron shower simulations are adequate for this purpose.

Future hadronic production and test beam data will provide a valuable tool to validating and improving the current simulation code. Hopefully this workshop will have stimulated further work into validating and improving the current hadron shower simulation codes.

REFERENCES

1. M. A. Thomson, Proc. of EPS-HEP 2003, Aachen. Topical Vol. of Eur. Phys. J. C Direct (2004).
2. ALEPH Collaboration, D. Buskulic et al., Nucl. Inst. Meth. **A360** (1995) 481.
3. I.G. Knowles and G.D. Lafferty, J. Phys. **G23** (1997) 731.
4. M. G. Green, S. L. Lloyd, P. N. Ratoff and D. R. Ward, "Electron-Positron Physics at the Z", IoP Publishing (1998).
5. GLD, A Large Detector Concept Study for the ILC, `http://ilcphys.kek.jp/gld/`; "The GLD Detector Outline Document", K. Abe et al., arXiv.org:physics/0607154.
6. LDC, The Large Detector Concept for the ILC, `http://www.ilcldc.org/`.
7. SiD, The Silicon Detector Design Study, `http://www-sid.slac.stanford.edu/`.
8. The 4th Concept, `http://www.4thconcept.org/`.
9. "First results of the DREAM project", R. Wigmans, Proc. of 11th International Conference on Calorimetry in High-Energy Physics (Calor 2004), Perugia, Italy, 28 Mar - 2 Apr 2004. Published in "Perugia 2004, Calorimetry in particle physics 241-257".
10. http://www.hep.phy.cam.ac.uk/ thomson/pandoraPFA/.
11. http://www-flc.desy.de/ilcsoft/ilcsoftware/.
12. O. Wendt, "Marlin and MarlinReco", to appear in Proc. of LCWS06, Bangalore, March 2006.
13. http://polywww.in2p3.fr/activites/physique/geant4/tesla/www/mokka/.
14. GEANT4 collaboration, S. Agostinelli et al., Nucl. Instr. and Meth. **A506** (2003) 3; GEANT4 collaboration, J. Allison et al., IEEE Trans. Nucl. Sci. 53 (2006) 1.
15. TESLA Technical Design Report, DESY 2001-011, ECFA 2001-2009 (2001).

Possibilities of new data in hadroproduction[1]

Rajendran Raja

Fermilab, P.O. Box 500, Batavia, Illinois 60510, U.S.A

Abstract. We describe experiments that can produce data that is of relevance to hadronic shower simulation. In particular we describe the possibilities to dramatically improve the quality of the data available for improving hadronic shower simulators made possible with the upgraded Main Injector Particle production Experiment (MIPP). We describe the status of MIPP which has to date acquired 18 million events of particle interactions using (5 GeV/c-120 GeV/c) $\pi^{\pm}, K^{\pm}$ and $p^{\pm}$ beams on various targets and the plans to upgrade the data acquisition speed of MIPP to make it run 100 times faster.

Keywords: Particle production, Calorimetry
PACS: 13.75.-n; 13.85.-t

INTRODUCTION

We describe experiments that will produce data in the near future that is of relevance to hadronic shower simulators. The HARP experiment [1] has acquired data using pion and proton beams in the momentum range (3-15 GeV/c) over a wide variety of targets ranging from beryllium to lead. In addition they also had cryogenic targets of hydrogen, deuterium, oxygen and nitrogen. They have recently published their results on pion production on p-Al collision [2]at 12.9 GeV/c. More publications from the large amount of data collected (420 million events) by HARP will add significantly to our knowledge of hadroproduction in their momentum range. In addition, the NA49 experiment [3] has proposed an upgrade, which if approved, will provide data on proton and pion interactions on Carbon in the momentum range 30-158 GeV/c. In what follows, we describe the MIPP experiment [4], and the data taken so far and the proposal to upgrade the experiment to make it acquire data 100 times faster.

CURRENT STATUS OF THE MIPP EXPERIMENT

We give a status report on the MIPP experiment and its performance to date. MIPP is situated in the Meson Center beamline at Fermilab. It received approval [5] in November 2001 and has installed and operated both the experiment and a newly designed secondary beamline in the interim. It received its first beams in March 2004, had an engineering run to commission the detector in 2004 and had its physics data-taking run in the period January 2005-March 2006. The experiment is currently busy analyzing its data.

---

[1] Work supported by Fermi Research Alliance, LLC, under contract No. DE-AC02-07CH11359 with the U.S. Department of Energy.

CP896, *Hadronic Shower Simulation Workshop*
edited by M. Albrow and R. Raja

MIPP is designed primarily as an experiment to measure and study in detail the dynamics associated with non-perturbative strong interactions. It has nearly 100% acceptance for charged particles and excellent momentum resolution. Using particle identification techniques that encompass dE/dx, time-of-flight [6], Multi-Cell Čerenkov [7] and a Ring Imaging Čerenkov (RICH)detector [8], MIPP is designed to identify charged particles at the 3σ or better level in nearly all of its final state phase space. MIPP has acquired data of high quality and statistics for beam momenta ranging from 5 GeV/c to 90 GeV/c for 6 beam species ($\pi^{\pm}, K^{\pm}$ and $p^{\pm}$) on a variety of targets as shown in Figure 1.

Data Summary 27 February 2006			Acquired Data by Target and Beam Energy Number of events, x 10^6									
Target			E									Total
Z	Element	Trigger Mix	5	20	35	40	55	60	65	85	120	
0	Empty	Normal		0.10	0.14			0.52			0.25	1.01
	K Mass	No Int.				5.48	0.50	7.39	0.96			14.33
	Empty LH	Normal		0.30				0.61		0.31		7.08
1	LH	Normal	0.21	1.94				1.98		1.73		
4	Be	p only									1.08	1.75
		Normal			0.10			0.56				
6	C	Mixed						0.21				1.33
	C 2%	Mixed		0.39				0.26			0.47	
	NuMI	p only									1.78	1.78
13	Al	Normal			0.10							0.10
83	Bi	p only									1.05	2.83
		Normal			0.52			1.26				
92	U	Normal						1.18				1.18
Total			0.21	2.73	0.86	5.48	0.50	13.97	0.96	2.04	4.63	31.38

FIGURE 1. The data taken during the first MIPP run as a function of nucleus. The numbers are in millions of events. During the last month of the run, the Jolly Green Giant magnet coils developed shorts. This time was used to acquire data without the TPC for exploring the feasibility of measuring the charged kaon mass using the RICH radii.

An important aspect of MIPP data-taking was the measurement of particle production off the NuMI [9] target in order to minimize the systematics in the near/far detector ratio in the MINOS [9] experiment. MIPP also made measurements with proton beams off various nuclei for the needs of proton radiography [5].

Another physics motivation behind MIPP is to restart the study of non-perturbative QCD interactions. Currently available data are of poor quality, and sparsely populate the beam momentum, p_T, and atomic weight phase space that makes comparisons between different experiments difficult. The MIPP TPC [10] digitizes the charged tracks in three dimensions, obviating the need for track matching across stereo views. Coupled with the particle identification capability of MIPP, the data from MIPP would add significantly to our knowledge base of non-perturbative QCD. This would help test inclusive scaling relations and also scaling nuclear reactions.

Experimental Setup

We designed a secondary beam [11] specific to our needs. The resonantly extracted protons from the Fermilab Main Injector are transported down the Meson Center line. They impinge on a 20 cm long copper target producing secondary beam particles. This target is imaged onto an adjustable momentum selection collimator which controls the momentum spread of the beam. This collimator is re-imaged on to our interaction target placed next to the TPC. The beam is tracked using three beam chambers and identified using two differential Čerenkovs [12] filled with gas, the composition and the pressure of which can be varied within limits as needed depending on the beam momentum and charge.

Figure 2 shows the layout of the apparatus. The TPC sits in a wide aperture magnet (the Jolly Green Giant) which has a peak field of 0.7 tesla. Downstream of the TPC are a 96 mirror multi-cell Čerenkov detector filled with C_4F_{10} gas, and a time of flight system. This is followed by a large aperture magnet (Rosie) which runs in opposite polarity (at -0.6 tesla) to the Jolly Green Giant to bend the particles back into the Ring Imaging Čerenkov counter. The RICH has CO_2 as the radiator and an array of phototubes of 32 rows and 89 columns [13]. Downstream of the RICH we have an electromagnetic calorimeter [14]and a hadron calorimeter [15] to measure forward-going photons and neutrons. The electromagnetic calorimeter provides a means of distinguishing forward neutrons from photons and will also serve as a device to measure the electron content of our beam at lower energies, which will be useful for measuring cross sections.

MIPP uses dE/dx in the TPC to separate pions, kaons and protons for momenta less than ≈ 1 GeV/c and the time of flight array of counters to do the particle identification for momenta less than 2 GeV/c. The multi-cell Čerenkov detector [7] contributes to particle identification in the momentum range ≈ 2.5 GeV/c-14 GeV/c and the RICH [8] for momenta higher than this. By combining information from all counters, we get the expected particle identification separation for K/p and π/K as shown in Figure 3. It can be seen that excellent separation at the 3σ or higher level exists for both K/p and π/p over almost all of phase space. Tracking of the beam particles and secondary beam particles is accomplished by a set of drift chambers [16] and proportional chambers [17] each of which have 4 stereo layers.

Some results from Acquired data

Figure 4 shows the pictures of reconstructed tracks in the TPC obtained during the data-taking run. The tracks are digitized and fitted as helices in three dimensions. Extrapolating three dimensional tracks to the other chambers makes the pattern recognition particularly easy.

MIPP

Main Injector Particle Production Experiment (FNAL-E907)

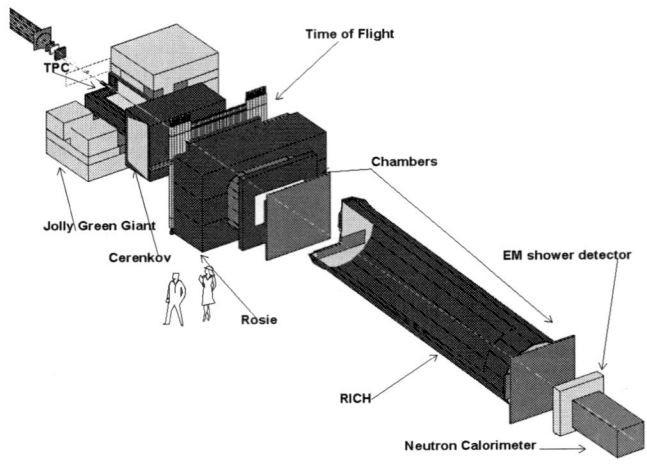

FIGURE 2. The experimental setup. The picture is a rendition in Geant3, which is used to simulate the detector.

Figure 5 shows the distribution of dE/dx of tracks measured in the TPC as a function of the track momentum in a preliminary analysis of p-Carbon data. The TPC is capable of separating pions, protons and kaons in the momentum range below ≈ 1 GeV/c. Figure 6 shows events with multiple rings in the RICH counter. Figure 7 shows the histogram of ring radii for a +40 GeV secondary beam. There is clean separation between pions, kaons and protons and their relative abundances [18] match expectations. Applying the particle identification trigger from the beam Čerenkovs enables us to separate the three particle species cleanly. The kaons which form 4% of the beam are cleanly picked out by the beam Čerenkov with very simple selection criteria. These can be made much more stringent with offline cuts to produce a very clean kaon beam.

The ring radius of the particle contains information on the mass of the particle. The pion and proton masses are very well known. The charged kaon mass, however, currently has measurement uncertainties of the order of 60 keV. Improving the precision of both charged kaon masses will pay dividends in rare K decay experiments involving charged kaons where the matrix elements depend on the kaon mass raised to large powers. Towards the end of our physics run, when the Jolly Green Giant magnet coils failed, we switched off the TPC and acquired data at the rate of 300 Hz to investigate how well

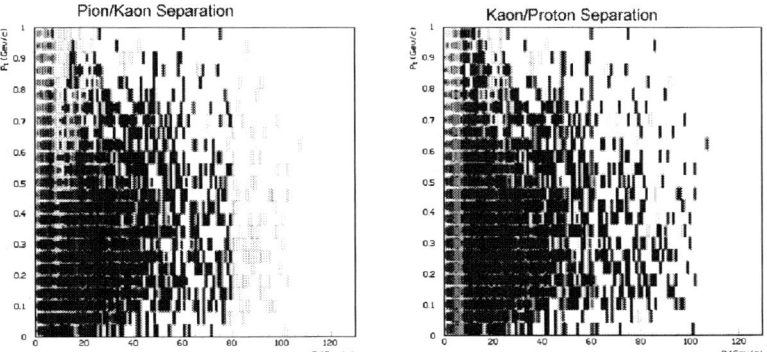

FIGURE 3. Particle identification plots for pion/kaon separation and for kaon/proton separation as a function of the longitudinal and transverse momentum of the outgoing final state particle. Black indicates separation at the 3σ level or better and grey indicates separation at the $1 - 3\sigma$ level. The boxes at largest values of the longitudinal momenta suffer from lack of kaon statistics.

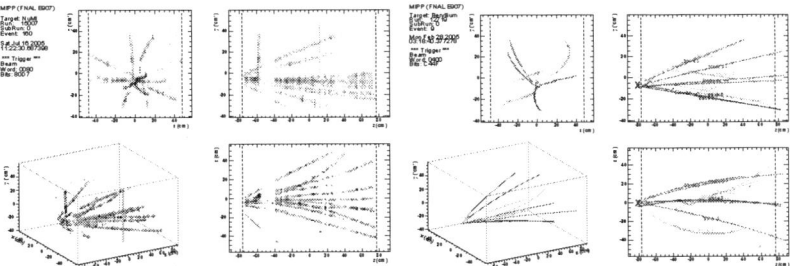

FIGURE 4. RAW and Reconstructed TPC tracks from two different events.

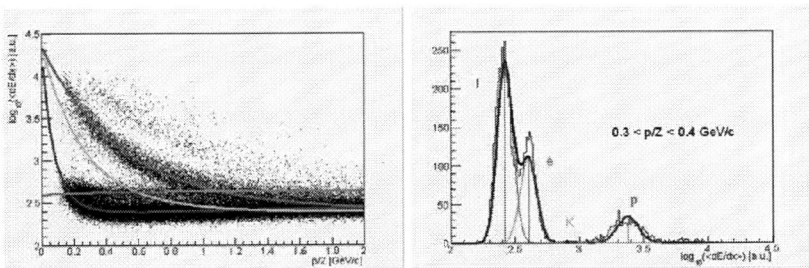

FIGURE 5. Preliminary dE/dx distributions in the TPC The scatter plot shows the electron, pion, kaon and proton peaks in the distribution as a function of the lab momentum for p-Carbon data. The second plot is the projection on the dE/dx axis for a momentum slice 0.3 GeV/c to 0.4 GeV/c.

we can measure the charged kaon mass. These events, whose statistics are indicated in

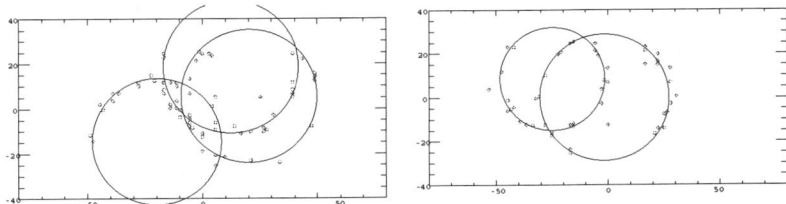

FIGURE 6. Examples of events with rings in the RICH counter for a 40 GeV/c beam. The x and y axes are in cm.

Figure 1, are currently being analyzed to evaluate the systematics involved in such a measurement.

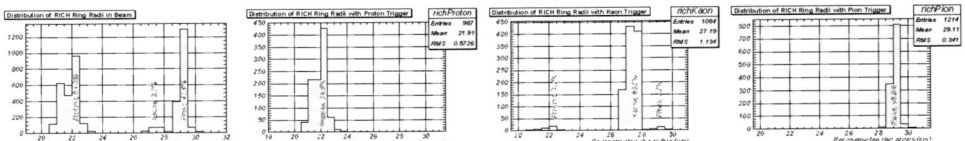

FIGURE 7. An example of a 40 GeV/c primary beam (non-interacting) trigger. The RICH identifies protons, kaons and pions by the ring radii. The beam Čerenkov detectors can be used to do the same. When the beam Čerenkov identification is used, one gets a very clean separation of pions, kaons and protons in the RICH.

NuMI target measurements

MIPP took 1.75 million events using 120 GeV/c primary beam protons impinging on the NuMI (spare) target. These events will play a crucial role in the prediction of neutrino fluxes in the NuMI beamline and will enable the MINOS experiment to control the systematics in the near/far detector ratios as well as helping them understand the near detector performance. Figure 8 shows a radiograph of the MIPP measurements of the MINOS target. The graphite slabs and cooling tubes can be seen. These events were obtained during the commissioning phase of this target measurement where the beam was not yet fully focused and aligned on the target. The 1.75 Million events on the NuMI target were obtained after the beam was aligned and centered on the target. Figure 9 shows the rich ring radii vs momentum of positive tracks originating from the NuMI target. Superimposed are the curves for known particles. This shows the excellent particle identification of the MIPP detector for forward going particles.

MIPP UPGRADE PROPOSAL

In October 2006, MIPP proposed to upgrade its detector by increasing the data acquisition speed of its TPC from 20 Hz to 3000 Hz using the ALTRO/PASA chips developed

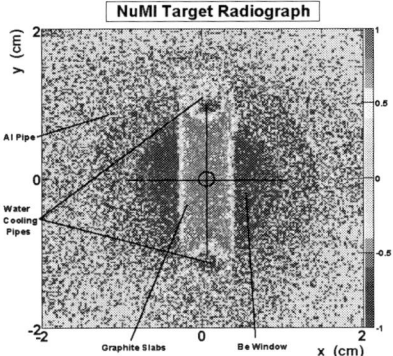

FIGURE 8. Radiograph of the MINOS target. The beam direction is perpendicular to the paper.

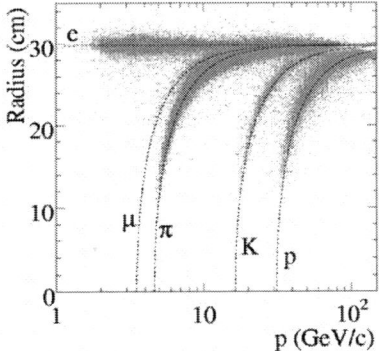

FIGURE 9. Preliminary data of RICH ring radii of positive tracks from the NuMI target vs momentum. Superimposed are the expected curves for e, μ, K and p particles.

for the ALICE collaboration [19]. With this large factor in data acquisition speed, it becomes possible to acquire 5 million events in a single day of running. The electronics of the rest of the detector will also be upgraded to run at this speed. We propose to use a hemisphere of the plastic ball detector [20] to measure the target recoil particles. This enables us to measure particle production on 30 of the most common nuclei found in particle physics detectors and improve the hadronic shower simulator programs. It will also permit us to study non-perturbative QCD in unprecedented detail. The baryon resonance spectrum can be investigated up to 3 GeV/c^2, using pion and kaon beams in the momentum range 1 GeV/c-5 GeV/c. The decision on the upgrade proposal has been deferred till analysis results from the present run are available and the collaboration has been strengthened further.

Upgrade Run Plan

We propose to conduct the upgrade running in three phases. In the first phase, we will acquire data on the neutrino targets, liquid nitrogen and take particle production measurements on 12 thin nuclear targets (5 million events per nucleus) which will improve the quality of hadronic shower simulation programs. The number of events acquired in each target is indicated in table 1.

TABLE 1.   Phase 1 Run Plan.

Target	Events (Millions)	Running Time (Days)
NuMI target 1	10	2
NuMI target 2	10	2
Liquid Hydrogen	20	4
Liquid Nitrogen	10	2
12 Nuclei		
D_2,Be,C,Al,Si,Hg,		
Fe,Ni,Cu,Zn,W,Pb	60	12
Total Events	110	22

TABLE 2.   Phase 2 Run Plan.

Target	Events (Millions)	Running Time (Days)
18 Nuclei		
Li,B,O_2,Mg,P,		
S,Ar,K,Ca,Ni,Nb,Ag,		
Sn,Pt,Au,Pb,Bi,U	90	18
10 Nuclei B-list		
Na,Ti,V, Cr,Mn,Mo,		
I, Cd, Cs, Ba	50	10
Total Events	140	28

During phase 2, we plan to complete the remaining 18 nuclei of the A-List, as detailed in the section on hadronic shower simulation and then proceed with the B-list if there is need. The second phase of running is detailed in table 2.

During phase 3, we plan to go into the tagged neutral beam mode, where we run the liquid hydrogen target and allow the ILC calorimetry to run simultaneously in place of the MIPP calorimeter to study the response of the ILC calorimeters to tagged neutral beams.

HADRONIC INTERACTION LIBRARY

With the upgrade, MIPP can change the landscape of data available for hadronic shower simulations by providing it in the form of a random access database which stores the four vectors and identities of the final state particles. With 6 million events per nucleus, distributed in ten beam momentum bins and six beam species and 100,000 events per bin, it is possible store this information in approximately 36 Gbytes of disk space. The data can be acquired in 36 calendar days of running. Such a random access library can be used directly by shower simulation programs or it can also be used to parametrize the models used in the programs currently. The former approach requires the use of a neutral particle algorithm which can be implemented assuming that the neutral particle rates are the average of the two charged particle states (e.g $\pi^0 = (\pi^+ + \pi^-)/2$ and $K^0 = (K^+ + K^-)/2$. Individual events can be picked in the appropriate reaction being simulated in the shower at a center of mass energy close to the one in the simulation and the center of mass longitudinal momenta of the events can then be scaled to match the event being simulated. It is possible to arrive at parametrisations of multiplicities as a function of $\sqrt{(s)}$ and this feature can be used to extrapolate the data to simulate showers at energies higher than those available in the data.

We plan to index each event by its missing mass (mass carried away by neutral particles). This will enable the simulation of neutral particles since we can look up events with small missing mass at the appropriate center of mass energy and turn the charged particles into neutral particles to simulate the neutral deposition.

Such a database of events can also act as a source of data for people formulating theoretical models of non-perturbative QCD interactions in the future.

CONCLUSIONS

MIPP has acquired high quality particle production data which it is busy analyzing. The MIPP upgrade will improve the quality and statistics of data available for hadronic shower programs by an order of magnitude.

REFERENCES

1. Information on the Harp experiment may be found at
   http://harp.web.cern.ch/harp/
   and also in review talks such as "HARP: A Hadron Production Experiment", E.Radicioni for the collaboration, hep-ex/0206028.
2. " Measurement of the production cross-section of positive pions in p-Al collisions at 12.9-GeV/c", HARP collaboration, Nucl. Phys. B732:1-45,2006.
3. "A New SPS programme", NA49 future collaboration, M.Gazdzicki for the collaboration, nucl-ex/0612007.
4. The main MIPP web page is at http://ppd.fnal.gov/experiments/e907/. Information on the MIPP collaboration may be found using links there.
5. The MIPP proposal and Addendum to the proposal may be found at http://ppd.fnal.gov/experiments/e907 /Proposal/E907_Proposal.html
6. The Time of Flight detector was fabricated by MIPP and consists of an array of 10 cm×10 cm scintillators and 5 cm×5 cm scintillators. See http://ppd.fnal.gov/experiments/e907 /TOF/TOF.html for a detailed description of the detector.
7. The multi-cell Čerenkov detector was first used in BNL experiment E766, then in Fermilab experiment E690 followed by other BNL experiments.
8. The details of the SELEX RICH construction and performance may be found at J. Engelfried *et al.*, *Nucl. Instr. and Meth.* **A43**,53(1999). We have replaced the front end electronics, and done extensive work on the safety systems. MIPP uses CO_2 gas as the radiator for the RICH.
9. NuMI stands for Neutrinos at the Main Injector and refers to the Fermilab Main Injector neutrino beam. MINOS is the first experiment to utilize this beam MINOS proposal may be found at "P-875: A

long baseline neutrino oscillation experiment at Fermilab", E. Ables *et al.*; FERMILAB-PROPOSAL-P875,(1995). See also their website at http://www-numi.fnal.gov/

10. The TPC was built by the BEVALAC group at Berkeley in the 1990's and used effectively at several Brookhaven experiments (e.g. E910) and then donated to Fermilab by LBNL for use in MIPP. See, G. Rai *et al.*, *IEEE Trans.Nucl.Sci.***37**,56(1990); LBL-28141.

11. "Beamline design for particle production Experiment, E907 at FNAL", C. Johnstone *et al.*, Proceedings of the PAC03 conference. So successful has the performance of this beamline been that the design has been adopted for use in the M-Test beamline upgrade for providing testbeams at Fermilab.

12. For more details on the beam Čerenkov system, see
http://ppd.fnal.gov/experiments/e907/Beam
/BeamCerenkov/BeamCerenkov.html

13. In 2004, we lost 20% of the phototubes in the RICH due to a fire in one of the phototube bases. This does not impact adversely on our pattern recognition, since the Čerenkov angle is large and there is plenty of light over most of our momentum range.

14. The EM calorimeter was fabricated by MIPP and uses lead as the radiator and an array of proportional tubes with 2.54 cm wire spacing as the readout. It has 10 radiation lengths and has 10 longitudinal segments.

15. The iron/scintillator fiber hadron calorimeter is recycled from the HyperCP collaboration. It has 9.7 interaction lengths and has 4 longitudinal and 2 transverse segmentations.

16. We have reused beam and drift chambers from the E690 collaboration. D.C.Christian *et al.*, *Nucl. Instr. and Meth.* **A345,62** (1994).

17. The large proportional chambers straddling the RICH find their use again after having been used by numerous previous experiments. M. De Palma *et al.*, *Nucl. Instr. and Meth.* **216** (1983) 393-397.

18. A. Malensek, Fermilab Technical Memo FN-341, 1981. (unpublished).

19. The details of the MIPP upgrade Proposal (P-960) can be found at
http://ppd.fnal.gov/experiments/e907/notes/
MIPPnotes/public/pdf/MIPP0138/MIPP0138.pdf

20. The plastic ball detector was constructed by GSI Darmstadt physicists in the early 1980's. See R. Albrecht *et. al.*, CERN Report CERN/SPSC/85-39 (august 1985), GSI Report GSI-85-32 (August 1985), for more details.

Summary of the 2006
Hadronic Shower Simulation Workshop

Laurie S. Waters

Los Alamos National Laboratory, MS K575, Los Alamos, New Mexico 87545

Abstract. The 2006 Hadronic Shower Simulation Workshop, held September 6-8, 2006 at Fermi National Laboratory brought together an international assembly of experts in the field of hadronic shower development. The overall goal was to present the current understanding of the physics of hadronic showers, and to study examples of how this is measured in particle-physics calorimetry. The modeling of such events is critical, and the major Monte Carlo codes, FLUKA, GEANT, MARS, MCNPX, and PHTS were represented at the workshop. A wide range of physics, much of which is used by the simulation codes was also discussed, ranging from the hadronic CEM, LAQGSM, and DTUJET models, down to low energy neutronics capabilities. Special purpose codes and methodologies used for specific applications such as muon and neutrino physics were also shown. The results of a code benchmarking exercises were presented and extensively discussed. This paper summarizes the key topics presented in the workshop.

Keywords: Monte Carlo, Radiation Transport, Particle Transport, Neutronics
PACS: 24.10.Lx

INTRODUCTION

The development of our ability to model hadronic and electromagnetic showers is largely driven by application. For example, the surge in research for large spallation neutron source facilities in the early 1990's, such as Accelerator Production of Tritium, PSI upgrades, and the Spallation Neutron source gave rise to improvements in shower physics initiated by protons below 2 GeV. One application for this community was how such showers affect material damage, and whether this is different from what one sees in a reactor. Research in this question led to improvements in the prediction of hydrogen and helium emission from the hadronic shower evaporation phase. Applications in space where shielding is at a minimum have stimulated research into shower impacts on crew dosimetry and electronics single event effects from cosmic rays. Homeland Security applications are driving research into deep penetration problems and delayed signals, involving new developments in modeling the decay of residual nuclei. Currently, the trend toward larger and more intense particle physics machines such as ILC and LHC increase the need for accurate predictions in calorimetry design. The intensities of these machines demand accurate simulations over a very broad range of energies in the shower, from the TeV incident particle to thermal neutron capture in calorimetry materials. Shower interactions in accelerator structures also affect both safety and background shielding for the experiments. Similar detail, although on a finer scale is needed to track single events in neutrino experiments. Finally, both the RHIC accelerator and proposals

CP896, *Hadronic Shower Simulation Workshop*
edited by M. Albrow and R. Raja
© 2007 American Institute of Physics 978-0-7354-0401-4/07/$23.00

for the Rare Isotope Accelerator in the US have resulted improvements in heavy ion shower modeling.

Simulation code developers and users have gathered on fairly regular bases in workshops such as the SARE (Simulating Accelerator Radiation Environments) [1], the SATIF (Shielding Aspects of Targets and Irradiation Facilities) [2] series. The present HSSW06 meeting has again involved both code developers and users, with a specific focus on particle physics calorimetry, and a goal of identifying areas in need of improvement. Developers of simulation codes were also invited to carryout a set of benchmark calculations. This paper summarizes the information presented in the workshop.

The day may arrive when simulation codes are so good that one may use them with complete blind faith. But, as Dr. Don Groom, our colloquium speaker reminded us, there is still much to be learned from using the physicist's innate intuition in the analysis of this problem. It is a message that he [3] and colleagues at CERN [4] have delivered in the past, and is certainly relevant today.

ALL-PARTICLE MONTE CARLO SIMULATION CODES

The workshop offered an excellent opportunity to formally compare the participating Monte Carlo simulation codes. The first section below will discuss the comparison topics which affect how problems are set up and run. Physics contained in the codes as presented in the workshop will be discussed in the following section. These are general-purpose 'all-particle' codes, which attempt to present as flexible and complete description of the shower process as possible, and including visualization of geometry and answers. They include FLUKA (FLUctuating KAscade) 2005 [5], GEANT4 (GEometry And Tracking) version 8.1 p1 [6], MARS (named after the planet) version 15 [7], MCNPX (Monte Carlo N-Particle eXtended) version 2.5.0 [8] and PHTS (multi-purpose Particle and Heavy Ion Transport code System) version 2.09 [9].

General Simulation Code Considerations

History and Structure

It is interesting in looking at the histories of the five codes to find how closely their development is tied to local needs, usually for a particular research accelerator.

FLUKA started development in 1962 and the FLUKA name was attached in 1970. Between 1970 and 1987 code development was a collaboration between CERN and groups at Leipzig and Helsinki. Since 1989 FLUKA has been developed at the National Institute of Nuclear Physics (INFN), Frascati, and is now developed, maintained and distributed in collaboration with CERN. The code is written in Fortran 77, and can be run standalone or in parallel. Source and binary are distributed with a 387 page manual. Workshops are held at least once per year. The user constructs an input file to control the problem execution.

GEANT first appeared in 1974, and has always been based at CERN. The last Fortran release of the code was GEANT 3.21 in 2000. The formal conversion of GEANT3 to C++ resulted in the GEANT4 code, which was first released in December, 1998.

GEANT4 is a international collaboration which includes CERN, ESA, IN2P3, INFN, KEK, LIP, PPARC, TRIUMF, and SLAC. A steering board provides overall direction. Source and binary are available with a 470 page on-line manual. Approximately 4 workshops are held each year. As an object-oriented C++ toolkit, GEANT4 is unique in that the user must code routines which control the selection of geometry, physics, tracking, and results. The code can be run in parallel if desired.

MARS has its origins in the FSU, and was first introduced in 1974, developing ideas from Feynman's inclusive approach to multi-particle reactions. MARS is now maintained at the Fermi National Laboratory. It is written in Fortran 95 and C, and a binary executable is distributed. The code can be run standalone or in parallel. A ~150 page manual is available and an average of 2 classes per year are held. Code execution is controlled by a user input file.

MCNPX is an extension of the MCNP code that increases the basic particle set, and adds models for use when evaluated nuclear data libraries are not available. The history of MCNP at Los Alamos is extensive, dating back to the Manhattan Project. The formal MCNPX project began in 1995, although predecessors date a few years earlier. The code is written in Fortran90 and C and can be run standalone or in parallel. Both source code and binary are released, along with a 470 page manual. Problem execution is controlled by a user-written input file.

Although the PHITS code was first released in 2003, it has an impressive list of predecessors, including NMTC, first released in Oak Ridge in 1951, and modified at JAERI in 1983. PHITS is well suited for the design of new accelerator facilities in Japan. Code development is a collaboration between JAEA, RIST, GSI and Chalmers University. It is written in Fortran 77, and both source and executable are distributed. The code can be run in parallel if desired. A 176 page manual is available, and the web site is in development. Workshops are held once per year. A user input file controls problem execution.

Code Distribution

The distribution of these codes and the assessment of the number of users must proceed cautiously. In the US, distribution of official releases of codes like this is handled by the Radiation Safety Information Computational Center (RSICC) in Oak Ridge. In Europe, the NEA fulfills a similar function. The export control laws of every country usually have something to say about distribution of various kinds of software, and each code must be in full compliance with appropriate regulations. It is also difficult to count the number of actual users, since one code release may be distributed to several users at an institution, without the developer's knowledge. At RSICC, codes are now given out to individuals on a sole-use basis for a fee per user which varies with the user's type of employer. Although this helps in counting the number of users, it is a disincentive for certain institutions to have a large number of individual users.

Some institutions still distribute codes at no cost outside of the more formal distribution centers. Beta test team distribution is often done this way, but still in strict compliance with local export control laws. Again, because of the practice of disseminating one copy of a code within an institution, the real number of users may be difficult to determine, unless users are registered in some way at the distribution point.

This is becoming a more common practice with all codes which still do free distribution from their own sites.

A more interesting, and not well studied question is exactly how many users might be counted as 'active'; actually using the code on a regular basis in their work. Users who take classes are usually have access to a code, however may never use it again after the class ends. People also change jobs, fail to keep their code version up to date, or use it only for a special project. Some have been intensive users in the past, but will have not used a code in recent years. How should these users be counted?

The best estimate of users, although no one was willing to say if all are 'active' among the code developers at the workshop is: MCNPX and GEANT4, ~2000 each, FLUKA ~1000, and MARS and PHITS ~300 each.

Geometry Capabilities

Geometry descriptions in Monte Carlo codes are usually either surface-based or combinatorial. In a surface-based code, the user defines a specific surface, then combines those surfaces into volumes. Occasionally a code will provide a 'macrobody' feature, in which a surface is pre-defined into a volume shape. With combinatorial geometry the user defines shapes, and combines them with Boolean algebra; intersection, union and compliment. MCNPX, MARS and PHITS use surface-based geometries, FLUKA and GEANT4 use combinatorial (PHITS also has a combinatorial option). Surfaces and shapes among the codes are fairly standard, although we note that GEANT4 has a unique twisted shape, and MCNPX is able to define a torus. In all codes, one can translate and rotate defined volumes and/or surfaces.

Duplication of volumes is essential in order to define objects such as CT scans, or highly segmented calorimetry. All the codes have some kind of rectangular lattice capability, MCNPX, MARS and PHITS have hexagonal lattices, and GEANT4 has a cylindrical lattice. All the codes can define nested volumes (MCNPX, PHITS and MARS through universe definitions). Repeated structures are also available in all codes.

It is often useful to have a capability to reflect particles at a boundary. For example, one can define a limited cylinder of air, and have particles specularly (mirror) reflected on the cylinder wall, which mimics an infinite volume – a useful device in modeling the atmosphere. All the codes have a reflection capability, with MCNPX also offering 'white' reflections with a cosine distribution, and a periodic boundary which simulates an infinite lattice. PHITS reflections are used for neutron albedo calculations.

All of the codes have viewing capabilities for geometry and plotting results, some built-in, some through external programs. Interactive Geometry viewers are contained in GEANT4 (3-D), MARS (2,3-D), MCNPX (2-D), and PHITS (2,3-D). GEANT4 has extensive interactive plotting and ray-tracing capabilities. MCNPX has a built-in plotter which can also overlay 'mesh tallies' on top of geometries, while other codes rely on outside custom viewing packages. To aid the user, outside Graphical User Interfaces are available for GEANT4 (GGE), MARS (Tcl/Tl), and MCNPX (VisEd and Moritz). GEANT4 and MCNPX have some CAD input capability, via their GUIs.

Electric and magnetic fields are available in all codes, although not formally released in MCNPX. Moving objects are available in FLUKA, GEANT4, and PHITS, and under development in MCNPX.

Source Capabilities

All of the codes can read a properly formatted external file of source particles. MCNPX can also write the results of a calculation on a surface, and read that file back in for sampling as a source in another problem. All codes offer the user a capability to write a specialized source subroutine, which must be compiled with the rest of the code.

All codes allow the user to designate an explicit source, specifying particle type, energy, position and angular distribution in the input file or source subroutine. Dependent distributions are also offered for MARS, MCNPX, and PHITS, where one variable is dependent on another (e.g., energy depends on particle type). GEANT4 has a specialized General Particle Source (GPS) element of their toolkit which allows dependent source specification.

MCNPX is the only code which considers neutron sources for calculation of k_{eff} eigenvalues and spatial eigenfunctions in criticality problems. Burnup capabilities for transmutation problems are also unique to MCNPX, stimulated by that code's use in reactor and accelerator transmutation of waste problems. MCNPX has included the CINDER'90 burnup code which allows transmutation of residual nuclides, from fission or from spallation, in the presence of a neutron flux.

Types of Answers

This is an area where some standardization of the definitions of various quantities calculated by the codes could benefit the community. The novice user often has little idea how to calculate quantities like flux or energy deposition, and benchmarking exercises between codes depend critically on the definitions used.

All the codes can calculate a volume flux for any particle. All can calculate a surface flux (Geant4 indicates the capability is 'limited'). Semi-deterministic calculations of flux as exemplified by 'point' and 'ring' detectors are contained in MCNPX for photons and neutrons with cross section libraries, and in MARS for neutrons only.

All codes can calculate surface current (counting particle weights), and again, GEANT4 indicates that this capability is 'limited'. All codes can calculate charge current.

Reaction rates are calculated in MCNPX, MARS and PHITS by weighting flux by number density and an interaction cross section. FLUKA uses the 'Star' approach, scaling the density of inelastic collisions to deduce various reaction rates.

All codes except PHITS indicate an energy deposition calculational capability. When using evaluated data libraries, codes will normally use heating factors contained in the library. Otherwise, energy deposition is a complex sum of ionization, residual recoil, local deposition from particles which fall below an energy cutoff, and other factors. Energy deposition is normally calculated on a particle-by-particle basis. For event-by-event calculations, such as one would see when looking at a signal from a detector, the 'pulse-height' calculation is appropriate. MCNPX, FLUKA and PHITS have this capability, while GEANT4 relies on user input, and MARS indicates no current capability.

All codes can calculate answers in terms of kinetic energy. Momentum and rapidity are calculated by GEANT4, FLUKA and MARS. This is a direct reflection of the origin of a code; those serving the high energy particle physics communities are familiar with these quantities, those which historically concentrate on the intermediate energy communities rely primarily on kinetic energy. It is assumed that the user can write a subroutine to cast the answer into any units they like for all codes.

Displacements per Atom (DPA), is a quantity needed for material damage studies. MCNPX has this capability with an auxiliary code, while MARS and PHITS do these calculations, and FLUKA indicates some capability. The user must define this in GEANT4.

Ever since their development at the Super Collider, 'Mesh" calculations have become quite popular, and all codes contain this capability. The user defines a mesh which is overlaid on top of the regular geometry and tracking proceeds through both geometry and mesh. Quantities such as flux, energy deposition and others can be scored in each element of the mesh and plotted as an extensive color map on top of the normal geometry. All codes now have rectangular and cylindrical mesh capabilities, while MARS and MCNPX also have spherical options.

The convergence of an answer in all codes is calculated by providing a variance. MCNPX has nine additional tests for answer convergence, and also provides confidence intervals for k_{eff} calculations.

Variance Reduction

For years, the primary reason cited for avoiding the use of Monte Carlo codes is the speed of the calculation. This has given rise to many different techniques to speed up the convergence of an answer (hence the name, 'variance reduction'). One broad class of techniques involves population control. These usually involve assigning a weight parameter to regions of the problem, which control 'splitting' and 'Russian Roulette' operations. In particle splitting, a particle crossing a boundary is split in two, each with half the weight. In Russian Roulette, particles are killed or survive when crossing a boundary, with appropriate adjustments in their weight. Region biasing, weight window meshes, and weight cutoffs are available in all codes. Energy biasing is available in all codes except GEANT4.

Modified sampling is another means of variance reduction. All codes have some means to bias source variables. Implicit capture, which involves killing particles based on the ratio of their absorption to total cross sections, is also present in all codes. Exponential transforms, which stretch the path length between collisions in a preferred direction by adjusting the total cross section, are present in MCNPX, FLUKA, and MARS. Angular biasing is also present in all codes. MCNPX further enhances the angular biasing feature by improving particle sampling in the vicinity of a point through the 'DXTRAN sphere'.

Code Physics

Basic physics processes in the codes are reviewed in this section, and the reader is referred to the individual code papers in this workshop for more detail.

Photons, Electrons, and Charged Particle Transport

Photon and electron physics are mature fields, and were not discussed much at the meeting, so will only be briefly reviewed here. The modeling of photoelectric, Compton, Rutherford, and pair/triplet production is present in all codes, and cross sections for these processes are often read from an evaluated data library such as Livermore's EADL and EPDL databases. FLUKA, GEANT4 and MCNPX use these libraries. Physical processes are modeled with the Integrated Tiger Series 3.0 coupled photon/electron transport code in MCNPX and PHITS. FLUKA, GEANT4 and MARS use custom models. Optical photon physics is offered by GEANT4 and FLUKA only, and is generally used to calculate light output from scintillation detectors.

Photonuclear physics is present in all codes except PHITS. In modeling photonuclear interactions, the codes differ as to how many processes they include as a function of energy. All capabilities model the Giant Dipole Resonance, either directly using tabulated resonance parameters, or read in from an evaluated nuclear data library, such as those available at the IAEA. Quasi-deuteron, delta and vector meson dominance processes are usually modeled with custom packages; CEM for MARS and MCNPX, CHIPS in GEANT4, and VMDM in FLUKA.

Electron physics packages used include ITS 3.0 for MCNPX and PHITS, and custom models for GEANT4, FLUKA and MARS. All the codes use a Continuous Slowing Down Approximation (CSDA) for charged particle transport, which determines step size based on mean energy loss in a pre-calculated path length. All codes use some variation of the Bethe-Bloch ionization formula for charged particles. For small angle scattering, FLUKA, MARS and PHITS use an improved Moliere model, while MCNPX uses a Rossi-Griesen algorithm, and GEANT4 uses Lewis theory. Energy straggling in MCNPX and PHITS uses Vavilov theory, while GEANT4 uses the Urban model. FLUKA and MARS have in-house models specifically for their own codes.

Neutronics

As high energy accelerators have increased in energy and current, the low energy neutron background problem has also increased, leading codes which previously specialized in high-energy physics to add a neutronics capability. Low energy neutron physics is a specialized field, and was the subject a dedicated talk at the workshop [10]. The capabilities of the codes in this area vary widely. All codes except FLUKA can read continuous-energy evaluated nuclear data libraries, favoring the ENDF format (nuclear data libraries typically have upper energies from 20-150 MeV). FLUKA has multigroup capability only. MCNPX inherits its neutron capabilities from the MCNP base code, which developed these techniques over decades of research in reactor physics. Both MARS and PHITS can couple neutrons produced in high energy interactions directly into the MCNP code.

Although it is tempting to use one data library for all applications, the user must keep in mind that various adjustments need to be made depending on problem conditions. Neutron thermal scattering can depend on material atomic structure in solids. Special $S(\alpha,\beta)$ cross sections have been evaluated for specific compounds which substitute for the regular evaluated libraries at thermal energies. Temperature affects the peak thermal cross section through Maxwell's equation E=3/2kT, and this can be adjusted directly in the code. Temperature also affects the widths of nuclear resonances, and one must use a cross section evaluated specifically for the temperature of interest.

Evaluated libraries may or may not contain information for secondary particle tracking. Fox example, in an (n,alpha) interaction, the energy and angular distribution of the emitted alpha particle must be available if that particle is to be tracked further. Whether such information is available in a library must be ascertained by the user. If not, it is usually possible to code in the particle emission parameters for simple cases.

The use of evaluated nuclear data libraries brings up the question of analog versus non-analog sampling, and this directly affects a code's ability to model a single particle throughout its entire track history. An analog code will simulate every interaction of that particle exactly, and energy will be conserved at the end of the event. Non-analog approximations can occur in many places in Monte Carlo neutronics transport, however the one that is of most concern leads to energy conservation on average over a large number of events, but not on an event-by-event basis. In MCNPX for inelastic reactions producing more than one secondary particle, the final state distribution of the secondaries is governed by various scattering laws. Often a cross section library will have more than one scattering law defined, and the laws are applied independently for each exiting particle. For example, it is possible for a 14.0 MeV neutron to produce two 12.0 MeV particles from a single interaction, with appropriate weight adjustments. This eliminates the correlation between the emitted particles, and is the basic difficulty in modeling individual events in library-based Monte Carlo codes. It is possible to recover the correlation through by modifying the libraries, or in post-processing, and this is an active area of research.

An interesting shower analogy to the non-analog neutronics problem occurs in the MARS code's option to use a model based on Feynman's inclusive approach to multi-particle reactions. At each interaction vertex, a particle cascade tree can be constructed using only a fixed number of representative particles and each particle carries a statistical weight, equal in the simplest case, to the mean particle multiplicity for that type of event. Energy is conserved on average over many collisions, but not on an event-by-event basis. The advantage of using non-analog approaches is primarily aimed at increasing the speed of the calculation.

Nuclear Interactions

There is a well-defined procedure for calculating 'spallation' hadronic interactions in Monte Carlo codes, which are considered appropriate for incident energies below a few GeV. Initially, an incoming hadron interacts with individual nucleons in a nucleus, which are considered to be a free Fermi gas limited by a potential describing the nuclear shape. Secondary particles are emitted according to resonance formation and decay using nucleon-nucleon cross sections. The is referred to as the Intranuclear Cascade (INC) phase. The Bertini model is the earliest and simplest example of an INC model, and is still included in many Monte Carlo packages. Most codes now include a pre-equilibrium phase just after INC, which accounts for nucleon emission on a time scale between the fast (10^{-22} sec) INC phase and the much slower (10^{-16} to 10^{-18} sec) compound nucleus reactions. An semi-classical, fast exciton pre-equilibrium model is often used which determines secondary production by calculating excitations in the nuclear Fermi sea. Quantum mechanical multistep models are also used. Early versions of pre-equilibrium models showed great improvement in code abilities to calculate large angle particle emission.

The residual nucleus then enters either a fission or evaporation stage. Evaporation models first emit light ions and neutrons, and finally, photons characteristic to the particular excited state of the residual. For light residuals the use of a Fermi-breakup model is common, which favors dissociation into light ions. Recent developments in Monte Carlo codes allow de-excitation of residual nuclei over very long periods of time, using nuclear half-lives and proper gamma emission line or multigroup spectra.

For higher energies, most codes move into quantum models, and several presentations were given in recent work in this area, on DPMJET [11], and the CEM/LAQGSM packages [12].

Overviews of options were given by all the speakers for their particular code packages. For interactions less than 5 GeV, FLUKA uses the PEANUT model, which includes a generalized INC model, exciton pre-equilibrium stage, and the FLUKA evaporation model. Emission of energetic light fragments, up to alphas, is described through a coalescence model. At higher energies, the Glauber-Gribov theory, a field theory formulation of the Glauber cascade is used. Recent work in nucleus-nucleus interactions uses BME (Boltzman Master Equation) for energies less than 0.1 GeV/n, an improved version of rQMD2.4 for energies between 0.1 and 5.0 GeV/n, and DPMJET-III for energies greater than 5 GeV.

GEANT4 includes a number of hadronic interaction options. For example, a classical approach is represented by the GHEISHA model, which was inherited from GEANT3, and re-engineered into C++. A Quark-Gluon string model is used for high energies, including pomeron exchange and a detailed nuclear model. String excitation follows the Fritiof approach. Low and high energy parametrized models are also available and were originally designed for fast shower studies. The initial interaction, fragmentation of hadrons, INC and de-excitation phases are all parameterized. There is no modeling of the target nucleus and no intra-nuclear tracking – and the method does not conserve energy or momentum event-by-event. Another option is the 'Binary Cascade', which is a hybrid between the classical cascade and full QMD model.

The inclusive approach in MARS has already been mentioned. For exclusive event generators below 5 GeV, MARS includes the CEM03.01 combined with Fermi breakup, coalescence, and an improved version of the Generalized Evaporation-fission model (GEM2). LAQGSM03 has recently been implemented for particle and heavy-ion projectiles from 10 MeV/A to 800 GeV/A. MARS has also long include the Dual-Parton model DPMJET3 for very high energy interactions.

MCNPX has several spallation options, including the Bertini INC, an exciton pre-equilibrium model, the Dresner evaporation, and Fermi-breakup models. The Rutherford-Appleton and ORNL fission models are both offered. This set of models was inherited from the predecessor LAHET code. MCNPX has since added and updated the CEM code. The INCL/ABLA package was also included, with special emphasis on hydrogen and helium gas production for materials damage studies. MCNPX can also read evaluated proton data libraries, which are available up to 150 MeV. An early version of FLUKA was phased in between 2.0 and 5.0 GeV for very high energy interactions. This has recently been supplemented with the LAQGSM03 code, which also models heavy ion interactions.

The predecessors of PHITS had their origins in the ORNL spallation models, and the code developers have continuously improved them over the years. The new JAM (Jet AA Microscopic Transport Model) code, as well as Bertini and JQMD models can be used up to 200 GeV for hadron-nucleus collisions. JQMD is used for nucleus-nucleus collisions. Much impressive work has been done by PHITS on the US proposals for the Rare Isotope Accelerator. GEM is used for evaporation processes.

Code Benchmarking Exercise

The benchmarking of these codes was discussed in detail during each presentation, but for the purposes of this summary, mention must be made of the formal workshop benchmark activity [13]. Seven problems were proposed, and offer the opportunity to compare the five codes in several different energy ranges. For example, Figure 1 show the results of part of benchmark Task 7, which had each code calculate energy deposition in a 10 cm long, 1 cm radius tungsten cylinder. This was a 'blind' test, with no actual data comparison.

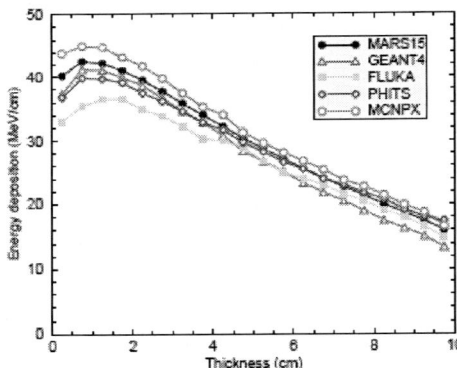

Figure 1. Energy deposition in a cylindrical tungsten target irradiated by 1 GeV protons.

Figure 2 shows the result of benchmark Task 5, which measures energy deposited in a copper absorber from 12 GeV protons incident on a copper target. Data is from a KEK experiment performed in 2004. It is difficult to discern detailed trends in these two examples; much depends on what options are set in each code. It is hoped that future meetings of the HSSW series will continue the benchmarking activities.

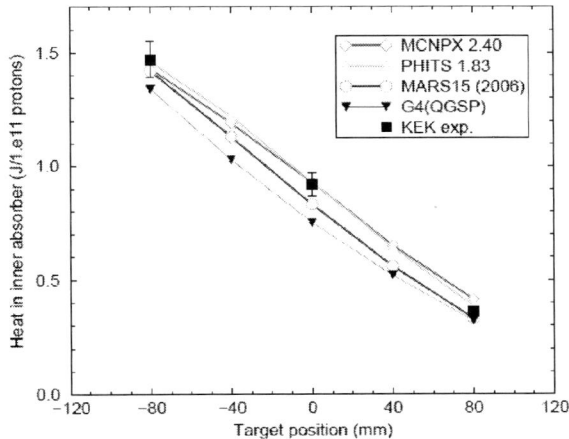

Figure 2. Energy deposition in a cylindrical absorber surrounding a copper target.

Applications

Several talks were given on specific applications of simulations in current experimental work, and many illustrated how the codes are used, and new simulation tools developed.

ATLAS End-Cap liquid Argon/copper calorimeter data was presented. Simulations were done with GEANT3 and GEANT4 (with the LHEP and QGSP physics lists) as a result of a long-standing validation program [13].

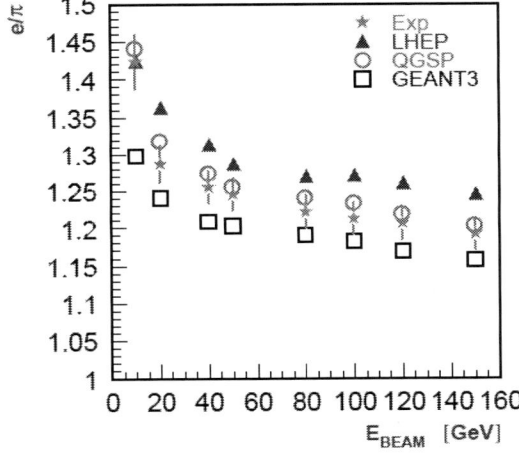

Figure 3. Comparison of GEANT calculations for ATLAS calorimeter e/π.

In general signal characteristics such as e/π and the energy dependence of the response in the non-compensating calorimeters is reasonably well described. Some problems remain in the spatial shower spread calculation. Figure 3 shows an example of GEANT3, GEANT4/LHEP and GEANT4/QGSP calculations for the e/π ratio.

Calculations again using the GEANT4 LHEP and QGSP lists were done for the CMS ECAL and HCAL. In general, the parameterized physics list LHEP show better agreement with data, while the model-based QGSP list predicts shorter showers than measured. Figure 4 shows an example of longitudinal shower profiles.

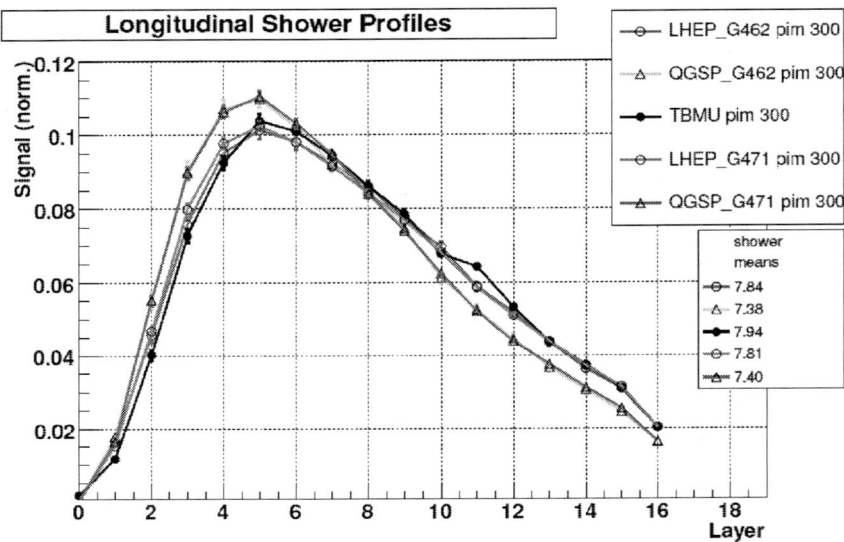

Figure 4. CMS longitudinal shower profiles compared with GEANT LHEP and QGSP physics.

MiniBooNE uses the decay in flight of a π^+ beam to produce muon neutrinos. The pion beam is produced via spallation in a beryllium target struck by 8.9 GeV protons. Pion production cross sections were measured in the Brookhaven E910 and HARP experiments and fit to a Sanford-Wang function. Subsequent neutrino production was computed, and compared to several Monte Carlo codes [15]. As seen in Figure 5, MARS shows the best match to the fitted cross sections.

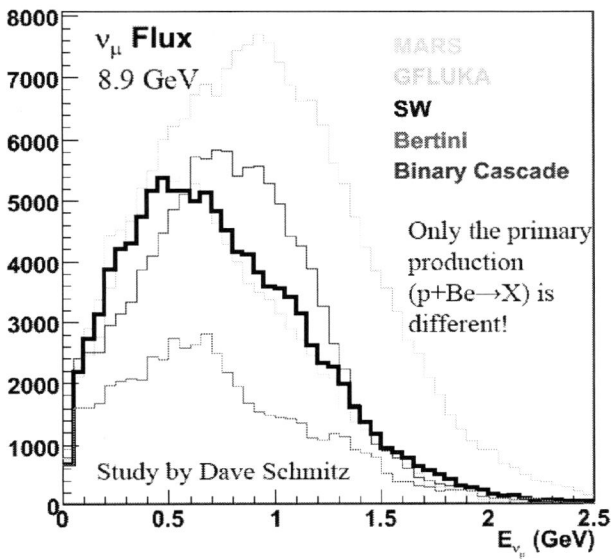

Figure 5. Calculation of muon neutrino production from pion decay. Pions were produced by 8.9 GeV protons incident on a Beryllium target.

In MINOS (Main Injector Neutrino Oscillation Search), a neutrino beam is provided by the 120 GeV/c main injector at Fermilab. The calorimeter detectors are located 735 km apart, one at Fermilab and one at the Soudan mine in Minnesota. Two talks were presented on simulation work for the Minos calorimeters, and on the need for neutrino event generators [16,17]. As part of the formulation of such event generators, cross sections for pion production were calculated from CDM03.01 (see Figure 6.)

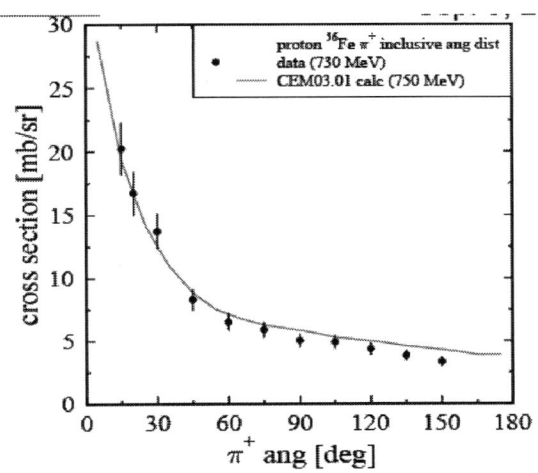

Figure 6. Comparison of CEM03.01 prediction of π+ angular distributions with data.

Atmospheric production of muons and neutrinos [19] was also discussed. It was emphasized that interpretation of cosmic ray data relies heavily on Monte Carlo simulations, particularly on the muon production. Measurements are underway at NA49, HARP and MIPP. Some HARP data was included in the benchmarking activity of this conference.

A number of other excellent points were made in the workshop, not all of which can be covered here. It was noted that the MIPP experiment has already acquired data of unprecedented quality on particle production using 6 beam species on several nuclei and the MINOS target. This data should be directly used by existing and future simulation codes, where it will be invaluable in improving neutrino production predictions [20].

ILC calorimeter design was presented [21], with a discussion of how jet energy resolution can be improved with increased longitudinal granularity. Excellent Monte Carlo simulations and PFA (Particle Flow Analysis) algorithms are needed to realize this vision.

In an interesting presentation on meaningful simulations of hadronic showers, crucial elements which must be present in any simulation code were identified [22]. Particularly important are π^0 production, the nuclear component and neutron transport. Benchmarks programs in these areas were encouraged.

SUMMARY

HSSW06 was an invaluable opportunity for experimentalists, code and model developers to interact on issues of importance to the high-energy community. The close examination of codes enabled the code developers to review their own progress, and share their most recent developments with others. Experimentalists had an opportunity to see the latest simulation capabilities, and to make known high priority requests for future development.

The colloquium speaker [23] presented a fascinating exploration of what can be learned through simple scaling laws, and reminded the audience of Moyer's old concept of a "universal" spectrum. It was shown that a calorimeter with two independent readouts (e.g., cherenkov and scintillator light), giving as much h/e contrast as possible, can give greatly improved signal resolution by examining the correlation between the two signals. It is hoped that such a methodology can be applied to some current experiment, and it is hoped that modern code developers will take heart the message that simple, fundamental rules underlie everything that we do.

HSSW06 was a great success, and the participants look forward to future meetings in this series.

REFERENCES

1. D. Filges and F. Goldenbaum, eds., "Proceedings of the Fifth Workshop on Simulating Accelerator Radiation Environments, (SARE-5), Models and Codes for Spallation Neutron Sources, July 17-18, 2000, OECD Headquarters, Paris, France", ISSN 1433-559X, ESS 112-01-T (April 2001).
2. "SATIF-8, Shielding Aspects of Accelerators, Targets and Irradiation Facilities, Pohang Accelerator Laboratory POSTECH, Pohang, Republic of Korea, 22-23 June 2002," Nuclear Energy Agency, Organization for Economic Co-operation and Development, in process.

3. Donald E. Groom, "A Minority Opinion", *Proceedings of the Workshop on Simulating Accelerator Radiation Environments, January 11-15, 1993, Santa Fe, New Mexico, USA,* Los Alamos National Laboratory, LA-2835C (October 1994), pp 36-40.
4. G. R. Stevenson, A. Fassò and M. Höfert, "Designing Accelerators without the Perfect Simulation Code," *Proceedings of the Workshop on Simulating Accelerator Radiation Environments, January 11-15, 1993, Santa Fe, New Mexico, USA,* Los Alamos National Laboratory, LA-2835C (October 1994), pp 41-47.
5. Paola Sala, these proceedings.
6.   John Apostolakis, Tatsumi Koi, and Dennis Wright, these proceedings.
7. Nikolai Mokhov, these proceedings
8. Gregg McKinney and Laurie Waters, these proceedings
9. Koji Niita and Reg Ronnigen, these proceedings.
10. Grady Hughes, these proceedings.
11. Johannes Ranft, these proceedings.
12. Stepan Mashnik, these proceedings.
13. D. Salihagić on behalf of the ATLAS LAr collaboration, these proceedings
14. Stefan Piperov for the CMS Collaboration, these proceedings.
15. Jonathan Link, these proceedings.
16. Mike Kordosky, these proceedings
17. Steve Dytman, these proceedings.
18. Christine Meurer for the HARP collaboration, these proceedings.
19. Giles Barr, these proceedings.
20. Rajendran Raja, these proceedings.
21. Mark Thompson, these proceedings.
22. Richard Wigmans, these proceedings.
23. Don Groom, these proceedings.

List of Participants

Adams	Todd	Florida State University	tadams@hep.fsu.edu
Albrow	Michael	Fermi National Accelerator Laboratory	albrow@fnal.gov
Alexa	Calin	IFIN	Calin.Alexa@cern.ch
Ansermet Tentindo	Silvia	EPFL, Switzerland	silvia.ansermet-tentindo@epfl.ch
Apostolakis	John	CERN	john.apostolakis@cern.ch
Atac	Muzaffer	Fermi National Accelerator Laboratory	matac@fnal.gov
Auer	Ralf	University of Erlangen-Nuremberg	ralf.auer@physik.uni-erlangen.de
Baek	Inseok	Michigan State University / NSCL	baek@nscl.msu.edu
Barr	Giles	Oxford University	g.barr1@physics.ox.ac.uk
Bazo Alba	Jose	Pontificia Universidad Catolica Del	jbazo@fnal.gov
Bhatti	Anwar	The Rockefeller University	bhatti@fnal.gov
Cammin	Jochen	University of Rochester	cammin@fnal.gov
Chakraborty	Dhiman	Northern Illinois University	dhiman@fnal.gov
Chandra	Avdhesh	University of California, Riverside	avdhesh@fnal.gov
Demarteau	Marcel	Fermi National Accelerator Laboratory	demarteau@fnal.gov
Denisov	Dmitri	Fermi National Accelerator Laboratory	denisovd@fnal.gov
Dytman	Steven	University of Pittsburgh	dytman@pitt.edu
Ellis	Malcolm	Fermi National Accelerator Laboratory	mellis@fnal.gov
Elvira	V. Daniel	Fermi National Accelerator Laboratory	daniel@fnal.gov
Gabriel	Tony	SID (ORNL/Retired)	jkgabrielta@earthlink.net
Gallmeier	Franz	Oak Ridge National Laboratory	gallmeierfz@ornl.gov
Garcia	Edmundo	University of Illinois, Chicago	ejgarcia@uic.edu
Gerbo	Davide	INFN	gerbo@fnal.gov
Gomes	Itacil	I.C.Gomes Consulting & Investment Inc.	icgomes@icgomes.com
Grachov	Oleg	University of Kansas	grachov@ku.edu
Groom	Donald	Lawrence Berkeley National Laboratory	deg@lbl.gov
Hagmann	Christian	Lawrence Livermore National Laboratory	hagmann1@llnl.gov
Handler	Thomas	University of Tennessee	thandler@utk.edu
Hennings-Yeomans	Raul	Case Western Reserve University	raul@casino.phys.cwru.edu
Huang	Xingtao	University of Puerto Rico	huangxt@fnal.gov
Hughes	H. Grady	Los Alamos National Laboratory	hgh@lanl.gov
Imrek	Jozsef	Brookhaven National Laboratory	jozsef.imrek@bnl.gov
James	Catherine	Fermi National Accelerator Laboratory	cjames@fnal.gov
Koi	Tatsumi	Stanford Linear Accelerator Center	tkoi@slac.stanford.edu
Kordosky	Michael	University College London	kordosky@fnal.gov
Kostin	Mikhail	NSCL Michigan State University	kostin@nscl.msu.edu
Kvita	Jiri	Charles University, Prague	kvita@fnal.gov
Lang	Karol	University of Texas, Austin	lang@hep.utexas.edu
Leitner	Rupert	Charles University	rupert.leitner@mff.cuni.cz
Leveling	Anthony	Fermi National Accelerator Laboratory	leveling@fnal.gov
Lima	Guilherme	Northern Illinois University	lima@fnal.gov
Link	Jonathan	Virginia Tech	link@fnal.gov
Lopez	Angel	University of Puerto Rico	angel@fnal.gov
Magill	Steve	Argonne National Laboratory	srm@anl.gov
Mal	Prolay	University of Notre Dame	prolay@fnal.gov
Mashnik	Stepan	Los Alamos National Laboratory	mashnik@lanl.gov
McKinney	Gregg	LANL	gwm@lanl.gov

Mendez	Hector	University of Puerto Rico	mendez@fnal.gov
Meurer	Christine	Forschungszentrum Karlsruhe	Christine.Meurer@ik.fzk.de
Meyer	Holger	Fermi National Accelerator Laboratory	hmeyer@fnal.gov
Micklich	Bradley	Argonne National Laboratory	bjmicklich@anl.gov
Mokhov	Nikolai	Fermi National Accelerator Laboratory	mokhov@fnal.gov
Monville	Maura E.	Fermi National Accelerator Laboratory	mauede@fnal.gov
Morfin	Jorge G.	Fermi National Accelerator Laboratory	morfin@fnal.gov
Nakao	Noriaki	Fermi National Accelerator Laboratory	nakao@fnal.gov
Neuffer	David	Fermi National Accelerator Laboratory	neuffer@fnal.gov
Nigmanov	Turgun	University of Michigan	nigmanov@fnal.gov
Niita	Koji	RIST, Japan	niita@tokai.rist.or.jp
Osiecki	Thomas	University of Texas, Austin	osiecki@mail.hep.utexas.edu
Para	Adam	Fermi National Accelerator Laboratory	para@fnal.gov
Piperov	Stefan	Fermi National Accelerator Laboratory	piperov@fnal.gov
Raja	Rajendran	Fermi National Accelerator Laboratory	raja@fnal.gov
Rakhno	Igor	Fermi National Accelerator Laboratory	rakhno@fnal.gov
Ranft	Johannes	Siegen University	Johannes.Ranft@cern.ch
Rangel	Murilo	CBPF, Brazil	rangel@fnal.gov
Reisetter	Angela	University of Minnesota	areisett@physics.umn.edu
Remec	Igor	Oak Ridge National Laboratory	remeci@ornl.gov
Repond	Jose	Argonne National Laboratory	repond@hep.anl.gov
Ribon	Alberto	CERN	Alberto.Ribon@cern.ch
Ronningen	Reginald	Michigan State University/NSCL	ronningen@nscl.msu.edu
Ronzhin	Anatoly	Fermi National Accelerator Laboratory	ronzhin@fnal.gov
Sala	Paola	INFN, Milan	paola.sala@mi.infn.it
Salihagic	Denis	Max-Planck-Institut fuer Physik	salihag@mppmu.mpg.de
Saoulidou	Niki	Fermi National Accelerator Laboratory	niki@fnal.gov
Solomey	Nickolas	E907-Fermilab	solomey@fnal.gov
Striganov	Sergei	Fermi National Accelerator Laboratory	strigano@fnal.gov
Thomson	Mark	University of Cambridge	thomson@hep.phy.cam.ac.uk
Tokar	Stanislav	Comenius University	tokar@fmph.uniba.sk
Vasileiou	Vlasios	University of Maryland	vlasisva@umdgrb.umd.edu
Waters	Laurie	Los Alamos National Laboratory	lsw@lanl.gov
Wigmans	Richard	Texas Tech University	wigmans@ttu.edu
Wright	Dennise	Stanford Linear Accelerator Center	dwright@slac.stanford.edu
Xia	Lei	ANL	lxia@hep.anl.gov
Yarba	Julia	Fermi National Accelerator Laboratory	yarba_j@fnal.gov
Yoo	Jonghee	Fermi National Accelerator Laboratory	yoo@fnal.gov
Yu	Jae	University of Texas at Arlington	jaehoonyu@uta.edu

AUTHOR INDEX

A

Apostolakis, J., 1

B

Barr, G. D., 150
Battistoni, G., 31
Blümer, J., 158

C

Cerutti, F., 31

D

Durkee, J. W., 81
Dytman, S., 178

E

Ellis, M., 168
Engel, R., 158

F

Fassò, A., 31
Fensin, M. L., 81
Ferrari, A., 31
Folger, G., 11, 21

G

Grady Hughes, H., 91
Groom, D. E., 137

H

Haungs, A., 158
Heikkinen, A., 11, 21

Hendricks, J. S., 81

I

Ivanchenko, V., 11
Ivantchenko, V., 21
Iwamoto, Y., 61
Iwase, H., 61

J

James, M. R., 81
Johns, R. C., 81

K

Kiryunin, A. E., 205
Koi, T., 11, 21
Kordosky, M., 185
Kossov, M., 11, 21

L

Lei, F., 21

M

Mancusi, D., 61
Matsuda, N., 61
McKinney, G. W., 81
Meurer, C., 158
Mokhov, N. V., 50
Muraro, S., 31

N

Nakashima, H., 61
Niita, K., 61